Point sources and multipoles in inverse scattering theory

Roland Potthast

Point sources and multipoles in inverse scattering theory

CRC Press
Taylor & Francis Group
Boca Raton London New York

CRC Press is an imprint of the
Taylor & Francis Group, an **informa** business

CRC Press
Taylor & Francis Group
6000 Broken Sound Parkway NW, Suite 300
Boca Raton, FL 33487-2742

Visit the Taylor & Francis Web site at
http://www.taylorandfrancis.com

and the CRC Press Web site at
http://www.crcpress.com

To my parents,
  who supported my way,
and to my wife, Regina,
  for sharing the joy and the work.

Contents

Preface

Books are written in different places and with various purposes motivating the authors and influencing style and content. Some works are mainly introductions into an area and arise from lectures and courses. Others summarize research and present results in a concise way.

In the present book, an introduction to recent research results is a main task. I consider research as a fascinating story in which many people take part and contribute to the whole of knowledge and technologies. It is important to realize that many different styles and characters are involved in this business. Many things, like the way of presenting results, are a matter of taste and personal strengths. I think that each person is important in his or her contributions, weaknesses, and strengths and should not be neglected by others.

Secondly, research is the story of important open problems and thrilling answers. Today's questions are for example: How can I succeed to simulate or calculate a natural phenomenon? Can I reconstruct a quantity from the knowledge of some others? Is it possile to get insight into the matter by mathematical analysis? Sometimes it is difficult to express what is missing and how to approach a particular problem. Then, it might be helpful to ask: Can we phrase or solve the question in a different way to get new insight into the matter? Often problems which have seemed insolvable with current techniques become accessible and solvable by new influences from technology or new mathematical tools.

For example, when solving inverse scattering problems I always hesitated to apply a general optimization scheme to a large nonlinear problem. I was convinced that it must be possible to get more concrete insight into the matter and to solve the inverse problem without multiply solving the direct problem. But for a long time we had no idea how this should be possible. Then, at the same time, the linear sampling method and the point-source method proved successful in reconstructing the shape of scatterers without using much optimization. Suddenly three or four different new methods (point-source method, singular-source method, linear sampling method, and modified linear sampling method) were alive and gave new input into the area.

A second story is the interpretation and understanding of Newton's method. During the doctoral dissertation, I spent a lot of time studying Fréchet derivatives. Since it is one main solution method to solve scattering problems, the lack of a differentiablity proof via the boundary integral approach was quite unsatisfactory. Then the possiblity to prove differentiability via boundary integral operators opened up before us. But the analysis of Newton's method in inverse scattering still remained incomplete: from the viewpoints of geometry and of convergence it was not clear what happened at each iteration step. Chapter 5 will present some recent results on the convergence and interpretation of Newton's method which answer these questions.

These are just examples of points where it was possible to find solutions to open questions. You will find more in this work, and other researchers tell *their* stories and experiences. But many questions are important and *not* answered until now. I think that we are just at the very beginning in inverse problems, remote sensing, image recognition, nondestructive testing, and medical imaging!

A large part of the scientific ideas in this book have been developed in interaction with Prof. Dr. Rainer Kreß, Göttingen, and Prof. Dr. David Colton, Delaware. Thank you for your encouragement and criticism. I also enjoyed many stimulating discussions in London with Dr. Simon Chandler-Wilde at Brunel University. Some work on the implementation of the methods is done in cooperation with Dr. Klaus Giebermann, Bonn. I would like to thank Dr. Tilo Arens for providing his code for both direct and inverse problems. Also, thanks are given to Dr. Imre Paulovits, ORGALOGIC GmbH/Systems Research Laboratories, Köln, for supporting the finish of my habilitation procedure in 1999 when I worked full-time in his company, and to Prof. R. Kress, Dr. S. Chandler-Wilde, Dr. Frank Hettlich and to Dr. Tilo Arens for reading and correcting parts of this book.

I am grateful to *Deutsche Forschungsgemeinschaft* (DFG), that financed my research over four years and paid for several trips to the U.S., including one year at the University of Delaware. I appreciate the hospitality of this institution, of Prof. Dr. Peter Monk and his family and of Heidi and Lothar Jeromin, Delaware.

Finally, I would like to express special thanks to my wife Regina, who strongly supported and encouraged me over the years.

Roland Potthast, Brunel University, London, 2001

Classics

As long as a branch of knowledge offers an abundance of problems, it is full of vitality.   *David Hilbert*

Before you generalize, formalize, and axiomatize, there must be mathematical substance.   *Hermann Weyl*

Our science, in contrast to others, is not founded on a single period of human history, but has accompanied the development of culture through all its stages. Mathematics is as much interwoven with Greek culture as with the most modern problems in engineering. It not only lends a hand to the progressive natural sciences but participates at the same time in the abstract investigations of logicians and philosophers.   *Felix Klein*

Visions

The youth will discover the joy and beauty of mathematics, its power and strength, the clear lines of thoughts, arguments, geometric vision, the joy to go beyond borders and exceed limits of old things, and, at the same time, to study concepts and proofs which are hundreds or thousands of years old and still are as young and amazing as a newborn baby.

Find the faith and courage to try something new and to go into thoughts which have not been thought before. There is the danger to come back with empty hands, to find that you are not gifted enough and not successful enough. But the thrilling presence of the new, the amazement when the fog slowly dissolves or when suddenly the long searched concept or truth opens up before you is worth the stake.

Appreciate the work of everyone, since every being has his or her place. Some rearrange long-known knowledge in a proper way, others go into unknown lands. We deeply need each other, even if we do not know this, to grasp the truth of this creation and to shape our lives. And we all need God, the creator, living spirit, holy love, the beginning and the end.

1 Introduction and tools

Scattering of acoustic or electromagnetic waves plays an important role in many fields of applied sciences. Acoustic and electromagnetic waves are used and investigated in such different areas as medical imaging, ultrasound tomography, material science, nondestructive testing, radar, remote sensing, aeronautics, and seismic exploration.

In the last twenty years the development of computational power has also had a strong impact on the classical fields of direct and inverse scattering. The computational simulation of scattering processes has become accessible using microcomputers and the field of inverse scattering problems arose. Inverse scattering is concerned with the reconstruction of scattering objects or their properties. It grew from its early beginnings in the middle of the century to a large and fast-developing area of applied mathematics and engineering.

In the first part of Chapter 1 we give a brief introduction into inverse scattering theory and outline main results. In the second part, we collect definitions and tools from functional analysis and the theory of integral equations, which are the basis for the further chapters.

1.1 A survey on inverse scattering theory

Scattering by obstacles and media. The classical area of acoustic and electromagnetic scattering is concerned mainly with two different problems, which are studied and applied in many different settings and applications.

The first problem is the scattering of acoustic or electromagnetic waves by an impenetrable scatterer, i.e., the waves do not significantly penetrate into the interior of the scattering obstacle D. In this case the scattering process is determined by the shape of D and and boundary conditions.

The second problem consists in the scattering of acoustic or electromagnetic waves by a penetrable scatterer, where the waves penetrate the obstacle and the interior structure of the obstacle strongly influences the scattering process. If the scatterer is homogeneous, the second problem leads to transmission problems; however if the scatterer is inhomogeneous, we speak of scattering by an inhomogeneous medium.

There exist two main approaches to the simulation of scattering processes. The first treats the problems in the time domain, the second uses a Fourier transform to map the differential equations into the frequency domain and investigate time-harmonic waves. We will follow the second approach.

We will use the letter $\mathcal{D}$ to denote the full scatterer with its physical properties and D to denote the interior of the support of the scatterer in

$\mathbb{R}^m$, $m = 2, 3$. We will always assume that the scatterer is bounded.
Mathematically the behavior of a time-harmonic acoustic wave

$$u(x)e^{-i\omega t}$$

in a homogeneous background medium is governed by the Helmholtz equation

$$\Delta u + \kappa^2 u = 0, \qquad (1.1.1)$$

where $\kappa = \omega/c_0 > 0$ is the wave number of the acoustic wave, w its frequency and c_0 the speed of sound. For scattering of an incident wave u^i by an impenetrable scatterer $\mathcal{D}$ a mathematical model also needs to take into account the behavior of the total field

$$u = u^i + u^s \qquad (1.1.2)$$

at the boundary ∂D of the scatterer. Here u^s denotes the scattered acoustic field. Different boundary conditions are used to model the underlying physical behavior. For a sound-soft scatterer the total field vanishes at the boundary, which leads to the Dirichlet boundary condition

$$u(x) = 0, \quad x \in \partial D. \qquad (1.1.3)$$

For a sound-hard scatterer the Neumann boundary condition

$$\frac{\partial u}{\partial \nu}(x) = 0, \quad x \in \partial D, \qquad (1.1.4)$$

is used, where ν denotes the exterior unit normal vector to the boundary ∂D. The model is completed by the Sommerfeld radiation condition

$$\lim_{r \to \infty} r^{\frac{m-1}{2}} \left(\frac{\partial u^s(x)}{\partial r} - i\kappa u^s(x) \right) = 0, \quad r = |x|, \qquad (1.1.5)$$

uniformly for all directions $x/|x|$ for the scattered field u^s. It physically implies that energy is transported to infinity and it is an important ingredient in ensuring that the physical correct solution of the scattering problem is selected.

To model scattering by inhomogeneous media the pairs of equations (1.1.1), (1.1.3) or (1.1.1), (1.1.4) are replaced by

$$\Delta u + \kappa^2 n(x) u = 0 \qquad (1.1.6)$$

in the whole space or $\mathbb{R}^m \setminus \partial D$. Here

$$n(x) := \frac{c_0^2}{c(x)^2} + i\sigma(x)$$

is the refractive index. The function $c(x)$ is the sound speed in the inhomogeneous medium, the constant c_0 is the sound speed in the homogeneous host medium and the term $\sigma(x) \geq 0$ models absorption. We assume that $c(x) = c_0$ and $\sigma(x) = 0$ outside of the inhomogeneous region D.

For scattering of electromagnetic waves in $\mathbb{R}^3$ the corresponding governing equations are the time-harmonic Maxwell equations

$$\text{curl } E - i\kappa H = 0, \qquad \text{curl } H + i\kappa E = 0 \qquad (1.1.7)$$

for the electric field E and the magnetic field H in a homogeneous medium, where $\kappa := \omega\sqrt{\epsilon_0\mu_0}$ is the wave number and ω the frequency of the time-harmonic wave, ϵ_0 the electric permittivity and μ_0 the magnetic permeability of the host medium. For scattering of an incident electromagnetic field E^i, H^i by a perfect conductor $\mathcal{D}$ the boundary condition

$$\nu(x) \times E(x) = 0, \quad x \in \partial D, \qquad (1.1.8)$$

on the total field

$$E := E^i + E^s \qquad (1.1.9)$$

models the behavior of the electric field at the boundary ∂D of D. The tangential components of the electric field E vanish at the boundary ∂D of the perfect conductor $\mathcal{D}$. The appropriate radiation condition is the Silver-Müller radiation condition

$$\lim_{r\to\infty} (H^s \times x - rE^s) = 0 \qquad (1.1.10)$$

for the scattered electromagnetic field E^s, H^s.

To describe scattering of electromagnetic waves by an inhomogeneous medium the equations (1.1.7) and (1.1.8) are replaced by

$$\text{curl } E - i\kappa H = 0, \qquad \text{curl } H + i\kappa n(x)E = 0, \qquad (1.1.11)$$

where the refractive index

$$n(x) := \frac{1}{\epsilon_0}\left(\epsilon(x) + i\frac{\sigma(x)}{\omega}\right)$$

is defined in terms of the permittivity $\epsilon(x)$ of the inhomogeneous medium, the permittivity ϵ_0 of the homogeneous background medium, the conductivity $\sigma(x)$ and the frequency ω of the wave. The magnetic permeability μ is assumed to be constant. We will also investigate anisotropic problems, where n is replaced by a matrix N.

The radiation conditions (1.1.5) or (1.1.10) together with the governing equations (1.1.1) or (1.1.7) imply the behavior

$$u^s(x) = \frac{e^{i\kappa r}}{r^{\frac{m-1}{2}}}\left\{u^\infty(\hat{x}) + O\left(\frac{1}{r}\right)\right\}, \quad r = |x| \to \infty, \qquad (1.1.12)$$

where $\hat{x} := x/|x|$ and

$$E^s(x) = \frac{e^{i\kappa r}}{r}\left\{ E^\infty(\hat{x}) + O\left(\frac{1}{r}\right)\right\}, \quad r = |x| \to \infty, \qquad (1.1.13)$$

of the scattered acoustic or electric field $u^s(x)$ or $E^s(x)$, respectively. Here u^∞ and E^∞ are known as the acoustic and electromagnetic far field pattern or scattering amplitude. In the acoustic case the far field pattern is a scalar function defined on the unit sphere or unit circle Ω, respectively. In the electromagnetic case the far field pattern is a tangential vector field on Ω.

Direct and inverse scattering problems. For a direct scattering problem the scatterer and the incident field are assumed to be given. The problem is to compute the scattered field or the far field pattern, respectively. Direct scattering problems have been studied for a long time and a number of different approaches for their solution have been developed (see [3], [7], [38], [57], [58], [62], [66], [80], [83], [90]). However, it is still an important problem of current research to develop efficient algorithms for the numerical computation of the scattered field, especially in three dimensions.

A whole range of different inverse problems are of interest in this framework. Given the far field pattern for scattering of plane waves we may try to reconstruct special properties of $\mathcal{D}$ or the full scatterer with all its properties. Different settings for the measurements lead to a variety of practically relevant mathematical and algorithmical problems.

In this work we will focus on the reconstruction of the shape D of the scatterer $\mathcal{D}$, i.e., the scattering domain for obstacle scattering and the support of the inhomogeneity for scattering by an inhomogeneous medium. For a wide range of applications it is not necessary to reconstruct the full behavior and actual values of the refractive index $n(x)$, but it is sufficient to approximately determine the support $\overline{D} := \text{supp}(n-1)$. For example, in nondestructive testing often this information is all that is needed.

As scattering data we will suppose that the far field pattern $u^\infty(\cdot, d)$ or $E^\infty(\cdot, d, p)$ is measured. These notations denote, respectively, the acoustic or electromagnetic far field patterns of incident plane waves, the incident fields given by

$$u^i(x, d) := e^{i\kappa x \cdot d}, \quad x \in \mathbb{R}^m, \qquad (1.1.14)$$

in the acoustic case or

$$\begin{aligned} E^i(x) &= i\kappa(d \times p) \times d\, e^{i\kappa \cdot d} \\ H^i(x) &= i\kappa d \times p\, e^{i\kappa x \cdot d} \end{aligned} \quad , x \in \mathbb{R}^3, \qquad (1.1.15)$$

in the electromagnetic case, where $d \in \Omega$ denotes the direction of incidence and $p \in \Omega$ the polarization of the electromagnetic wave. We will investigate the cases where the far field pattern is given for one, a finite number or all incident plane wave directions, with the far field pattern measured either on the whole unit sphere Ω or at a given finite number n of observation points $\hat{x}_j \in \Omega$, $j = 1, ..., n$.

Basic mathematical problems. The investigation of an inverse problem starts with several basic mathematical questions, which are strongly related to the inverse nature of the task.

1. *Uniqueness.* It has to be asked which data sets uniquely determine the object. Without at least local uniqueness there is no possibility to reconstruct unknown objects by stable algorithms.

2. *Existence.* One can ask whether a solution of the inverse problem exists for a given data set. Errors in the measurements may lead to data for which there is no solution to the inverse problem.

3. *Stability.* Because of errors in the measurements or in the numerical storage of data we need to investigate stability for the reconstruction of D from the given far field data. Mathematically this is the problem of continuity of the nonlinear inverse operator under appropriate assumptions.

4. *Algorithms.* Efficient and stable reconstruction algorithms need to be developed. This leads us to the numerical and algorithmical analysis.

We will now place important results in a historical context and give a brief introduction to the contributions to the above questions which are discussed later in the book in more detail. This includes a sketch of related approaches and a description of differences and similarities.

The ill-posedness of inverse scattering problems. First, consider one of the main features of inverse scattering problems. Let B denote a ball with fixed radius R_e around the origin. Given the *a priori* information that B contains the scatterer $\mathcal{D}$ in its interior, by

$$\mathcal{F} : u^s|_{\partial B} \mapsto u^\infty \qquad (1.1.16)$$

we denote the operator which maps the scattered field $u^s(x)$, $x \in \partial B$, onto its far field pattern $u^\infty(\hat{x})$, $\hat{x} \in \Omega$. Computing $\mathcal{F}$ explicitly, it can be seen to be compact in any reasonable function space, for example from $C(\partial B)$ into $L^2(\Omega)$ (see Chapter 2 for further details). Thus, by functional analytic arguments, the range of the operator $\mathcal{F}$ cannot be the whole space and the inverse $\mathcal{F}^{-1}$ of the operator $\mathcal{F}$ cannot be bounded. This indicates that the inverse scattering problem is an ill-posed problem in the sense of

Hadamard [22], i.e., that in this case the demands of existence and stability are violated.

The problem of existence for given measured data. Consider the problem of existence. We already pointed out that, in general, we cannot expect existence, i.e., we cannot expect that a far field pattern, which has been polluted by measurement error, will any longer be the exact far field pattern corresponding to a scattering problem. The most we can expect is the denseness or completeness of the far field patterns for a given set of incident waves or of scatterers $\mathcal{D}$ in some measurement space, for example in $L^2(\Omega)$. Completeness of the set of far field patterns for the above acoustic and electromagnetic scattering problems for a set of incident plane waves $\{u^i(\cdot, d_n) : n \in \mathbb{N}\}$, where $\{d_n : n \in \mathbb{N}\}$ is dense in the unit sphere Ω, has been investigated in detail by Colton, Kirsch, Kress, Blöhbaum and Päivärinta between 1984 and 1990 (see [8] for further references). Necessary conditions for a function in $L^2(\Omega)$ to be a far field pattern can be given in terms of the decay of its Fourier coefficients with respect to spherical harmonics (see [8], Theorem 2.16). This can be obtained by an expansion of the scattered field outside of the ball B. Further necessary conditions have been given by Müller [63] (see also Colton and Kress [7]) using the theory of entire functions of exponential type. More recently Kirsch ([41], [42]) obtained a characterization of the set of far field patterns for a given scatterer in terms of its series representation with respect to the eigenfunctions of the corresponding far field operator

$$(F\varphi)(\hat{x}) := \int_\Omega u^\infty(\hat{x}, d)\, \varphi(d) ds(d), \quad \hat{x} \in \Omega. \qquad (1.1.17)$$

We will describe these results in Section 7.2. A corresponding method for the reconstruction of the support D of the scatterer $\mathcal{D}$ is described below. To the author's knowledge no general characterization of the set of far field patterns for arbitrary scatterers $\mathcal{D}$ is known.

In this work, we will assume that the given data are either the exact far field data u^∞ for scattering by a scatterer $\mathcal{D}$ or some measured data

$$u_\delta^\infty(\cdot) \in L^2(\Omega) \quad \text{or} \quad u_\delta^\infty(\cdot, \cdot) \in L^2(\Omega \times \Omega)$$

with

$$\|u^\infty(\cdot) - u_\delta^\infty(\cdot)\|_{L^2(\Omega)} \leq \delta \quad \text{or} \quad \|u^\infty(\cdot, \cdot) - u_\delta^\infty(\cdot, \cdot)\|_{L^2(\Omega \times \Omega)} \leq \delta. \quad (1.1.18)$$

We will also study the finite data case, where a finite number of measured data $u^\infty_{(n_o,n_i),\delta} \in L^2(\Omega_{n_o} \times \Omega_{n_i})$ are given with

$$\left(\frac{c_m}{n_o} \frac{c_m}{n_i} \sum_{\hat{x} \in \Omega_{n_o}} \sum_{d \in \Omega_{n_i}} \left| u^\infty(\hat{x}, d) - u^\infty_{(n_o,n_i),\delta}(\hat{x}, d) \right|^2 \right)^{\frac{1}{2}} \leq \delta \qquad (1.1.19)$$

for $\delta \geq 0$ with some constant c_m. Here we assume $(\Omega_n)_{n \in \mathbb{N}}$ to be a sequence of finite subsets Ω_n of Ω, such that Ω_n consists of n elements and for given ϵ we can find n such that the distance

$$d(\hat{x}, \Omega_n) := \inf_{d \in \Omega_n} |\hat{x} - d|$$

is smaller than ϵ for all $\hat{x} \in \Omega$. The left-hand side of (1.1.19) defines a norm $\| \cdot \|_{L^2(\Omega_{n_o} \times \Omega_{n_i})}$. The positive real number δ in (1.1.18) and (1.1.19) is referred to as the data error.

Uniqueness results for reconstructions. The origin of uniqueness results for inverse obstacle scattering problems can be found in the works of Rellich in the 40's. He proved that the far field pattern uniquely determines the (analytic) scattered field in the exterior of the scatterer $\mathcal{D}$ (which result we refer to as Rellich's Lemma). Then, Schiffer (see [57]) showed for the inverse acoustic obstacle scattering problem with Dirichlet boundary condition, that the far field pattern $u^\infty(\hat{x}, d)$, $\hat{x}, d \in \Omega$, for all incident plane waves and for one fixed wave number κ uniquely determines the domain of the scatterer. The corresponding result for the reconstruction of the acoustic refractive index in three dimensions was obtained by Nachman [65], Novikov [67] and Ramm [81], and considerably simplified by Hähner [23], (see also [24]), in 1996. Analogous results for the electromagnetic problems were first obtained by Colton and Päivärinta [11] in 1990, Colton and Kress [8] in 1993, and by Ola, Päivärinta and Somersalo [68] in 1993.

In 1983, Colton and Sleeman [15] investigated the case where the sound-soft scatterer is known to be a subset of a ball with given radius R_e. They showed that the support is determined by a finite number N of incident plane waves depending on R_e. If R_e is small enough, one wave is sufficient to determine the scatterer.

So far it has not been possible to extend Schiffer's approach or the ideas of Colton and Sleeman to the sound-hard scatterer or to the case of an inhomogeneous medium. In 1992, Isakov [34] obtained uniqueness results for penetrable obstacles using different techniques, which were simplified and applied to impenetrable sound-soft and sound-hard scatterers by Kirsch and Kress [47] in 1993. The results could also be successfully transferred to the case of electromagnetic obstacle scattering [8]. Since these ideas will be the starting point of a large part of this work (with contributions to uniqueness, stability, the finite data problem, and reconstruction algorithms), we briefly want to describe the main ingredients.

Consider the scattered acoustic field $\Phi^s(\cdot, z)$ for scattering of a point-source $\Phi(\cdot, z)$ with source point $z \in \mathbb{R}^m \setminus \overline{D}$, where Φ is the standard fundamental solution of the Helmholtz equation in dimension m, $m = 2, 3$. From the sound-soft boundary condition and the singularity of the incident point-source we derive that for a point $x \in \partial D$ we have

$$\Phi^s(x, z) \to \infty, \quad z \to x. \tag{1.1.20}$$

Kirsch and Kress used (1.1.20) to show that if the far field patterns of two scatterers $\mathcal{D}_1$ and $\mathcal{D}_2$ for scattering of plane waves coincide for all directions of incidence $d \in \Omega$, then the domains D_1 and D_2 are the same.

In Chapter 3 we will further develop the techniques of Kirsch and Kress to derive uniqueness of the support D of an inhomogeneous medium n, if n has a jump in one of its derivatives at the boundary of the scatterer $\mathcal{D}$. For the three-dimensional case this also can be obtained from the results of Nachman [65], Novikov [67], and Ramm [81] for the acoustic and from Colton and Päivärinta [12] in the electromagnetic case. In two dimensions, Sun and Uhlmann [86] proved uniqueness of the support of n, if n has a jump at the boundary. They use Fourier techniques, which are different from our approach. Our techniques will be constructive and allow the development of reconstruction methods in Chapter 6.

ϵ-Uniqueness for finite data. Uniqueness results for inverse scattering problems usually assume the full far field pattern on the unit sphere to be given. Often it is assumed that the far field pattern is known for all or a full open set of directions $d \in \Omega$ of the incident plane waves. Since most proofs use Rellich's lemma, the knowledge of the far field pattern at least in an open subset of Ω seems to be necessary to uniquely determine the scattered field u^s.

From a practical perspective it is reasonable to ask the question: What can be said if the far field pattern is given only at a finite number of measurement points and for a finite number of waves? In Chapter 3, we develop a technique to answer this question avoiding the use of Rellich's uniqueness results. It leads to a relaxed concept of uniqueness, a preliminary version of which was first proposed in 1998 (see [73]). We will prove ϵ-uniqueness for the reconstruction of the shape of a scatterer, i.e., that given $\epsilon > 0$ there are $n_o, n_i \in \mathbb{N}$ such that, if for two scatterers $\mathcal{D}_1$ and $\mathcal{D}_2$ the far field patterns for all n_i directions of incidence $d \in \Omega_{n_i}$ coincide at the n_o observations points $\hat{x} \in \Omega_{n_o}$, the Hausdorff distance

$$d(D_1, D_2) := \max \left\{ \sup_{x \in D_1} d(x, D_2), \sup_{x \in D_2} d(x, D_1) \right\} \tag{1.1.21}$$

between the scatterers D_1 and D_2 satisfies the estimate

$$d(D_1, D_2) < \epsilon.$$

Since the concept of ϵ-uniqueness is close to stability, let us postpone further considerations until after discussing the stability question.

Stability estimates. We have already pointed out that inverse scattering problems are ill-posed problems, i.e., that for the inverse of the non-linear scattering operator in general we do not have stability. There exist two main approaches to restore stability results for this ill-posed problem.

The first approach consists of a modification of the norm used for the far field pattern. In inverse scattering it was used for example in 1990 by Stefanov [85] to study stability for inverse scattering by a medium and has been extended by Hähner [24] in 1998 to electromagnetic and elastic wave scattering. In principle they consider a space X of functions on Ω with a very strong norm $\| \cdot \|_X$ involving all derivatives of functions $\varphi \in C^\infty(\Omega)$, such that the inverse $\mathcal{F}^{-1} : X \to C(\partial B)$ of (1.1.16) becomes a bounded operator. More recently these ideas have been extended by Hähner and Hohage to prove logarithmic stability estimates for the reconstruction of the refractive index with respect to L^2-norms in the space of the far field patterns, see [25]. But note that so far these results do not include the stability estimates for the support of inhomogeneous medium scatterers as discussed below.

Another approach is the use of *a priori* bounds on the set of fields or objects to be determined (see for example [37]). This approach has been applied to scattering theory by Isakov [34], [35] (see also [36]). We will use the well known fact that the inverse of a continuous mapping is continuous if it is defined on a compact subset of a Banach space. Thus with appropriate restrictions on the set of scatterers, stability can be restored and stability estimates obtained.

Isakov's restrictions mainly consist of a uniform bound on the $C^{2,\alpha}$-norm of all boundaries in a special parametrization. For the reconstruction of the shape of a sound-soft scatterer from the knowledge of the far field pattern for one incident wave, Isakov derives a double-logarithmic estimate

$$d(D_1, D_2) \; < \; C \Big(\ln \Big| \ln \|u_1^\infty(\cdot) - u_2^\infty(\cdot)\|_{C(\Omega)} \Big| \Big)^{-\gamma} \qquad (1.1.22)$$

for the Hausdorff distance $d(D_1, D_2)$ of the domains D_1, D_2 with positive constants C, γ depending on a bound for the $C^{2,\alpha}$-norm of the boundary. So far Isakov's techniques could not be used to treat other boundary conditions or electromagnetic scattering problems.

In Chapter 3 we will pursue the idea of imposing appropriate restrictions on the set of scatterers under consideration. With techniques different from Isakov we will be able to derive stability estimates for the reconstruction of the shape of either a penetrable or impenetrable scatterer from the knowledge of the far field patterns for all incident plane waves. More explicitly we prove an estimate of the form

$$d(D_1, D_2) \leq F\Big(\|u_1^\infty(\cdot, \cdot) - u_2^\infty(\cdot, \cdot)\|_{L^2(\Omega) \times L^2(\Omega)}\Big), \qquad (1.1.23)$$

where F is a function with the property

$$F(\delta) \to 0, \quad \delta \to 0, \qquad (1.1.24)$$

which can be computed according to some *a priori* knowledge on the class of scatterers. For the convex hulls $\mathcal{H}(D)$ of the shape D of scatterers $\mathcal{D}$ we derive a logarithmic estimate

$$d\Big(\mathcal{H}(D_1), \mathcal{H}(D_2)\Big) \leq C \left| \ln \|u_1^\infty(\cdot, \cdot) - u_2^\infty(\cdot, \cdot)\|_{L^2(\Omega) \times L^2(\Omega)} \right|^{-\gamma} \quad (1.1.25)$$

with constants C and γ.

Our technique is inspired by the uniqueness proof of Isakov, Kirsch and Kress. The first step is the explicit estimation of the behavior of $\Phi^s(z, z)$ for $z \to \partial D$. As a second step we develop a method for the approximate reconstruction of $\Phi^s(z, z)$ in the exterior of the domain

$$D_\rho := \{y \in \mathbb{R}^m : d(y, D) < \rho\} \qquad (1.1.26)$$

with some small parameter $\rho > 0$ from the scattering data $u^\infty(\hat{x}, d)$ for all $\hat{x}, d$ in Ω. Estimating the bound of the approximate reconstruction operator $Q : L^2(\Omega) \times L^2(\Omega) \to C(B \setminus D_\rho)$ and using the fact that $\Phi^s(z, z)$ is large only in a neighborhood of the boundary, stability estimates will be obtained in Chapter 3. For the acoustic sound-soft and sound-hard scatterer the results can be found in [70]. Similar stability results for the reconstruction of the support of media and for electromagnetic scattering problems will be derived in Chapter 3.

At this point we would like to relate these results to some demands on the degree of ill-posedness of an inverse problem formulated by Fritz John [37] in 1960. For purposes of computation John demands Hölder continuous dependence of a problem on the data. Here, for the reconstruction of the domains, we obtained logarithmic continuity, which is a typical type of estimate for continuing solutions of the wave equation in space-like directions. As shown in [70], for the reconstruction of the scattered field u^s on fixed compact subsets U of the open exterior of the convex hull $\mathcal{H}(D)$ of D

$$\left\| u_1^\infty - u_2^\infty \right\|_{L^2(\Omega)} \leq \delta$$

yields the estimate

$$\left|u_1^s(x) - u_2^s(x)\right| \le \alpha\delta^{\overline{|\ln(-\gamma\ln(\delta))|}^{\frac{\beta}{}}}, \quad x \in U,$$

with constants $\alpha, \beta, \gamma > 0$. This can be proven with the same techniques which we will use in Chapter 3. These estimates come close to Hölder continuity demanded by John and are reflected by the numerical results of Chapters 5 and 6.

Stability for the case of finite data: ϵ-stability. With the help of the stability estimates it is not difficult to derive related statements for the case of finite data.

We will work with the same assumptions as for stability or ϵ-uniqueness, i.e., a uniform bound on the C^2-norm of the boundaries (and corresponding assumptions on the uniform smoothness of the refractive index n), to derive a uniform bound for the far field patterns in $C^1(\Omega \times \Omega)$. This bound can be used to relate the distance of two far field patterns at a finite number of points to the distance of the full far field patterns. Then for the case of finitely many measurements we derive ϵ-uniqueness and a modified stability statement, which we will refer to as ϵ-stability.

Consider a simple example for the derivation of ϵ-uniqueness. Given $\epsilon > 0$ we can use (1.1.25) to obtain a $\delta > 0$ such that

$$\|u_1^\infty(\cdot, \cdot) - u_2^\infty(\cdot, \cdot)\|_{L^2(\Omega)\times L^2(\Omega)} \le \delta \tag{1.1.27}$$

implies

$$d\Big(\mathcal{H}(D_1), \mathcal{H}(D_2)\Big) \le \epsilon. \tag{1.1.28}$$

Now given δ, we choose $n \in I\!\!N$ sufficiently large such that with the help of the bound on $\|u_j^\infty(\cdot, \cdot)\|_{C^1(\Omega\times\Omega)}$ for $j = 1, 2$, the equation

$$u_1^\infty(\hat{x}, d) = u^\infty(\hat{x}, d) \text{ for all } \hat{x}, d \in \Omega_n \tag{1.1.29}$$

yields (1.1.27). Thus given ϵ we may choose $n \in I\!\!N$ such that (1.1.29) implies (1.1.28), i.e., we have proven the statement of ϵ-uniqueness for the convex hulls of the scattering domains as a simple consequence of stability.

Now, we describe the concept of ϵ-stability. Since in general we do not have uniqueness, for a finite data set we will not be able to obtain stability. More explicitly, we cannot obtain a function $F(\delta)$ which satisfies (1.1.24) and an estimate of the type (1.1.23) when the data error is considered at finitely many points. But for given $\epsilon > 0$ it is possible to find $n_o, n_i \in I\!\!N$ and a function $F_{(n_o, n_i)}(\delta)$, such that F has the behavior

$$\limsup_{\delta \to 0} F_{(n_o, n_i)}(\delta) \le \epsilon \tag{1.1.30}$$

and the domains D_1 and D_2 satisfy the estimate

$$d(D_1, D_2) \leq F_{(n_o, n_i)}\Big(\|u^\infty_{1,(n_o,n_i)} - u^\infty_{2,(n_o,n_i)}\|_{L^2(\Omega_{n_o} \times \Omega_{n_i})}\Big) \qquad (1.1.31)$$

with $\| \cdot \|_{L^2(\Omega_{n_o} \times \Omega_{n_i})}$ given by (1.1.19) and

$$u^\infty_{j,(n_0,n_i)} := u^\infty_j\Big|_{\Omega_{n_o} \times \Omega_{n_i}} \in L^2(\Omega_{n_o} \times \Omega_{n_i})$$

for $j = 1, 2$. We call a statement of this form ϵ-stability. We will prove ϵ-stability for the reconstruction of the shape of the scatterer for the scattering problems described above and explicitly study the behavior of the function $F_{(n_o, n_i)}(\delta)$ for the reconstruction of the convex hull of a scatterer.

Three main categories of reconstruction methods. Consider now the problem of the actual reconstruction of the shape of the unknown scatterer. Basically three different types of reconstruction methods have been developed. We will summarize their main features and use them as a background to explain our results.

The first category and Newton's method. The first approach to consider is the inverse problem as a nonlinear ill-posed operator equation and adapt iterative methods of gradient- or Newton-type to solve this equation. For inverse obstacle scattering this approach mainly relies on the Fréchet differentiability or domain derivative of the scattered field with respect to variations of the boundary of the scatterer and a characterization of the derivative as a corresponding boundary value problem. Results for interior boundary value problems were obtained in 1980 by Simon [84] with the help of the implicit function theorem, for the scattering problems in 1993 by Kirsch [44] using variational methods, and in 1994 by the author [78] by means of integral equations.

For a discussion of the large number of papers on the numerical implementation of these type of methods, (for example Murch, Tan and Wall [64], Roger [82], Tobocman [88], Wang and Chen [89], Kirsch [44] and [45], Kress and Rundell [54], [55] and [56], Kress [50], [52] and [53], Mönch [61], Hohage [30], Hanke, Hettlich and Scherzer [26], and Hettlich [27]) we refer to [28] and [31], see also [8]. In Section 5.4 we will study the convergence of a regularized Newton scheme in inverse acoustic scattering and relate the Newton method to the point-source method, which belongs to the second category of inversion schemes.

The second category and the point-source method. The second category in principle splits the inverse scattering problem into a linear ill-posed part to reconstruct the scattered field in the exterior of the scatterer and a nonlinear well-posed part to find the boundary of the scatterer or the refractive index using the boundary condition or the partial differential equation, respectively.

In Chapter 5, with the point-source method, we describe a method of this second category developed by the author since 1995. Different steps in this development can be found in [74], [75] and [76]. Other examples of methods which belong to the second category are the optimization methods proposed by Colton and Monk in 1985 and by Kirsch and Kress in 1986, both described in [8].

The main aim of the point-source method is the explicit construction of a kernel $g(z, d)$, such that in the domain $B \setminus D_\rho$, where D_ρ is given by (1.1.26) and B by (1.1.16), the scattered field u^s is approximated in the form $A u^\infty$ with a linear integral operator

$$(A\varphi)(z) := \int_\Omega g(z, \hat{x})\varphi(\hat{x}) \, ds(\hat{x}), \quad z \in B. \tag{1.1.32}$$

To this end the far field pattern $\Phi^\infty(\cdot, z)$ for incident point-sources $\Phi(\cdot, z)$ with source-point z is considered. Given some *a priori* knowledge on the size of the scatterer, the kernel g will be constructed in the following three steps.

1. An approximation for a point-source by a superposition of plane waves

$$\Phi(x, z) \approx \int_\Omega e^{i\kappa x \cdot d} \tilde{g}(z, d) \, ds(d), \quad x \in \overline{D}, \ z \in B \setminus D_\rho \tag{1.1.33}$$

is computed.

2. Passing to the far field patterns, an approximation for the far field pattern due to point-sources by a superposition of the far field patterns of plane waves

$$\Phi^\infty(\hat{x}, z) \approx \int_\Omega u^\infty(\hat{x}, d)\tilde{g}(z, d) \, ds(d), \quad \hat{x} \in \Omega, \ z \in B \setminus D_\rho \tag{1.1.34}$$

is obtained.

3. Using the mixed reciprocity relation $\Phi^\infty(\hat{x}, z) = \gamma u^s(z, -\hat{x})$, the far field reciprocity relation $u^\infty(\hat{x}, d) = u^\infty(-d, -\hat{x})$ and the substitution $d \to -d$, an approximation

$$u^s(z, -\hat{x}) \approx \int_\Omega u^\infty(d, -\hat{x}) \left\{ \frac{1}{\gamma} \tilde{g}(z, -d) \right\} \, ds(d), \quad \hat{x} \in \Omega, \ z \in B \setminus D_\rho \tag{1.1.35}$$

for the scattered field u^s from its far field pattern u^∞ is derived. Here, $u^s(\cdot, d)$ denotes the scattered field for an incident plane wave with direction of incidence $d \in \Omega$ and $u^\infty(\cdot, d)$ is the corresponding far field pattern.

Using rotations and translations of the approximating functions, the computation of g can be performed efficiently. Given the reconstruction of the scattered field u^s, parts of the unknown boundary ∂D of D can be found using the total field $u = u^i + u^s$ and the boundary condition. For electromagnetic waves a corresponding operator will be constructed (see also [75]).

Note that some similarities exist between the construction procedure of the point-source method and the Backus-Gilbert method or mollifier methods [16], [43], [59].

The third category, the method of singular sources and linear sampling methods. Since 1996 methods for the reconstruction of the shape of a scatterer have been developed, which are based on characterizations of the boundary of the scatterer independent of its physical properties. For an algorithm of category III the boundary condition or physical properties of the scatterer do not need to be known. The independence of a reconstruction method on the physical properties of the scatterers is of great practical importance, since in many cases knowledge about these properties of the objects to be reconstructed is not available. We will introduce two different approaches: the method of singular sources and linear sampling methods.

A linear sampling method was proposed in 1996 by Colton and Kirsch [6]; see also Colton and Monk [10] and Colton, Piana, and Potthast [13] for a formulation of the mathematical basis and numerical realization of the method. The idea is to characterize the boundary ∂D of a scatterer D by the behavior of the solution $g = g(z, \cdot) \in L^2(\Omega)$ of a linear integral equation of the first kind

$$(Fg)(\hat{x}) = e^{-i\kappa z \cdot \hat{x}}, \quad \hat{x} \in \Omega, \qquad (1.1.36)$$

for $z \in D$, where F is the far field operator

$$(Fg)(\hat{x}) := \int_\Omega u^\infty(\hat{x}, d)g(d)\, ds(d), \quad \hat{x} \in \Omega. \qquad (1.1.37)$$

Examining either an interior boundary value problem or an interior transmission problem, in [13] (see also [8]) it is shown that there exists an approximate solution of (1.1.36) with

$$\|g(z, \cdot)\|_{L^2(\Omega)} \to \infty \quad \text{and} \quad \|v_g(z, \cdot)\|_{L^2(D)} \to \infty \quad \text{for} \quad z \to \partial D, \qquad (1.1.38)$$

where v_g denotes the Herglotz wave function

$$v_g(x) := \int_\Omega e^{i\kappa x \cdot d} g(d) \, ds(d), \quad x \in \mathbb{R}^m. \qquad (1.1.39)$$

Colton and Kirsch propose to compute a regularized approximate solution of (1.1.36) on a grid containing D. The domain D then can be found as the set of points where $\|g(z,\cdot)\|_{L^2(\Omega)}$ is large. Numerical experiments can be found in [5], [6], [10], and [13].

Colton, Piana, and Potthast [13] also applied Morozov's discrepancy principle for the solution of (1.1.36) and used either $\|g(z,\cdot)\|$ or the values of the regularization parameter $\alpha(z)$ to determine the shape D of the scatterer $\mathcal{D}$. Numerical results using this method can be found in [5].

We will introduce the linear sampling method and present the main results in Chapter 7, studying the method for the reconstruction of impenetrable obstacles and inhomogeneous isotropic, orthotropic, and anisotropic media.

In 1998 and 1999, Kirsch [41], [42] was able to derive a characterization of the shape D of the scatterer $\mathcal{D}$ for acoustic scattering in the case of obstacles or non-absorbing media, i.e., a real-valued refractive index n. Kirsch showed that the domain D is the set of points z, where the equation

$$(F^*F)^{1/4} g(\hat{x}) = e^{-i\kappa z \cdot \hat{x}}, \quad \hat{x} \in \Omega, \qquad (1.1.40)$$

is solvable. This characterization holds both for scattering by obstacles [42] or for a non-absorbing medium [41]. The support of the scatterer can then be found as above, computing approximate regularized solutions of (1.1.40) and using either the norm $\|g(z,\cdot)\|$ of the solution or the size of the regularization parameter $\alpha(z)$, which is chosen according to Morozov's discrepancy principle. A comparison of numerical results for the method of Colton and Kirsch (1.1.36) and the version proposed by Kirsch (1.1.40) can be found in [5]. More recently, Kirsch managed to extend his approach to more general refractive indices by a reformulation into an optimization problem, see [40]. We will introduce the main ideas of Kirsch in Section 7.2 and outline a proof for his result in the case of obstacle scattering.

In Chapter 6, we propose a method of singular sources for the reconstruction of the support of obstacles or scattering media from the knowledge of the far field pattern $u^\infty(\hat{x}, d)$, $\hat{x}, d \in \Omega$. It is based on the ideas used in the uniqueness and stability proofs. For the reconstruction of obstacles in the case of acoustic waves theoretical and numerical results can be found in [70].

We want to sketch the main ideas of this method. As described above, the boundary ∂D of an obstacle D is the set of points where the scattered field $\Phi^s(z, z)$ for incident point-sources becomes singular. We construct a kernel $g_\tau(x, d)g_\eta(z, \tilde{d})$, such that, for $x, z \in B \setminus D_\rho$, an approximation for the field $\Phi^s(x, z)$ is obtained in the form $(Qu^\infty)(x, z)$ with the bounded linear integral operator Q defined by

$$(Q\varphi)(x, z) := \int_\Omega \int_\Omega g_\tau(x, d)g_\eta(z, \tilde{d})\varphi(d, \tilde{d}) \, ds(d)ds(\tilde{d}), \quad x, z \in B. \quad (1.1.41)$$

The boundary ∂D is found as the set of points where $(Qu^\infty)(z, z)$ is large. We investigate the method both for the reconstruction of obstacles and the support of media and give numerical examples.

The methods of the second and the third category have important differences at a fundamental level. Methods of the second category use the scattered field and the boundary condition to determine the scatterer. The boundary condition does not need to be known for the reconstructions with methods of the third category. The missing knowledge, as we show and discuss in Chapter 6, leads to a greater ill-posedness of the inverse scattering problem.

Contents. We split our presentation into seven chapters.

- Section 1.1 of this chapter has already been used for a survey about inverse acoustic and electromagnetic scattering theory. Section 1.2 serves to present basic definitions and tools for further use.

- In Chapter 2 we study the solutions to the direct scattering problems and derive properties on which our investigation of the inverse problems will be based. The main results of Chapter 2 will be uniform bounds for integral operators and scattering maps and estimates for the behavior of the scattered field $\Phi^s_{\mu,q}(z, z)$ for incident multipoles of order μ and polarization $q \in \Omega$.

- The themes of Chapter 3 are uniqueness and stability for the reconstruction of the shape of a scatterer. We first derive uniqueness results from the knowledge of the full far field patterns $u^\infty(\hat{x}, d)$, $\hat{x}, d \in \Omega$. In a second part we develop a technique to derive stability estimates for the reconstruction of the shape of both impenetrable and penetrable scatterers from the knowledge of the far field patterns for all incident plane waves.

- The finite data case is investigated in Chapter 4. We investigate the question of uniqueness if the far field patterns are known only for a finite number of observation points and a finite number of incident plane waves. For this situation we propose a concept which we call ϵ-uniqueness: given ϵ there are numbers n_o, n_i such that the far field patterns for n_i directions of incidence measured at n_o observations directions determine the unknown shape up to an error ϵ in the Hausdorff distance between the domains.

In a second part stability for a finite set of measured data is studied. For this case we propose a concept of ϵ-stability: given ϵ there are numbers n_o, n_i and a function $F_{(n_i, n_o)}$, such that (1.1.30) and (1.1.31) are satisfied.

- In Chapter 5 a point-source method is introduced for the reconstruction of a scattered field u^s from its far field pattern u^∞ and the construction of the shape of an unknown impenetrable scatterer $\mathcal{D}$. We explicitly construct a family of bounded linear integral operators (1.1.32) for the reconstruction of u^s and prove error estimates and convergence to the true scattered field. Numerical examples for reconstructions in two and three dimensions are given. We also relate the point-source method to Newton's method for domain reconstruction and prove convergence of a regularized Newton scheme.

- A method for the reconstruction of the shape of both impenetrable and penetrable scatterers is proposed in Chapter 6. We call it the method of singular sources, since it uses the singular behavior of the scattered fields $\Phi^s_{\mu,q}(z, z)$ of multipoles, if the source point z of the incident multipole $\Phi_{\mu,q}(\cdot, z)$ tends to the boundary of the scatterer. Some numerical examples in two dimensions demonstrate the applicability of the ideas.

- Chapter 7 is dedicated to linear sampling methods. We introduce the method including full proofs for the reconstruction of inhomogeneous isotropic, orthotropic and anisotropic media. In Section 7.2 we also present the spectral theory of the far field operator which was developed by Kirsch to prove convergence of a modified linear sampling method for domain reconstructions in inverse scattering.

1.2 Basic definitions and tools.

In this part we introduce basic definitions, notations, and theorems from analysis, functional analysis, scattering, and potential theory for later use. An introduction to these areas can be found in [8], [29], [43], [51], and [87].

Direct and inverse scattering theory investigates scattering of acoustic, electromagnetic or elastic waves from scatterers in $I\!\!R^m$, $m = 2, 3$. We will restrict our attention to bounded scatterers with a sufficiently smooth boundary to introduce and study properties of the solutions and inversion schemes.

We consider our assumptions of smoothness and boundedness as a first and important step towards more complicated and realistic problems. Here, we can study all characteristic problems and difficulties of inverse scattering problems in a well-defined setting. Note that even with rather strong assumptions we cover a wide range of applications.

Domains, balls, cylinders.
By D we denote a bounded open set in $I\!\!R^m$, $m \in I\!\!N$, with boundary ∂D and closure $\overline{D}$, such that the exterior of D in $I\!\!R^m$ is connected. $B_r(x)$ is the open ball with radius r and center x in $I\!\!R^m$. The lower half plane is given by

$$H := \{x = (x_1, ..., x_m) : x_m \leq 0\}$$

and we define $H_r := B_r(0) \cap H$.
Let $Z_{a,r}$ be the open finite cylinder

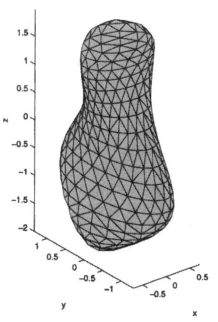

Figure 1.1: Triangulation of scatterer with smooth boundary.

$$Z_{a,r} := \Big\{ x = (x_1, ..., x_m) \in I\!\!R^m, \sqrt{x_1^2 + ... + x_{m-1}^2} < r, |x_m| < a \Big\}. \tag{1.2.1}$$

We use the notation $Z_{a,r}(x, p)$ for the cylinder defined by (1.2.1) in the coordinate system with origin x and the x_m-axis given by $p \in \Omega$.

Function spaces. For the treatment of scattering problems using boundary integral equations we will need to work both with Hölder spaces and with Sobolev spaces. Sobolev spaces provide a Hilbert-space setting which is of advantage for many parts of our analysis, whereas the Hölder-space analysis is a powerful tool for the treatment of boundary integral operators on the boundary of smooth scattering domains.

The spaces of continuous or l-times continuously differentiable functions on D or ∂D are denoted by $C(D)$, $C^l(D)$ or $C(\partial D)$, $C^l(\partial D)$, respectively. The space of l-times Hölder continuously differentiable functions with Hölder coefficient α is $C^{l,\alpha}(D)$ or $C^{l,\alpha}(\partial D)$, respectively. For a multi-index

$$\gamma := (\gamma_1, ..., \gamma_m)$$

we define

$$|\gamma| := \sum_{j=1}^m \gamma_j. \tag{1.2.2}$$

The l-th derivatives of a function $f \in C^l(D)$ are given by

$$f^{(\gamma)} := \frac{\partial^{|\gamma|} f}{\partial^{\gamma_1} x_1 ... \partial^{\gamma_m} x_m}$$

for all $\gamma \in \mathbb{N}_0^m$ with $|\gamma| = l$. We will need the space $L^2(D)$ of square-integrable functions on D and the Sobolev spaces $H^l(D)$, which are defined as the closure of $C^l(\overline{D})$ with respect to the norm

$$\|f\|_{H^l(D)} := \sum_{|\gamma| \le l} \|f^{(\gamma)}\|_{L^2(D)}. \tag{1.2.3}$$

Parametrizations of boundaries. Given $m \in \mathbb{N}_0$ and $\alpha \in [0,1]$ the boundary ∂D is said to be of class $C^{l,\alpha}$, if for each point $x \in \partial D$ there is an open set $V \in \mathbb{R}^m$ with $x \in V$ and a bijective mapping $\psi \in C^{l,\alpha}(B_1(0))$ such that $\psi(B_1(0)) = V$ and $\psi(H_1) = V \cap D$. Since ∂D is compact, we can always find a finite number of such domains V which cover ∂D. A parametrization of $V \cap \partial D$ is given by $\psi|_{U_1}$ with

$$U_1 := \{x \in B_1(0) : x_m = 0\}. \tag{1.2.4}$$

Then $\{U_1, \psi, V \cap \partial D\}$ is called a set of local coordinates for ∂D. For domains of class $C^{l,\alpha}$ with $l \ge 1$ and $\alpha \in [0,1]$ let $\nu(x)$ be the exterior unit normal vector to the boundary ∂D of D at the point $x \in \partial D$. We use special local coordinates for ∂D. For a point $x \in \partial D$ we can find a coordinate system K_x with origin x and the x_m-axis given by $\{x + h\nu(x) : h \in \mathbb{R}\}$. In this special coordinate system in a neighborhood $Z_{a,r}$ of 0 with $r, a > 0$, a parametrization of $\partial D \cap Z_{a,r}$ is given by

$$\partial D \cap Z_{a,r} = \left\{ \left(t_1, ..., t_{m-1}, f(t_1, ..., t_{m-1}) \right) : (t_1, ..., t_{m-1}) \in B_r(0) \right\}$$
(1.2.5)

with a mapping $f \in C^{l,\alpha}(B_r(0))$ defined on the open set $B_r(0) \subset \mathbb{R}^{m-1}$. The function $f \in C^{l,\alpha}(B_r(0))$ is uniquely determined up to rotation and we have

$$D \cap Z_{a,r} = \{(t_1, ..., t_m) \in Z_{a,r} : t_m \leq f(t_1, ..., t_{m-1})\}. \tag{1.2.6}$$

Classes of domains. In Chapters 3 and 4 we will investigate stability of reconstructions and the case of finite data. To this end we need various properties of special functions and integral operators to be valid uniformly for special sets of domains under consideration. Here, we specify classes of domains with mainly three different appropriate conditions. First, we need the domains to be a subset of a fixed large ball (boundedness). Second, we assume that the smoothness of the domain is uniform in the sense that we obtain uniform parametrizations. Third, we need an infinite cone condition which enables us to reach each point of the boundary ∂D of a scatterer by a fixed infinite cone which is a subset of the exterior of D. This cone will be used to obtain uniform reconstructions of the scattered fields and also excludes several bad cases of sequences of domains where different parts of the boundaries converge towards each other.

DEFINITION 1.2.1 (**Classes of domains.**) *Given constants R_e, r_0, a_0, $l \in \mathbb{N}$, $\alpha \in [0,1]$, $C_0 > 0$ and $\beta_e > 0$ we define the class*

$$\mathcal{A} = \mathcal{A}(R_e, r_0, a_0, l, \alpha, C_0) \tag{1.2.7}$$

of domains in $\mathbb{R}^m$, which satisfy the following conditions:

  *1. (**Boundedness.**) For all $D \in \mathcal{A}$ we have*

$$D \subset B_{R_e}(0). \tag{1.2.8}$$

  *2. (**Uniform smoothness.**) For each point $x \in \partial D$ and the special coordinate system K_x defined above there exists $r = r_0$, $a = a_0$ and some function $f \in C^{l,\alpha}(B_{r_0}(0))$, such that the equations (1.2.5) and (1.2.6), and the estimate*

$$\|f\|_{C^{l,\alpha}(B_r(0))} \leq C_0 \tag{1.2.9}$$

are satisfied. Without loss of generality we will assume that $a_0 \geq 2r_0$.

3. **(Exterior cone condition.)** *For each $x \in \mathbb{R}^m \setminus D$ there is a cone*

$$\mathrm{co}(x, p, \beta_e) := \left\{ y \in \mathbb{R}^m : \frac{y-x}{|y-x|} \cdot p \geq \cos(\beta_e) \right\} \tag{1.2.10}$$

with direction $p \in \Omega$, and opening angle β_e in the exterior $\mathbb{R}^m \setminus D$ of D.

In contrast to the theory of partial differential equations, where cone conditions are used to describe the regularity of the boundary of a domain, here the infinite cone condition is a geometrical condition.

It can be seen as a condition to limit non-convexity and, simultaneously, allow scatterers consisting of several separate components. It guarantees that each point x on the boundary ∂D can be reached by an exterior infinite cone $\mathrm{co}(x, p, \beta_e)$ with a direction $p \in \Omega$ depending on x and a fixed given opening angle β_e. We use the notation $B := B_{R_e}(0)$.

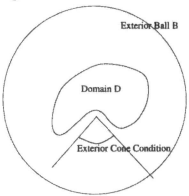

Figure 1.2: Boundedness and exterior cone condition.

To work with the class $\mathcal{A}$ of domains we need to note some of its properties.

THEOREM 1.2.2 *For parameters $R_e, r_0, a_0 > 0$, $l \geq 2$, $\alpha \in [0, 1]$ and $C_0 > 0$ we obtain for the class of domains $\mathcal{A} = \mathcal{A}(R_e, r_0, a_0, l, \alpha, C_0)$ the following properties.*

1. *There is a radius $r_i = r_i(r_0, a_0, C_0)$, such that for each domain $D \in \mathcal{A}$ and each point $x \in \partial D$ we have*

$$B_{r_i}(x + \nu(x)r_i) \subset \mathbb{R}^m \setminus \overline{D} \tag{1.2.11}$$

and

$$B_{r_i}(x - \nu(x)r_i) \subset D. \tag{1.2.12}$$

2. *Each domain $D \in \mathcal{A}$ is a union of balls with radius r_i.*

3. *For each domain $D \in \mathcal{A}$ and each point $x \in \partial D$ we have*

$$x + \nu(x)s \notin D, \quad s \in [0, r_i] \tag{1.2.13}$$

and

$$x - \nu(x)s \in D, \quad s \in (0, r_i]. \tag{1.2.14}$$

4. *There is a number $L_1 = L_1(R_e, r_0, a_0, C_0) \in \mathbb{N}$, such that for each domain D in $\mathcal{A}$ the boundary ∂D of D is piecewise parametrized by at most L_1 twice continuously differentiable injective mappings*

$$\psi_j : B_{r_0}(0) \to \mathbb{R}^m, \quad j = 1, ..., L_1, \tag{1.2.15}$$

of the form (1.2.5) with a norm (1.2.9) bounded uniformly for all domains $D \in \mathcal{A}$.

5. *There is a number $L_2 = L_2(R_e, r_0, a_0, C_0)$, such that for every scatterer D in $\mathcal{A}$ the domain D is parametrized by L_2 injective mappings*

$$\psi_j : Z_{a, r_0} \to \mathbb{R}^m, \quad j = 1, ..., L_2, \tag{1.2.16}$$

where ψ_j is an element of $C^{l, \alpha}(Z_{a, r_0})$, $j = 1, ..., L_2$, with a norm bounded uniformly for all domains $D \in \mathcal{A}$.

Proof. Given $x \in \partial D$ the property 1 is obtained with the help of the special coordinate system K_x and the bound C_0 on the norm

$$\|f\|_{C^{2,0}(B_{r_0}(0))}$$

by elementary calculations as follows. We only need to consider the line $x_2 = 0$ in a neighborhood of 0. Here the boundary of $B_{r_i}(0 + \nu(0)r_i)$ is given by

$$g(x_1) := r_i - \sqrt{r_i^2 - x_1^2}, \quad 0 \le x_1 < r_i.$$

For $r_i < \frac{1}{C_0}$ we estimate the derivative of g by

$$g'(x_1) = \frac{x_1}{\sqrt{r_i^2 - x_1^2}} \ge \frac{x_1}{r_i} > C_0 x_1, \quad 0 \le x_1 < r_i. \tag{1.2.17}$$

From

$$\frac{\partial f(t, 0)}{\partial t}(x_1, 0) = \int_0^{x_1} \frac{\partial^2 f(t, 0)}{\partial t^2} dt$$

for the derivative of f we derive

$$\left| \frac{\partial f(t, 0)}{\partial t}(x_1, 0) \right| \le C_0 x_1, \quad 0 \le x_1 < r_i. \tag{1.2.18}$$

Thus for the derivatives of g and f from (1.2.17) and (1.2.18) we obtain the estimate

$$g'(x_1) > \left| \frac{\partial f(x_1, 0)}{\partial x_1}(x_1) \right|, \quad 0 \leq x_1 < r_i,$$

which yields $g(x_1) > f(x_1, 0)$, $0 \leq x_1 < r_i$, and thus property 1.

Properties 2 and 3 are immediate consequences of property 1. Property 4 can be obtained by compactness of

$$\overline{B_{R_*}(0)} \subset \mathbb{R}^m, \tag{1.2.19}$$

since there is a finite number L_1 of balls $B_{r_0}(x)$, $x \in B_{R_*}(0)$, which cover $B_{R_*}(0)$ and for each ball $B_{r_0}(x)$ the set $\partial D \cap B_{r_0}(x)$ is parametrized by special local coordinates of the form (1.2.5) with norm bounded by (1.2.9).

To prove 5, we first proceed as in 4 and obtain L_1 local coordinate systems K_{x_j} and corresponding local coordinates of the form (1.2.5) covering ∂D. For the local coordinates K_{x_j} we obtain a mapping $\psi_{1j} : Z_{a,r_0} \to Z_{a,r_0} \cap D$ by

$$(t_1, ..., t_m) \mapsto \left(t_1, ..., t_{m-1}, \frac{t_m + a}{2a} \Big[f(t_1, ..., t_{m-1}) + a \Big] - a \right).$$

For this mapping from (1.2.9) we derive $\|\psi_{1j}\|_{C^{l,\alpha}(Z_{a,r_0})} \leq C$ with some constant C uniformly for $D \in \mathcal{A}$. The set

$$G := D \setminus \left(\bigcup_{j=1}^{L_1} \psi_{1j}(Z_{a,r_0}) \right)$$

is a compact subset of D with

$$d(G, \partial D) \geq \tau > 0 \tag{1.2.20}$$

uniformly for all domains $D \in \mathcal{A}$.

Second, there is a finite number $\tilde{L}$ of cylinders $Z_{\tau/3, \tau/3}(y_j, p_j)$, $y_j, p_j \in \mathbb{R}^m$, covering $B_{R_*}(0)$. We choose those cylinders which are contained in D. Because of (1.2.20) they cover G. Using translation, rotation and multiplication with a diagonal matrix with diagonal terms $\tau/(3r_0)$ and $\tau/(3a)$ we obtain continuously differentiable parametrizations $\psi_{2j} : Z_{a,r_0} \to Z_{\tau/3, \tau/3}(y_j, p_j)$, $j = 1, ..., \tilde{L}$, such that

$$G \subset \left(\bigcup_{j=1}^{L_2} \psi_{2j}(Z_{a,r_0}) \right).$$

The proof is now completed with $L_2 := L_1 + \tilde{L}$ by combining steps one and two.   □

When working with the exterior cone condition the following technical lemma will be useful.

LEMMA 1.2.3 *Given the class $\mathcal{A}$ of domains, there is $0 < \beta_0 \leq \beta_e$ and $\rho_0 > 0$ such that all domains*

$$D_\rho := \{ y \in \mathbb{R}^m : d(y, D) < \rho \} \qquad (1.2.21)$$

with $D \in \mathcal{A}$ and $0 \leq \rho \leq \rho_0$ satisfy the exterior cone condition with angle β_0.

Proof. We will show that there is $0 < \beta_0 < \beta_e$, such that for each point x in $\mathbb{R}^m \setminus D_\rho \subset \mathbb{R}^m \setminus D$ and each cone $\text{co}(x, p, \beta_e) \in \mathbb{R}^m \setminus D$ the cone $\text{co}(x, p, \beta_0)$ satisfies

$$\text{co}(x, p, \beta_0) \subset \mathbb{R}^m \setminus D_\rho.$$

According to Theorem 1.2.2 the domain D is the union of balls with radius r_i. Thus D is a subset of

$$G := \bigcup_{B_{r_i}(y) \subset \mathbb{R}^m \setminus (B_\rho(x) \cup \text{co}(x,p,\beta_e))} B_{r_i}(y).$$

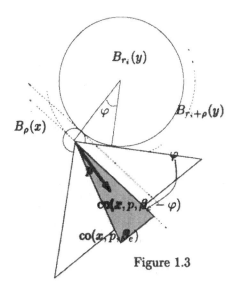

Considering balls $B_{r_i}(y)$, which touch both $B_\rho(x)$ and $\text{co}(x, p, \beta_e)$, we obtain by elementary geometric arguments from Figure 1.3 that for angles

$$\varphi := \arccos(\frac{r_i}{r_i + \rho}) \qquad (1.2.22)$$

with

$$\varphi < \beta_e \qquad (1.2.23)$$

we have

$$\begin{aligned} &B_{r_i+\rho}(y) \cap \\ &\text{co}(x, p, \beta_e - \varphi) = \emptyset. \end{aligned}$$

Figure 1.3

Thus we have proven

$$\text{co}(x, p, \beta_e - \varphi) \cap D_\rho \subset \text{co}(x, p, \beta_e - \varphi) \cap G_\rho = \emptyset,$$

i.e., we have

$$\text{co}(x, p, \beta_e - \varphi) \subset \mathbb{R}^m \setminus D_\rho.$$

The proof is complete by observing that $\arccos(r_i/(r_i + \rho)) \to 0, \rho \to 0$, and choosing ρ_0, β_0 such that for all $\rho < \rho_0$ and φ defined by (1.2.22) the estimate (1.2.23) is satisfied. $\qquad \square$

Spherical harmonics, Bessel and Hankel functions. Indispensable tools for the investigation of both direct and inverse scattering problems are special solutions to the Helmholtz equation. For later use we now introduce Legendre polynomials, associated Legendre polynomials, spherical harmonics, Bessel and Neumann functions and note some of their properties. If not pointed out otherwise, for a proof of these properties we refer to [8].

Let Y_n^l for $l = -n, ..., n$ and $n = 0, 1, 2, ...$ be a complete orthonormal system of spherical harmonics, as for example given by

$$Y_n^l(\theta, \varphi) := \sqrt{\frac{2n+1}{4\pi} \frac{(n-|l|)!}{(n+|l|)!}} \, P_n^{|l|}(\cos(\theta)) e^{il\varphi} \qquad (1.2.24)$$

for $l = -n, ..., n$, $n = 0, 1, 2, ...$ Here, P_n^l are the associated Legendre functions, which can be derived from the Legendre polynomials P_n by

$$P_n^l(t) := (1 - t^2)^{l/2} \frac{d^l P_n(t)}{dt^l}, \quad l = 0, 1, ..., n. \qquad (1.2.25)$$

The Legendre polynomials

$$P_n(x) := \frac{1}{2^n \, n!} \frac{d^n}{dx^n} \left\{ (x^2 - 1)^n \right\}, \, n = 0, 1, ... \qquad (1.2.26)$$

form an orthogonal system in $L^2[-1, 1]$; more explicitly we have

$$\int_{-1}^{1} P_n(t) P_l(t) dt = \frac{2}{2n+1} \, \delta_{nl}, \, n, l = 0, 1, 2, ... \qquad (1.2.27)$$

They satisfy the inequality

$$|P_n(t)| \leq 1, \, -1 \leq t \leq 1, \, n = 0, 1, 2, ... \qquad (1.2.28)$$

For $2n+1$ orthonormal spherical harmonics of order n the addition theorem

$$\sum_{l=-n}^{n} Y_n^l(\hat{x}) \overline{Y_n^l(\hat{y})} = \frac{2n+1}{4\pi} P_n(\cos(\theta)) \qquad (1.2.29)$$

holds for $\hat{x}, \hat{y} \in \Omega$, where θ is the angle between $\hat{x}$ and $\hat{y}$. For the surface gradient of spherical harmonics we note the estimate

$$\left| \text{Grad } Y_n^l(\hat{x}) \right| \leq C \, n^{3/2} \|Y_n^l\|_{L^2(\Omega)}, \, \hat{x} \in \Omega \qquad (1.2.30)$$

(see Section X, Lemma 6.1 of [60]).

The spherical Bessel functions and spherical Neumann functions of order n are given by the series

$$j_n(t) := \sum_{p=0}^{\infty} \frac{(-1)^p t^{n+2p}}{2^p p! 1 \cdot 3 \cdots (2n + 2p + 1)} \qquad (1.2.31)$$

and

$$y_n(t) := -\frac{(2n)!}{2^n n!} \sum_{p=0}^{\infty} \frac{(-1)^p t^{2p-n-1}}{2^p p! (-2n + 1)(-2n + 3) \cdots (-2n + 2p - 1)}, \qquad (1.2.32)$$

where the first coefficient in the series (1.2.32) has to be set equal to one. The linear combinations

$$h_n^{(1,2)} := j_n \pm i y_n \qquad (1.2.33)$$

are known as spherical Hankel functions of the first and second kind of order n. By straightforward calculation from the series (1.2.31) and (1.2.32) it is possible to derive the differentiation formula

$$t^{n+1} f_{n-1}(t) = \frac{d}{dt} \left\{ t^{n+1} f_n(t) \right\} \qquad (1.2.34)$$

for both $f_n = j_n$ and $f_n = y_n$, and together with Stirlings formula

$$n! = \sqrt{2\pi n} \left(\frac{n}{e} \right)^n (1 + o(1)), \quad n \to \infty \qquad (1.2.35)$$

we obtain the behavior

$$j_n(t) = \frac{1}{2n+1} \left(\frac{et}{2n} \right)^n \left(1 + O\left(\frac{1}{n} \right) \right), \quad n \to \infty, \qquad (1.2.36)$$

and

$$h_n^{(1)}(t) = \frac{-\sqrt{2}}{t} \left(\frac{2n}{et} \right)^n \left(1 + O\left(\frac{1}{n} \right) \right), \quad n \to \infty, \qquad (1.2.37)$$

uniformly on compact subsets of $(0, \infty)$.

Point-sources and multipoles. One can use the spherical harmonics and the spherical Bessel or Hankel functions to construct special solutions of the Helmholtz equation. Given a spherical harmonic Y_n of order n, the function

$$u_n(x) := j_n(\kappa |x|) Y_n(\hat{x}) \qquad (1.2.38)$$

is an entire solution of the Helmholtz equation. The multipole of order n

$$v_n(x) := h_n^{(1)}(\kappa |x|) Y_n(\hat{x}) \qquad (1.2.39)$$

is a radiating solution to the Helmholtz equation in $\mathbb{R}^m \setminus \{0\}$.

Modulo a constant, the three-dimensional multipole of order zero is the point-source

$$\Phi(x, z) := \frac{1}{4\pi} \frac{e^{i\kappa|x-z|}}{|x-z|}, \quad (1.2.40)$$

the multipole of order one is a dipole, the multipole of order two a quadrupole, etc.

Figure 1.4

Multipole expansions, i.e., expansions of solutions of the Helmholtz equation with respect to the functions v_n, are used both in direct and inverse scattering. For the proof of explicit stability estimates in Chapter 3, we will need the multipole-expansion of the fundamental solution

$$\Phi(x, y) = i\kappa \sum_{n=0}^{\infty} \sum_{l=-n}^{n} h_n^{(1)}(\kappa|x|) Y_n^l(\hat{x}) j_n(\kappa|y|) \overline{Y_n^l(\hat{y})}, \quad (1.2.41)$$

where $\hat{x} = x/|x|$ and $\hat{y} = y/|y|$. Here, the series and its term-by-term derivatives with respect to x and y are absolutely and uniformly convergent on compact subsets of $|x| > |y|$. Further tools are the Funk-Hecke formula

$$\int_\Omega e^{-i\kappa x \cdot \hat{z}} Y_n(\hat{z}) \, ds(\hat{z}) = \frac{4\pi}{i^n} j_n(\kappa|x|) Y_n(\hat{x}), \quad x \in \mathbb{R}^m \quad (1.2.42)$$

for spherical harmonics Y_n of order n and the Jacobi-Anger expansion

$$e^{i\kappa x \cdot d} = \sum_{n=0}^{\infty} i^n (2n+1) j_n(\kappa|x|) P_n(\cos(\theta)), \quad x \in \mathbb{R}^m, \quad (1.2.43)$$

where d is a unit vector, θ denotes the angle between x and d and the convergence is uniform on compact subsets of $\mathbb{R}^m$.

The two-dimensional case. For scattering in $\mathbb{R}^2$ the multipoles and some constants have to be modified. The functions j_n and y_n are replaced by the Bessel function of order n

$$J_n(t) := \sum_{p=0}^{\infty} \frac{(-1)^p}{p!(n+p)!} \left(\frac{t}{2}\right)^{n+2p} \quad (1.2.44)$$

and the Neumann function of order n

$$
\begin{aligned}
Y_n(t) \quad &:= \frac{2}{\pi} \left\{ \ln \frac{t}{2} + C \right\} J_n(t) - \frac{1}{\pi} \sum_{p=0}^{n-1} \frac{(n-1-p)!}{p!} \left(\frac{2}{t} \right)^{n-2p} \\
&- \frac{1}{\pi} \sum_{p=0}^{\infty} \frac{(-1)^p}{p!(n+p)!} \left(\frac{t}{2} \right)^{n+2p} \{ \psi(n+p) + \psi(p) \}
\end{aligned}
\tag{1.2.45}
$$

for $n = 0, 1, 2, ...$, where we define $\psi(0) := 0$,

$$
\psi(p) := \sum_{l=1}^{p} \frac{1}{l}, \quad p = 1, 2, ...,
\tag{1.2.46}
$$

and

$$
C := \lim_{p \to \infty} \left\{ \sum_{l=1}^{p} \frac{1}{l} - \ln p \right\}
\tag{1.2.47}
$$

denotes Euler's constant, and if $n = 0$ the finite sum in (1.2.45) is set equal to zero. The linear combinations

$$
H_n^{(1,2)} := J_n \pm iY_n
\tag{1.2.48}
$$

are called Hankel functions of the first and second kind of order n, respectively. The multipoles in $\mathbb{R}^2$ are given by the functions

$$
V_n(x) := H_n^{(1)}(\kappa r) e^{\pm in\varphi}
\tag{1.2.49}
$$

with the polar coordinates (r, φ). For the two-dimensional fundamental solution

$$
\Phi(x, z) := \frac{i}{4} H_0^{(1)}(\kappa |x - z|)
\tag{1.2.50}
$$

we obtain the multipole-expansion

$$
\Phi(x, y) = \frac{i}{4} H_0^{(1)}(\kappa |x|) J_0(\kappa |y|) + \frac{i}{2} \sum_{n=1}^{\infty} H_n^{(1)}(\kappa |x|) J_n(\kappa |y|) \cos(n\theta), \tag{1.2.51}
$$

where θ denotes the angle between x and y. It is valid uniformly on compact subsets of $|x| > |y|$. The Jacobi-Anger expansion (1.2.43) in $\mathbb{R}^2$ assumes the form

$$
e^{i\kappa x \cdot d} = J_0(\kappa |x|) + 2 \sum_{n=1}^{\infty} i^n J_n(\kappa |x|) \cos(n\theta), \quad x \in \mathbb{R}^2.
\tag{1.2.52}
$$

Single- and double-layer potentials, jump relations. We will use the boundary-layer approach to investigate the properties of the solutions to scattering problems for impenetrable scatterers. With the help of boundary-layer potentials the scattering problems are reduced to integral equations on the boundary of the scatterer.

For a domain $D \subset I\!\!R^m$ with boundary of class C^2 consider the single-layer potential

$$u(x) := \int_{\partial D} \Phi(x, y)\varphi(y) ds(y), \quad x \in I\!\!R^m, \tag{1.2.53}$$

and the double-layer potential

$$v(x) := \int_{\partial D} \frac{\partial \Phi(x, y)}{\partial \nu(y)} \varphi(y) ds(y), \quad x \in I\!\!R^m \setminus \partial D. \tag{1.2.54}$$

Later we will also use $P_1\varphi$ and $P_2\varphi$ for the single-layer or double-layer potential, respectively. The behavior of the single- and double-layer potentials at the boundary ∂D is described by the following jump relations.

THEOREM 1.2.4 (**Jump relations.**) *The single-layer potential* u *with continuous density* φ *is continuous throughout* $I\!\!R^m$. *On the boundary we have*

$$u(x) = \int_{\partial D} \Phi(x, y)\varphi(y) \, ds(y), \quad x \in \partial D, \tag{1.2.55}$$

and

$$\frac{\partial u_\pm}{\partial \nu}(x) = \int_{\partial D} \frac{\partial \Phi(x, y)}{\partial \nu(x)} \varphi(y) \, ds(y) \mp \frac{1}{2}\varphi(x), \quad x \in \partial D, \tag{1.2.56}$$

where

$$\frac{\partial u_\pm}{\partial \nu}(x) := \lim_{h \to +0} \nu(x) \cdot \operatorname{grad} u(x \pm h\nu(x)) \tag{1.2.57}$$

is to be understood in the sense of uniform convergence on ∂D. *The double-layer potential* v *with density* φ *can be continuously extended from* D *to* $\overline{D}$ *and from* $I\!\!R^m \setminus \overline{D}$ *to* $I\!\!R^m \setminus D$ *with limiting values*

$$\frac{\partial v_\pm}{\partial \nu}(x) = \int_{\partial D} \frac{\partial \Phi(x, y)}{\partial \nu(y)} \varphi(y) \, ds(y) \pm \frac{1}{2}\varphi(x), \quad x \in \partial D, \tag{1.2.58}$$

where

$$v_\pm(x) := \lim_{h \to +0} v(x \pm h\nu(x))$$

and where the integral exists as an improper integral. For a density $\varphi \in L^2(\partial D)$ *the jump relations (1.2.55) to (1.2.58) have to be replaced by*

$$\lim_{h \to +0} \int_{\partial D} \left| u(x \pm h\nu(x)) - \int_{\partial D} \Phi(x, y)\varphi(y) ds(y) \right|^2 ds(x) = 0, \tag{1.2.59}$$

$$\lim_{h\to+0}\int_{\partial D}\left|\tfrac{\partial u}{\partial \nu}(x\pm h\nu(x))-\int_{\partial D}\tfrac{\partial \Phi(x,y)}{\partial \nu(x)}\varphi(y)ds(y)\pm\tfrac{1}{2}\varphi(x)\right|^2 ds(x)=0 \tag{1.2.60}$$

and

$$\lim_{h\to+0}\int_{\partial D}\left|\tfrac{\partial u}{\partial \nu}(x\pm h\nu(x))-\int_{\partial D}\tfrac{\partial \Phi(x,y)}{\partial \nu(y)}\varphi(y)ds(y)\mp\tfrac{1}{2}\varphi(x)\right|^2 ds(x)=0. \tag{1.2.61}$$

Proof. The proof for continuous densities can be found in [7], Theorems 2.12, 2.13 and 2.19; the proof for $\varphi \in L^2(\partial D)$ is due to Kersten [39]. □

For the treatment of point-sources with source-point on the boundary of a domain, in Chapter 5 jump-relations with L^p-densities will be investigated.

Riesz and Fredholm theory. With the help of boundary-layer potentials and jump relations the acoustic and electromagnetic scattering problems can be reduced to boundary integral equations of the second kind, i.e., operator equations of the form

$$(I-A)\varphi=f \tag{1.2.62}$$

with a compact linear operator $A : X \to X$ defined on a normed space X. Integral equations of this kind can be solved using the following theorem of Riesz.

THEOREM 1.2.5 (**Riesz Theorem.**) *Let X be a normed space and $A : X \to X$ a compact linear operator. If the homogeneous equation*

$$(I-A)\varphi=0$$

only has the trivial solution $\varphi=0$, then for all $f \in X$ the inhomogeneous equation

$$(I-A)\varphi=f$$

has a unique solution $\varphi \in X$ and this solution depends continuously on f.

Proof. See Corollary 1.17 of [7]. □

According to the Riesz theorem the injectivity of an operator $I-A$ yields its continuous invertibility. Usually the injectivity of an integral operator corresponding to a scattering problem is obtained from the uniqueness of the solution to this scattering problem.

For the investigation of special scattered fields for scattering from inhomogeneous medium scatterers we will need to study the integral equations of the scattering problems both in the spaces of continuous and square-integrable functions. The injectivity of the integral operators in $L^2(D)$ will be obtained from the results in $C(D)$ with the help of dual systems, defined on subspaces of $L^2(D)$ by the sesquilinear form

$$\langle \varphi, \psi \rangle := \int_D \varphi(y)\overline{\psi(y)}dy \tag{1.2.63}$$

for $\varphi, \psi \in L^2(D)$.

THEOREM 1.2.6 (**Fredholm Alternative Theorem**) *Let X and Y be normed spaces, $\langle X, Y \rangle$ a dual system and $A : X \to X$, $B : Y \to Y$ compact adjoint operators. We have either*

$$N(I - A) = \{0\} \qquad and \qquad N(I - B) = \{0\}$$

and

$$(I - A)(X) = X \qquad and \qquad (I - B)(Y) = Y$$

or

$$\dim N(I - A) = \dim N(I - B) \in I\!N$$

and

$$(I - A)(X) = N(I - B)^{\perp} \qquad and \qquad (I - B)(Y) = N(I - A)^{\perp}.$$

Proof. For a proof we refer the reader to [51]. □

For operators A with $\|A\| < 1$ the invertibility of $I - A$ is obtained by more elementary arguments (generalizing the geometric series).

THEOREM 1.2.7 (**Neumann series.**) *Let $A : X \to X$ be a bounded linear operator on a Banach space X with $\|A\| < 1$. Then $I - A$ has a bounded inverse operator on X which is given by the Neumann series*

$$(I - A)^{-1} = \sum_{j=0}^{\infty} A^k \tag{1.2.64}$$

and which satisfies

$$\|(I - A)^{-1}\| \leq \frac{1}{1 - \|A\|}. \tag{1.2.65}$$

Proof. We refer to [51]. □

Singular systems and Riesz basis. A main object of inverse scattering theory is the operator $\mathcal{F}$ which maps the incident waves u^i defined on the support of the scatterer onto the far field pattern of the scattered fields. It is a linear operator and it is compact in all reasonable spaces. An important part of the study of the inverse problems will be the spectral theory of compact linear operators.

Let $A : X \to Y$ be a linear compact operator. The nonnegative square roots of the eigenvalues of the self-adjoint compact operator $A^*A : X \to X$ are called singular values of A. It is well known from elementary functional analysis that the singular values form an at most countable set accumulating only at zero. All nonzero eigenvalues have finite multiplicity, i.e., the dimension of the nullspaces $N(\mu_j^2 I - A^*A)$ is finite. Eigenvectors corresponding to different eigenvalues are orthogonal.

THEOREM 1.2.8 (**Singular System.**) *Let (μ_j) denote the sequence of the nonzero singular values of the compact linear operator $A \neq 0$ ordered as a decreasing sequence and repeated according to their multiplicity. Then there exist orthonormal sequences φ_j in X and $\tilde{\varphi}_j$ in Y such that*

$$A\varphi_j = \mu_j\tilde{\varphi}_j, \qquad A^*\tilde{\varphi}_j = \mu_j\varphi_j \qquad (1.2.66)$$

for all $j \in \mathbb{N}$. For each $\varphi \in X$ we have the singular value decomposition

$$\varphi = \sum_{j=1}^{\infty}(\varphi, \varphi_j)\varphi_j + Q\varphi \qquad (1.2.67)$$

with the orthogonal projection operator $Q : X \to N(A)$ and

$$A\varphi = \sum_{j=1}^{\infty}\mu_j(\varphi, \varphi_j)\tilde{\varphi}_j. \qquad (1.2.68)$$

Each system with these properties is called a singular system of A.

Proof. We refer to [8] or [51].   □

Orthonormal basis as obtained from the singular value decomposition form a general tool which is widely used in functional analysis and for the treatment of inverse problems. For the spectral theory of scattering problems we will need a more general type of expansion. A set $\{\varphi_j : j \in \mathbb{N}\}$ is called Riesz basis of a complex Hilbert space X, if every $\varphi \in X$ is of the form

$$\varphi = \sum_{j=1}^{\infty}\alpha_j\varphi_j \qquad (1.2.69)$$

with coefficients α_j which satisfy

$$\sum_{j=1}^{\infty} |\alpha_j|^2 < \infty, \tag{1.2.70}$$

and every sequence $(\alpha_j), j \in I\!\!N$ in $\mathbb{C}$ with (1.2.70) converges in X towards an element φ such that equation (1.2.69) is true.

THEOREM 1.2.9 *Let X be a complex Hilbert space, $K : X \to X$ a linear compact operator with $\text{Im}(K\varphi, \varphi) \neq 0$ for all $\varphi \in X, \varphi \neq 0$. Let $\{\varphi_j : j \in I\!\!N\}$ be a linearly independent and complete set in X and assume that we have the orthogonality relation*

$$((I + K)\varphi_j, \varphi_l) = s_j \delta_{jl}, \quad j, l \in I\!\!N, \tag{1.2.71}$$

where $s_j \in \mathbb{C}$ satisfies the following conditions: There exists $r > 0$ with

$$|s_j| = r \ \forall j \in I\!\!N, \qquad \text{Im}(s_j) \to 0, \ j \to \infty. \tag{1.2.72}$$

Then $\{\varphi_j : j \in I\!\!N\}$ is a Riesz basis of X.

Proof. We refer to [42]. □

Ill-posed problems and regularization. In general, inverse problems are ill-posed in the sense of Hadamard [22], i.e., the demands of uniqueness, existence and stability are violated. Ill-posed equations of the type

$$A(\varphi) = f \tag{1.2.73}$$

with a compact (linear or nonlinear) operator $A : X \to Y$ are usually solved approximately by a family of bounded (linear or nonlinear) regularization operators

$$R_\alpha : Y \to X, \quad \alpha > 0, \tag{1.2.74}$$

with the property

$$\lim_{\alpha \to 0} R_\alpha(A(x)) = x \quad \text{for all } x \in X, \tag{1.2.75}$$

i.e., the operators $R_\alpha A$ converge pointwise to the identity for $\alpha \to 0$. Here, α is called the regularization parameter. If R_α satisfies (1.2.75), then the family of operators R_α is called a regularization strategy (see [43]). In the presence of data error of size δ we calculate the solution of (1.2.73) with f replaced by f^δ, i.e., we calculate

$$\varphi^\delta := R_{\alpha(\delta)} f^\delta \tag{1.2.76}$$

with a regularization parameter $\alpha(\delta)$ depending on $\delta > 0$. Of course, we would like to choose the regularization parameter in a way such that our approximate solution tends towards the true solution if the data error tends to zero. This motivates the following definition.

A strategy for the choice of the parameter α depending on the error level δ is called regular, if for all $f \in A(X)$ and all $f^\delta \in Y$ with $\|f^\delta - f\| \leq \delta$ there holds

$$R_{\alpha(\delta)}f^\delta \to A^{-1}f, \quad \delta \to 0. \tag{1.2.77}$$

We will call a set of regularization operators R_α with a regular strategy for the inversion of (1.2.73) a convergent regularization.

We already discussed the different aspects of the ill-posedness of an equation or problem in the preceding survey. For compact linear operators in Hilbert spaces these aspects can be investigated using the singular system.

THEOREM 1.2.10 (Picard.) *Let $A : X \to Y$ be a compact linear operator with singular system $(\mu_j, \varphi_j, \tilde{\varphi}_j)$. The equation of the first kind*

$$A\varphi = f \tag{1.2.78}$$

is solvable if and only if f belongs to the orthogonal complement $N(A^*)^\perp$ and satisfies

$$\sum_{j=1}^{\infty} \frac{1}{\mu_j^2} |(f, \tilde{\varphi}_j)|^2 < \infty. \tag{1.2.79}$$

In this case a solution is given by

$$\varphi = \sum_{j=1}^{\infty} \frac{1}{\mu_j} (f, \tilde{\varphi}_j) \varphi_j. \tag{1.2.80}$$

Proof. See [8].   □

In Picard's Theorem the ill-posedness of the inverse operator A^{-1} is due to the convergence $\mu_j \to 0$ of the sequence of singular values. The faster the decay of the singular values, the more ill-posed is the equation (1.2.78). A natural regularization method for A^{-1} is obtained by taking a finite sum instead of the series (1.2.80).

THEOREM 1.2.11 (Spectral cut-off.) *Let $A : X \to Y$ be an injective compact linear operator with singular system $(\mu_j, \varphi_j, \tilde{\varphi}_j)$. Then the spectral cut-off*

$$R_n f := \sum_{\mu_j \geq \mu_n} \frac{1}{\mu_j} (f, \tilde{\varphi}_j) \varphi_j \tag{1.2.81}$$

describes a regularization scheme with regularization parameter $m \to \infty$ and $\|R_m\| = 1/\mu_n$. The spectral cut-off is regular if

$$\alpha(\delta) \to 0, \quad \frac{\delta^2}{\alpha(\delta)} \to 0, \quad \text{for } \delta \to 0. \tag{1.2.82}$$

Proof. See [8] and [51]. □

Another well-known regularization strategy (compare for example [19]) for the approximate solution of equation (1.2.73) for linear operators $A :$ $X \to Y$ on Hilbert spaces X, Y is given by the Tikhonov regularization scheme, which computes an approximate solution φ_α by

$$\varphi_\alpha := (\alpha I + A^* A)^{-1} A^* f. \tag{1.2.83}$$

We need to collect some of its properties.

THEOREM 1.2.12 (**Tikhonov regularization.**) *The Tikhonov regularization (1.2.83) defines a regularization scheme with $\|R_\alpha\| \leq \frac{1}{2\alpha}$. It is regular provided the parameter $\alpha(\delta)$ is chosen such that (1.2.82) is satisfied.*

Proof. See [51]. □

A further possibility to approximately solve (1.2.73) is minimum norm solutions. For a bounded linear operator $A : X \to Y$ between two normed spaces X and Y, $\delta > 0$ and $f \in Y$ an element $\varphi_0 \in X$ is called minimum norm solution of $A\varphi = f$ with discrepancy δ, if $\|A\varphi_0 - f\| \leq \delta$ and

$$\|\varphi_0\| = \inf \{ \|\varphi\| : \|A\varphi - f\| \leq \delta \}.$$

THEOREM 1.2.13 (**Minimum norm solutions.**) *We consider two Hilbert spaces X and Y. If $A : X \to Y$ has dense range in Y, then for each $f \in Y$ there is a unique minimum norm solution φ_0 of $H\varphi = f$ with discrepancy δ. The minimum norm solution φ_0 can be calculated by*

$$\varphi_0 = (\alpha I + A^* A)^{-1} A^* f, \tag{1.2.84}$$

where α is a zero of the function

$$G(\alpha) := \left\| (\alpha I + A^* A)^{-1} A^* f - f \right\|^2 - \delta^2. \tag{1.2.85}$$

For an injective bounded linear operator A minimum norm solutions define a regular strategy for the choice of the regularization parameter α for the Tikhonov regularization.

Proof. A proof is given in [51].   □

The preceding theorem provides an *a posteriori* strategy for the choice of the parameter α in the Tikhonov regularization scheme for the approximate solution of $A\varphi = f$.

Next, we will introduce a fourth well-known regularization method for later use. For a bounded linear injective operator $A : X \to Y$, $\rho > 0$ and $f \in Y$ we call an element $\varphi_0 \in X$ a quasi-solution of (1.2.73) with constraint ρ, if $\|\varphi_0\| \leq \rho$ and

$$\|A\varphi_0 - f\| = \inf\{\|A\varphi - f\| : \|\varphi\| \leq \rho\}. \tag{1.2.86}$$

For quasi-solutions we obtain regularity only if the constraint is chosen such that $\|A^{-1}f\| = \rho$ (see [51]). In Chapter 5 we will use quasi-solutions as smoothing operators for an application where we do not need regularity.

So far we only treated regularization for the case where the data have errors which might strongly influence the solution of an ill-posed equation. In Chapter 7 we will be faced with the case where the ill-posed operator itself is given by measured data. For this situation the Tikhonov-Morozov regularization is adequate.

THEOREM 1.2.14 (**Tikhonov-Morozov regularization.**) *Let $A_\delta : X \to Y$ be a family of injective and compact operators between Hilbert spaces X and Y with dense ranges $R(A_\delta)$ such that $\|A_0 - A_\delta\| \leq \delta$ for all $\delta > 0$. Given $f \in Y$ with $f \notin N(A_\delta^*)$ for all $\delta \geq 0$ the Tikhonov-Morozov solution of the equation $A_\delta \varphi = r$ is the solution of the system*

$$(\alpha I + A_\delta^* A_\delta)\varphi_\alpha = A_\delta^* r, \tag{1.2.87}$$

where α is chosen such that

$$\|A_\delta \varphi_\alpha - r\| = \delta \|\varphi_\alpha\|. \tag{1.2.88}$$

If the unperturbed equation is solvable by some $g \in X$, then

$$g_{\alpha(\delta)} \to g, \quad \delta \to 0. \tag{1.2.89}$$

If the unperturbed equation $A_0 g = r$ is not solvable, then $\|g_{\alpha(\delta)}\| \to \infty$ for $\delta \to 0$.

Proof. We refer to [42].   □

Scatterers, domains, boundary conditions. For the different scattering objects under consideration we need to clarify our notation for scatterers and scattering domains. An impenetrable acoustic or electromagnetic scatterer is given by a domain D and a boundary condition. We will use the letter $\mathcal{D}$ for the full scatterer with all its properties. The type of a scatterer is either sound-soft or sound-hard for the acoustic problems or perfect-conductor for electromagnetic scattering. Thus, an impenetrable scatterer $\mathcal{D}$ can be viewed as a pair

$$\mathcal{D} = (D, type) \tag{1.2.90}$$

of its domain D and its boundary condition.

For penetrable scatterers the situation is slightly different. Again, we use $\mathcal{D}$ for the full scatterer. The scatterer $\mathcal{D}$ is located in a domain D, defined as the interior of the support of the inhomogeneity given by a refractive index n (or N for the cases of orthotropic or anisotropic media) with $n|_{\mathbb{R}^m \setminus \overline{D}} = 1$ and $n|_{\overline{D}} \in C^{0,\alpha}(\overline{D})$. We write the full scatterer $\mathcal{D}$ as a pair

$$\mathcal{D} = (D, n). \tag{1.2.91}$$

We will study uniqueness, stability and algorithms for the reconstruction of the domain D of impenetrable and penetrable scatterers $\mathcal{D}$ for both acoustic and electromagnetic scattering problems.

2 Direct scattering problems

For the investigation and solution of inverse scattering problems a good knowledge about the direct scattering problems is essential. Thus in this chapter on direct scattering problems we collect and derive definitions and results for further use in the following sections.

2.1 Acoustic obstacle scattering

We consider acoustic scattering from a bounded sound-soft or sound-hard impenetrable scatterer $\mathcal{D}$. The scatterer $\mathcal{D} = (D, type)$ consists of a domain $D \subset \mathbb{R}^m$ for $m = 2, 3$ and a boundary condition *type* for the total field on the boundary ∂D. We always assume the boundary of ∂D of D to be of class C^2 and the open exterior $\mathbb{R}^m \setminus \overline{D}$ of D to be connected. An incident field u^i is a solution to the Helmholtz equation

$$\Delta u + \kappa^2 u = 0 \tag{2.1.1}$$

with wave number $\kappa > 0$ on a domain containing D in its interior.

DEFINITION 2.1.1 *Given an incident field u^i and a scatterer $\mathcal{D}$, the direct acoustic obstacle scattering problem is to find a scattered field*

$$u^s \in C^2(\mathbb{R}^m \setminus \overline{D}) \cap C(\mathbb{R}^m \setminus D),$$

which solves the Helmholtz equation (2.1.1) in $\mathbb{R}^m \setminus \overline{D}$ and satisfies the Sommerfeld radiation condition

$$r^{\frac{m-1}{2}}\left(\frac{\partial u^s}{\partial r} - i\kappa u^s\right) \to 0, \ \ r = |x| \to \infty, \tag{2.1.2}$$

uniformly in all directions $\hat{x} = x/|x|$, such that the total field

$$u = u^i + u^s \tag{2.1.3}$$

satisfies the sound-soft boundary condition

$$u^i + u^s = 0 \ on \ \partial D \tag{2.1.4}$$

or the sound-hard boundary condition

$$\frac{\partial}{\partial \nu}\left(u^i + u^s\right) = 0 \ on \ \partial D. \tag{2.1.5}$$

Here ν denotes the unit outward normal vector to ∂D and the normal derivative in (2.1.5) is understood in the sense of (1.2.57). A solution u of the Helmholtz equation in the exterior of some ball B satisfying (2.1.2) is called radiating.

The main tools for the investigation and solution of the direct scattering problem are Green's integral theorems. In particular, for $u, v \in C^2(\overline{G})$ we have Green's second theorem

$$\int_G \{u\Delta v - v\Delta u\}\, dx = \int_{\partial G} \left(u\frac{\partial v}{\partial \nu} - v\frac{\partial u}{\partial \nu} \right) ds, \qquad (2.1.6)$$

where G denotes a domain of class C^1 and ν the unit normal vector to the boundary ∂G directed into the exterior of G. Green's integral theorems can be used to derive Green's formula for a radiating solution u^s of the direct acoustic scattering problem

$$u^s(x) = \int_{\partial D} \left\{ u^s(y)\frac{\partial \Phi(x,y)}{\partial \nu(y)} - \frac{\partial u^s}{\partial \nu}(y)\Phi(x,y) \right\} ds(y), \quad x \in \mathbb{R}^m \setminus \overline{D}$$

$$\qquad (2.1.7)$$

and the asymptotic behavior

$$u^s(x) = \frac{e^{i\kappa|x|}}{|x|^{\frac{m-1}{2}}} \left\{ u^\infty(\hat{x}) + O\left(\frac{1}{|x|}\right) \right\}, |x| \to \infty, \qquad (2.1.8)$$

where $\hat{x} := x/|x| \in \Omega$ and $\Omega := \{x \in \mathbb{R}^m : |x| = 1\}$, see [8], Theorem 2.5. The function u^∞ is called the far field pattern of the scattered acoustic wave.

As incident fields u^i plane waves and point-sources are of special interest. We denote the scattered field for an incident plane wave

$$u^i(x, d) := e^{i\kappa x \cdot d}, \quad x \in \mathbb{R}^m,$$

with direction $d \in \Omega$ by $u^s(x, d)$, $x \in \mathbb{R}^m \setminus D$, and the corresponding far field pattern by $u^\infty(\hat{x}, d)$, $\hat{x} \in \Omega$. An incident point-source $\Phi(\cdot, z)$ with source point $z \in \mathbb{R}^m$ is given by the fundamental solution to the Helmholtz equation (1.2.40). The scattered field for an incident point-source $\Phi(\cdot, z)$ with source point z is denoted by $\Phi^s(\cdot, z)$ and the corresponding far field pattern by $\Phi^\infty(\hat{x}, z)$, $\hat{x} \in \Omega$.

Several approaches have been developed to solve the direct scattering problem. We will use integral equations to obtain a representation of the solution in terms of boundary-layer potentials and to study properties of the scattered fields. To this end, let us introduce the classical boundary integral operators. We use the single-layer operator

$$(S\varphi)(x) := 2 \int_{\partial D} \Phi(x, y)\varphi(y) ds(y), \quad x \in \partial D, \qquad (2.1.9)$$

the double-layer operator

$$(K\varphi)(x) := 2 \int_{\partial D} \frac{\partial \Phi(x, y)}{\partial \nu(y)}\varphi(y) ds(y), \quad x \in \partial D, \qquad (2.1.10)$$

the L^2-adjoint of the double-layer operator

$$(K^*\varphi)(x) := 2\int_{\partial D} \frac{\partial\Phi(x,y)}{\partial\nu(x)}\varphi(y)ds(y), \quad x \in \partial D, \qquad (2.1.11)$$

and the normal derivative of the double-layer operator

$$(T\varphi)(x) := 2\frac{\partial}{\partial\nu(x)}\int_{\partial D} \frac{\partial\Phi(x,y)}{\partial\nu(y)}\varphi(y)ds(y), \quad x \in \partial D. \qquad (2.1.12)$$

For a derivation of the following uniqueness and existence results for the direct acoustic scattering problems we refer to [8], Chapter 3. Here we only summarize the results.

THEOREM 2.1.2 *The direct acoustic scattering problem with sound-soft or sound-hard boundary condition has a unique solution and the solution depends continuously on the boundary data $u^i|_{\partial D}$ or $\frac{\partial u^i}{\partial\nu}|_{\partial D}$, respectively, of the incident field in $C(\partial D)$ with respect to uniform convergence of the solution and all its derivatives on closed subsets of $\mathbb{R}^m \setminus \overline{D}$.*

In particular, for the case of the sound-soft boundary condition, the solution can be represented as a combined acoustic potential

$$u^s(x) = \int_{\partial D}\left\{\frac{\partial\Phi(x,y)}{\partial\nu(y)} - i\Phi(x,y)\right\}\varphi(y)ds(y), \quad x \in \mathbb{R}^m \setminus \overline{D}, \quad (2.1.13)$$

where the density $\varphi \in C(\partial D)$ is a solution of the boundary integral equation

$$(I + K - iS)\varphi = -2u^i|_{\partial D}. \qquad (2.1.14)$$

Here I stands for the identity operator. For the case of the sound-hard boundary condition, a representation of the solution is given by the modified acoustic single- and double-layer potential

$$u^s(x) := \int_{\partial D}\left\{\Phi(x,y)\varphi(y) + i\frac{\partial\Phi(x,y)}{\partial\nu(y)}(S_0^2\varphi)(y)\right\} ds(y), \quad x \in \mathbb{R}^m \setminus \partial D,$$
$$(2.1.15)$$

where S_0 denotes the operator S in the case $\kappa = 0$ and the density $\varphi \in C(\partial D)$ solves the boundary integral equation

$$(I - K^* - iTS_0^2)\varphi = 2\frac{\partial u^i}{\partial\nu}. \qquad (2.1.16)$$

Proof. We refer to Theorems 3.7, 3.9 and 3.10 of [8].   □

Far field and mixed reciprocity relations. Before we investigate more details of the behavior of integral operators for the solution to the direct scattering problem, we introduce some symmetry properties of the scattered fields or far field patterns, respectively, which are called reciprocity relations.

Due to reciprocity relations, the role of source and receiver in the scattering process can be exchanged. Reciprocity relations play an important role for the investigation of both direct and inverse scattering problems.

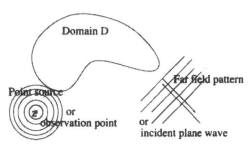

Figure 2.1: Reciprocity relation.

We will use reciprocity relations in nearly all further sections, for example for a proof of the Isakov-Kirsch-Kress uniqueness theorem, to prove uniqueness for the support of scattering media, to obtain stability estimates, to introduce the point-source method and also to derive the method of singular sources.

For further argumentation we distinguish between different reciprocity relations according to the location of the source and receiver in the near field, the far field or with either the source in the near field and the observations in the far field or vice versa.

THEOREM 2.1.3 (**Far field reciprocity relation.**) *The far field patterns for scattering of plane waves by a sound-soft or sound-hard scatterer satisfy*

$$u^\infty(\hat{x}, d) = u^\infty(-d, -\hat{x}), \quad \hat{x}, d \in \Omega. \qquad (2.1.17)$$

Proof. We refer to [8], Theorem 3.13 for the sound-soft scatterer. The sound-hard boundary condition can be treated analogously, see for example the proof of the mixed reciprocity relation below.   □

For the mixed reciprocity relations we need the constant

$$\gamma_m = \begin{cases} \dfrac{e^{i\pi/4}}{\sqrt{8\pi\kappa}}, & m = 2 \\[3mm] \dfrac{1}{4\pi}, & m = 3, \end{cases} \qquad (2.1.18)$$

depending on the dimension $m = 2, 3$.

THEOREM 2.1.4 (**Mixed reciprocity relation.**) *For acoustic scattering of plane waves $u^i(\cdot, d)$, $d \in \Omega$ and point-sources $\Phi(\cdot, z)$, $z \in \mathbb{R}^m \setminus \overline{D}$ from a sound-soft and a sound hard scatterer D we have*

$$\Phi^\infty(\hat{x}, z) = \gamma_m \, u^s(z, -\hat{x}), \quad z \in \mathbb{R}^m \setminus \overline{D}, \ \hat{x} \in \Omega. \tag{2.1.19}$$

Proof. The proof for the sound-soft scatterer is due to Kress [50]. By Green's second theorem (2.1.6) we have that

$$\int_{\partial D} \left(\Phi^s(y, z) \frac{\partial u^s(y, d)}{\partial \nu(y)} - \frac{\partial \Phi^s(y, z)}{\partial \nu(y)} u^s(y, d) \right) ds(y) = 0, \tag{2.1.20}$$

for $z \in \mathbb{R}^m \setminus \overline{D}$, $d \in \Omega$. Passing to the limit $|x| \to \infty$ in Green's formula (2.1.7) we obtain the representation

$$\Phi^\infty(\hat{x}, z) = \gamma_m \int_{\partial D} \left(\Phi^s(y, z) \frac{\partial e^{-i\kappa \hat{x} \cdot y}}{\partial \nu(y)} - \frac{\partial \Phi^s}{\partial \nu}(y, z) e^{-i\kappa \hat{x} \cdot y} \right) ds(y) \tag{2.1.21}$$

for $z \in \mathbb{R}^m \setminus \overline{D}$, $\hat{x} \in \Omega$. Let $u(., d)$ denote the total field for the sound-soft or the sound-hard scattering problem with incident plane wave of direction d. Adding γ_m times (2.1.20) with d replaced by $-\hat{x}$ to equation (2.1.21), with the help of the boundary conditions we obtain

$$\Phi^\infty(\hat{x}, z) = \gamma_m \int_{\partial D} \Phi^s(y, z) \frac{\partial u(y, -\hat{x})}{\partial \nu(y)} ds(y), \quad z \in \mathbb{R}^m \setminus \overline{D}, \ \hat{x} \in \Omega, \tag{2.1.22}$$

for the sound-soft scatterer and

$$\Phi^\infty(\hat{x}, z) = -\gamma_m \int_{\partial D} \frac{\partial \Phi^s(y, z)}{\partial \nu(y)} u(y, -\hat{x}) ds(y), \quad z \in \mathbb{R}^m \setminus \overline{D}, \ \hat{x} \in \Omega, \tag{2.1.23}$$

for the sound-hard scatterer. Again from Green's theorem we have the representation formula

$$u^s(x, d) = -\int_{\partial D} \Phi(x, y) \frac{\partial u(y, d)}{\partial \nu} ds(y), \quad x \in \mathbb{R}^m \setminus \overline{D}, \ d \in \Omega \tag{2.1.24}$$

for the sound-soft boundary condition and

$$u^s(x, d) = \int_{\partial D} \frac{\partial \Phi}{\partial \nu}(x, y) u(y, d) ds(y), \quad x \in \mathbb{R}^m \setminus \overline{D}, \ d \in \Omega \tag{2.1.25}$$

for the sound-hard boundary condition. Now from (2.1.22), (2.1.24) and (2.1.23), (2.1.25) using the boundary condition for Φ^s we obtain (2.1.19) both for the sound-soft and sound-hard boundary condition. □

Classes of impenetrable scatterers. Before starting to look for any object we need at least some rough idea about what we are looking for and where it should be. In the framework of scattering theory this means that we need to formulate appropriate assumptions on the scatterers. For example bounds on the size of the scatterer occur in uniqueness theorems (see for example [8], Theorem 5.2). As indicated in the introduction, bounds on the curvature of the scatterer can be used to obtain results on stability. To prove convergence of reconstruction algorithms, it is usually required to know some geometric properties of the scatterers under consideration. For example, it is required to know a part of the interior of the unknown domain for the method of Kirsch and Kress (Section 5.4 of [8]).

We now define classes of scatterers for further investigation. They do not describe the weakest possible restrictions for the different statements and algorithms, but they play the role of some simple limitations, which are adequate for the behavior of the inverse scattering problems under consideration.

DEFINITION 2.1.5 *Given positive constants $R_e, r_0, a_0, C_0, \beta_e$, we define the class*

$$C_{ss} = C_{ss}(R_e, r_0, a_0, C_0, \beta_e)$$

as the set of sound-soft scatterers with domain

$$D \in \mathcal{A}(R_e, r_0, a_0, l, \alpha, C_0, \beta_e) \qquad (2.1.26)$$

for $l = 2$, $\alpha = 0$ and $\mathcal{A}$ given by (1.2.7). The class

$$C_{sh} = C_{sh}(R_e, r_0, a_0, C_0, \beta_e)$$

is the set of sound-hard scatterers $\mathcal{D}$ with domain D satisfying (2.1.26). The classes C_{ss} and C_{sh} together form the class

$$C_{obst} := C_{ss} \cup C_{sh}.$$

To define the convergence of a sequence of domains or boundaries, respectively, we use the parametrizations $(\psi_j)_{j=1,\ldots,L_1}$, which we constructed in Theorem 1.2.2.

DEFINITION 2.1.6 *The convergence*

$$\partial \tilde{D} \to \partial D$$

is understood as a convergence of the parametrizations $\tilde{\psi}_j \to \psi_j$, $j = 1,\ldots,L_1$ in the norm of $C^{l,\alpha}(B_{r_0}(0))$.

We first derive a compactness property of the class $\mathcal{C}_{obst}$.

LEMMA 2.1.7 *Given a sequence $(\mathcal{D}_j)_{j\in I\!\!N}$ of scatterers $\mathcal{D}_j \in \mathcal{C}_{obst}$, there is a subsequence $(\mathcal{D}_{j_k})_{k\in I\!\!N}$ of $(\mathcal{D}_j)_{j\in I\!\!N}$, for which the sequence of domains $(D_{j_k})_{k\in I\!\!N}$ converges in the $C^{1,\alpha}$-norm to a domain $D \subset B$.*

Proof. Consider a sequence $(\mathcal{D}_j)_{j\in I\!\!N}$ of scatterers. Due to Theorem 1.2.2 we obtain mappings $\psi_{jl} : B_{r_i}(0) \to I\!\!R^m$ with $\|\psi_{jl}\|_{C^2(B_{r_i}(0))} \leq C_0$, such that the sets

$$V_{jl} := \psi_{jl}(B_{r_i}(0)), \quad l = 1, ..., L_1,$$

cover ∂D_j. Since a bounded subset of $C^2(B_{r_i}(0))$ is a relatively compact subset of $C^{1,\alpha}(B_{r_i}(0))$, we can find a convergent subsequence $(\psi_{j_k 1})_{k\in I\!\!N}$ of $(\psi_{j1})_{j\in I\!\!N}$ in $C^{1,\alpha}(B_{r_i}(0))$. Then, we consider the corresponding subsequence $(\psi_{j_k 2})_{k\in I\!\!N}$ of $(\psi_{j2})_{j\in I\!\!N}$ and again choose a convergent subsequence $(\psi_{j_{(k_n)}2})_{n\in I\!\!N}$ of $(\psi_{j_k 2})_{k\in I\!\!N}$. We proceed in the same way with $l = 3, ..., L_1$ and after L_1 steps we obtain a subsequence $(\mathcal{D}_{j_k})_{k\in I\!\!N}$ of $(\mathcal{D}_j)_{j\in I\!\!N}$ with a convergent sequence $(D_{j_k})_{k\in I\!\!N}$ of domains. This completes the proof. $\square$

Properties of boundary integral operators and their dependence on the boundary. We now collect some results on the integral operators S, K, K^* and T. Care has to be taken, as we need most bounds and constants to hold uniformly for scatterers $\mathcal{D} \in \mathcal{C}_{obst}$. We first summarize the classical mapping properties of the potential operators. As for S_0 we use the notation T_0 for the scatterer T in the case $\kappa = 0$.

THEOREM 2.1.8 *For $\alpha \in (0,1)$ the operators S, K, K^* and $T - T_0$ are bounded operators from $C(\partial D)$ into $C^{0,\alpha}(\partial D)$. The operators S and K are also bounded from $C^{0,\alpha}(\partial D)$ into $C^{1,\alpha}(\partial D)$. The operator T is bounded from $C^{1,\alpha}(\partial D)$ into $C^{0,\alpha}(\partial D)$. The double-layer potential defines a bounded operator from $C^{0,\alpha}(\partial D)$ into $C(B \setminus \overline{D})$. All bounds hold uniformly for scatterers $\mathcal{D} \in \mathcal{C}_{obst}$.*

Proof. The proofs for the mapping properties of the operators can be found in the Theorems 2.12, 2.15, 2.30 and 2.31 of [7]. Using the properties of the class $\mathcal{A}$ it has been worked out in [79] that the estimates are satisfied uniformly for scatterers $\mathcal{D} \in \mathcal{C}_{obst}$. $\square$

For the use of continuity and compactness arguments, we need to investigate the dependence of the operators S, K, K^* and $T - T_0$ on the domain ∂D. We investigate the operators as bounded linear operators on $C(\partial D)$.

To treat functions and operators on ∂D in dependence on ∂D we need to define appropriate reference spaces. This will be the spaces

$$X := \left[C^{1,\alpha}(B_{r_0}(0)) \right]^{L_1},$$

and for $l \in \{0,1\}$

$$Y := [C^{l,\alpha}(B_{r_0}(0))]^{L_1}.$$

For each scatterer $\mathcal{D} \in \mathcal{C}_{obst}$ there is a parametrization $(\psi_j)_{j=1,...,L_1} \in X$. This defines a mapping

$$\Pi : \mathcal{C}_{obst} \to X, \quad \partial D \mapsto (\psi_j)_{j=1,...,L_1}.$$

Later we will need the mapping

$$\Psi_{\partial D} : C^{l,\alpha}(\partial D) \to Y, \quad \varphi \mapsto (\varphi \circ \psi_j)_{j=1,...,L_1}.$$

To study operators on the boundary of domains we will define an appropriate reference space in a neighbourhood V of a reference domain D_0. Given a domain $D_0 \in \mathcal{A}$ with normal vectors $\nu_0(x), x \in \partial D_0$, and a $C^{1,\alpha}$-vectorfield v_0 on the boundary ∂D_0 with $v_0(x) \cdot \nu_0(x) \geq \frac{1}{2}$, all boundaries in a neighbourhood V of ∂D_0 are given uniquely by some function $r \in C^{1,\alpha}(\partial D_0)$ in the form $\{\tilde{x} = x + v_0(x)r(x), x \in \partial D\}$. We define $W := C^{l,\alpha}(\partial D_0)$, $l = 0, 1$, and mappings

$$\tilde{\Psi}_{\partial D} : C^{l,\alpha}(\partial D) \to W, \quad \varphi \mapsto \varphi(x + v_0(x)r(x)), \quad x \in \partial D_0. \qquad (2.1.27)$$

Thus, for ∂D in V the mapping

$$\tilde{\Psi}_{\partial D}^{-1} : W \to C^{l,\alpha}(\partial D)$$

is well defined and bounded.

DEFINITION 2.1.9 *For functions* $\varphi \in C^{l,\alpha}(\partial D)$ *and* $\varphi_n \in C^{l,\alpha}(\partial D_n), n \in \mathbb{N}$, *we say that*

$$\varphi_n \to \varphi, \quad n \to \infty,$$

if $\partial D_n \to \partial D, \ n \to \infty$, *and*

$$\Psi_{\partial D_n}(\varphi_n) \to \Psi_{\partial D}(\varphi), \quad n \to \infty,$$

in Y. *A bounded linear operator* J *on* $C^{l,\alpha}(\partial D)$, $l \in \{0,1\}, \alpha \in [0,1]$, *is said to depend continuously on* ∂D, *if the mapping*

$$\hat{J}(\partial D) := \tilde{\Psi}_{\partial D} J(\partial D) \tilde{\Psi}_{\partial D}^{-1} \ \in BL(W, W)$$

depends continuously on ∂D.

For the investigation of continuity properties of boundary integral operators with a weakly singular kernel, we need some further technical tools.

LEMMA 2.1.10 *For scatterers* $D \in C_{obst}$ *we uniformly have the estimate*

$$|\nu(z_0) \cdot (y - z_0)| \leq L\,|y - z_0|^2, \quad y, z_0 \in \partial D, \qquad (2.1.28)$$

with a constant L and there are constants $\tau, C > 0$ such that

$$\left| \frac{1}{|y - z_h|} - \frac{1}{|y - z_{-h}|} \right| \leq C, \quad y, z_0 \in \partial D, 0 < h \leq \tau, \qquad (2.1.29)$$

where $z_h := z_0 + \nu(z_0)h$.

Proof. First note, that for each point $z_0 \in \partial D$ the statements have to be shown only for y in a neighbourhood $B_{r_i}(z_0)$ of z_0, since for $y \notin B_{r_i}(z_0)$ the terms are bounded by a constant uniformly for all domains $D \in C_{obst}$. For $y \in B_{r_i}(z_0)$ we will give a proof by choosing the special coordinate system K_{z_0} introduced in Section 1.2. The origin of this system is z_0 and the direction of the third axis e_3 is given by $\nu(z_0)$. By rotation around the third axis we can obtain 0 for the second coordinate of y. According to Theorem 1.2.2 in a neighbourhood of z_0 the intersection of the boundary ∂D with the $e_1 - e_3$-plane of the new coordinate system is given in the form $(t, 0, f(t))$ with a function f with $|f(t)| \leq C_0 t^2$ for all t with $|t| \leq r_0$.

To prove the first part of the theorem we estimate

$$\begin{aligned} |\nu(z_0) \cdot (y - z_0)| &= \left| (0, 0, 1) \cdot \big((t, 0, f(t)) - (0, 0, 0) \big) \right| \\ &= |f(t)| \\ &\leq C_0 t^2 \\ &\leq C_0\,|y - z_0|^2. \end{aligned} \qquad (2.1.30)$$

For the second statement we calculate and estimate with the help of the mean value theorem

$$\begin{aligned} & \left| \frac{1}{|y - z_h|} - \frac{1}{|y - z_{-h}|} \right| \\ &= \left| \left| (t, 0, f(t)) - (0, 0, h) \right|^{-1} - \left| (t, 0, f(t)) - (0, 0, -h) \right|^{-1} \right| \\ &= \left| \left(t^2 + [f(t)]^2 + h^2 - 2f(t)h \right)^{-\frac{1}{2}} - \left(t^2 + [f(t)]^2 + h^2 + 2f(t)h \right)^{-\frac{1}{2}} \right| \\ &\leq \frac{2f(t)h}{(t^2 + [f(t)]^2 + h^2 - 2|f(t)|h)^{\frac{3}{2}}} \\ &\leq \frac{2C_0 t^2 h}{((1 - 2C_0h)t^2 + h^2)^{\frac{3}{2}}}. \end{aligned} \qquad (2.1.31)$$

We now use spherical coordinates (r, φ) for (t, h), i.e., we insert $t = r \sin(\varphi)$ and $h = r \cos(\varphi)$ into (2.1.31). With the help of the estimate

$$\left(\frac{1}{2} r^2 \sin^2(\varphi) + r^2 \cos^2 \varphi \right)^{\frac{3}{2}} \geq \sqrt{\frac{1}{8}} r^3$$

we derive

$$\frac{2 C_0 t^2 h}{((1 - 2C_0 h)t^2 + h^2)^{\frac{3}{2}}} \leq 2\sqrt{8}\, C_0 \sin^2(\varphi) \cos(\varphi) \leq 2\sqrt{8}\, C_0 \qquad (2.1.32)$$

for all $0 < h \leq \frac{1}{4C_0}$ and all $0 \leq t \leq r_0$. This ends the proof. $\square$

LEMMA 2.1.11 *Consider normed spaces X, Y, a subset $V \subset X$ and a function $f : V \to Y$. We assume that for each $\epsilon > 0$ we have the decomposition*

$$f = f_{1,\epsilon} + f_{2,\epsilon},$$

with functions $f_{1,\epsilon} : V \to Y$ and $f_{2,\epsilon} : V \to Y$. If for each fixed $\epsilon > 0$ the function $f_{2,\epsilon}$ depends continuously on x at a point $x_0 \in V$ and if the family $(f_{1,\epsilon})_{\epsilon > 0}$ satisfies the estimate

$$\|f_{1,\epsilon}(x)\| \leq c\epsilon \qquad (2.1.33)$$

with a constant c in a neighbourhood V' of x_0, then the function f depends continuously on x at the point x_0.

Proof. Given $\tau > 0$ we have to find $\delta > 0$, such that $\|f(x) - f(x_0)\| \leq \tau$ for all $x \in B_\delta(x_0) \subset V \subset X$. We first choose $\epsilon := \tau/(4c)$ and obtain

$$\|f_{1,\epsilon}(x)\| \leq \tau/4$$

for all $x \in V'$. From the continuity of $f_{2,\epsilon}$ we obtain a $\delta > 0$ such that

$$\|f_{2,\epsilon}(x) - f_{2,\epsilon}(x_0)\| \leq \tau/2$$

for all $\|x - x_0\| \leq \delta$. Adding $f_{1,\epsilon}$ and $f_{2,\epsilon}$ from

$$\|f(x) - f(x_0)\| \leq \|f_{1,\epsilon}(x)\| + \|f_{1,\epsilon}(x_0)\| + \|f_{2,\epsilon}(x) - f_{2,\epsilon}(x_0)\|$$

we derive the statement of the lemma. $\square$

THEOREM 2.1.12 *The operators S, K, K^* and $T - T_0$ as elements of the space $BL(C(\partial D), C(\partial D))$ depend continuously on the boundary ∂D of the scatterer $\mathcal{D} \in C_{obst}$ with respect to the $C^{1,\alpha}$-norm for ∂D.*

Proof. The continuity statement for C^2-boundaries is a consequence of more general results on the Fréchet differentiability of the operators with respect to the domain, see [78] and [79]. Here, we will give a proof for boundaries of class $C^{1,\alpha}$, which does not use Fréchet derivatives.

We consider the operator K^* in the three dimensional case $m = 3$. The operator K^* has the kernel

$$\nu(x) \cdot \nabla_x \Phi(x,y), \quad x, y \in \partial D.$$

Let Φ_0 denote the fundamental solution in the case $\kappa = 0$ and K_0^* the corresponding operator K^*. First we note that by (2.1.28) the difference

$$
\begin{aligned}
\nu(x) \cdot \nabla_x & \left(\Phi(x,y) - \Phi_0(x,y)\right) \\
&= \nu(x) \cdot \nabla_x \left(\frac{e^{i\kappa|x-y|}}{4\pi|x-y|} - \frac{1}{4\pi|x-y|}\right) \\
&= \nu(x) \cdot \nabla_x \left(\sum_{n=1}^{\infty} \frac{(i\kappa)^n \, |x-y|^{n-1}}{4\pi \, n!}\right) \qquad (2.1.34) \\
&= \frac{\nu(x) \cdot (x-y)}{4\pi \, |x-y|} \left(\sum_{n=2}^{\infty} \frac{(i\kappa)^n \, (n-1) \, |x-y|^{n-2}}{n!}\right)
\end{aligned}
$$

is a continuous function in $x, y \in \partial D$. For such kernels the continuity statements are straightforward to prove, i.e., the difference $K^* - K_0^*$ depends continuously on the boundary ∂D. We investigate the weakly singular part K_0^* with the kernel

$$\nu(x) \cdot \nabla_y \Phi_0(x,y) = \frac{\nu(y) \cdot (x-y)}{|x-y|^3}, \quad x, y \in \partial D. \qquad (2.1.35)$$

By definition of the class $\mathcal{A}$ we can choose a local coordinate system with $x = 0$ and $\nu(x) = (0, 0, 1)$. Then the boundary ∂D in a neighbourhood of x is represented in the form $(s, t, f(s,t))$, $s, t \in \mathbb{R}$, with $|f(s,t)| \le C_0|(s,t)|^2$. We obtain the estimate

$$
\begin{aligned}
\left|\frac{\nu(x) \cdot (x-y)}{|x-y|^3}\right| &= \left|\frac{f(s,t)}{(s^2 + t^2 + [f(s,t)]^2)^{\frac{3}{2}}}\right| \\
&\le C_0(s^2 + t^2)^{-\frac{1}{2}} \\
&\le C_0|x-y|^{-1} \qquad (2.1.36)
\end{aligned}
$$

in a neighbourhood $|x-y| \le \frac{1}{\sqrt{2}\,C_0}$ of x. Thus for all scatterers $\mathcal{D} \in \mathcal{C}_{obst}$ the kernel (2.1.35) of the integral operator K_0^* has a weak singularity, which is bounded by the weakly singular function (2.1.36). Decomposing the domain of integration into $B_\epsilon(x) \cap \partial D$ and $\partial D \setminus B_\epsilon(x)$ for all sufficiently small $\epsilon > 0$ we obtain

$$(K_0^*\varphi)(x) \;=\; \int_{B_\epsilon(x)\cap\partial D} k_0(x,y)\varphi(y)ds(y) \qquad (2.1.37)$$

$$+ \int_{\partial D\setminus B_\epsilon(x)} k_0(x,y)\varphi(y)ds(y)$$

with the kernel k_0 given by (2.1.35). We can use the bound (2.1.36) for the singularity of the kernel to estimate the first term of (2.1.37) uniformly for all scatterers $\mathcal{D} \in \mathcal{C}_{obst}$ and all $x \in \partial D$ by

$$\left| \int_{B_\epsilon(x)\cap\partial D} k_0(x,y)\varphi(y)ds(y) \right| \leq c\epsilon\|\varphi\|_{C(\partial D)} \qquad (2.1.38)$$

with some constant c. For each fixed $\epsilon > 0$ the second integral depends continuously on the boundary ∂D and it is bounded uniformly for $\epsilon > 0$ and $\mathcal{D} \in \mathcal{C}_{obst}$ by

$$\left| \int_{B_\epsilon(x)\cap\partial D} k_0(x,y)\varphi(y)ds(y) \right| \leq c\|\varphi\|_{C(\partial D)} \qquad (2.1.39)$$

with some constant c. Let V be a neigbourhood of some reference surface ∂D_0 as introduced above. We define the mappings $K_{j,\epsilon}^* : V \to BL(W,W)$ for $j = 1,2$ by

$$(K_{1,\epsilon}^*(\partial D)\varphi)(x) := \int_{B_\epsilon(x)\cap\partial D} k_0(x,y)\varphi(y)ds(y), \;\; x \in \partial D,$$

and

$$(K_{2,\epsilon}^*(\partial D)\varphi)(x) := \int_{\partial D\setminus B_\epsilon(x)} k_0(x,y)\varphi(y)ds(y), \;\; x \in \partial D.$$

From (2.1.37) we derive $K_0^* = K_{1,\epsilon}^* + K_{2,\epsilon}^*$. Now we apply Lemma 2.1.11 to the mapping

$$V \to BL(W,W), \;\; \partial D \mapsto K^*(\partial D)$$

to obtain with the help of (2.1.38) and (2.1.39) the continuous dependence of K_0^* on ∂D. Thus we have proven that $K^* = (K^* - K_0^*) + K_0^*$ depends continuously on the boundary ∂D with respect to $C^{1,\alpha}$-norm for ∂D.

The statement for S, K and $T - T_0$ and for the two-dimensional case $m = 2$ can be proven analogously.   □

Uniform bounds for classes of domains. To obtain uniform bounds for the mapping of the incident field onto the scattered field or its far field pattern, respectively, we have to study the inverse of the integral operators $I + K - iS$ for the sound-soft and $I - K' - iTS_0^2$ for the sound-hard boundary condition. Their existence and boundedness is obtained from the Riesz theory. But the Riesz theory does not give a possibility to control the bounds. To derive uniform boundedness in $\mathcal{C}_{obst}$ we use compactness and continuity arguments in the following theorem.

As a preparation we consider the strongly singular integral operator T used for the solution of the sound-hard scattering problem. Strongly singular operators are more difficult to handle than operators with weakly singular kernels. Due to the following relations, for the products ST and TS it is possible to obtain a representation in terms of weakly singular integral operators K and K^*. The integral operators S and T satisfy the relations

$$ST = K^2 - I \tag{2.1.40}$$

and

$$TS = (K^*)^2 - I. \tag{2.1.41}$$

A proof can be found in [8], equations (3.12) and (3.13).

THEOREM 2.1.13 *In the space* $BL(C(\partial D), C(\partial D))$ *the integral operators*

$$(I + K - iS)^{-1} \tag{2.1.42}$$

and

$$(I - K^* - iTS_0^2)^{-1} \tag{2.1.43}$$

are bounded by a constant c uniformly for scatterers $\mathcal{D} \in \mathcal{C}_{obst}$.

Proof. First we consider $(I + K - iS)^{-1}$. The boundary integral operators S and K depend continuously on the boundary with respect to the $C^{1,\alpha}$-norm. Since the inverse A^{-1} of an operator A depends continuously on A, the operators (2.1.42) and (2.1.43) depend continuously on the boundary ∂D with respect to the $C^{1,\alpha}$-norm. Using the compactness of the imbedding of $C^2(B_{r_i}(0))$ into $C^{1,\alpha}(B_{r_i}(0))$, the statement is a consequence of the fact that continuous functions on compact sets are bounded.

To treat the operator $(I - K^* - iTS_0^2)^{-1}$ we use (2.1.41) to obtain

$$\begin{aligned} (I - K^* - iTS_0^2)^{-1} &= (I - K^* - i(T - T_0)S_0^2 + T_0 S_0^2)^{-1} \\ &= (I - K^* - i(T - T_0)S_0^2 + [(K_0^*)^2 - I]S_0)^{-1} \end{aligned}$$

and proceed in the same way as for $(I + K - iS)^{-1}$. $\qquad\square$

THEOREM 2.1.14 *The mappings of the incident field $u^i \in C(\partial D)$ onto the far field pattern $u^\infty \in C^1(\Omega)$ of the scattered field u^s for scattering by a sound-soft scatterer and of the normal derivative $\frac{\partial u^i}{\partial \nu} \in C(\partial D)$ onto the far field pattern $u^\infty \in C^1(\Omega)$ of u^s for scattering by a sound-hard scatterer are bounded uniformly for $D \in \mathcal{C}_{obst}$ by a constant c_∞.*

Proof. We use the representations (2.1.13) and (2.1.15) for the solution of the scattering problems. Then the far field patterns of the scattered fields u^s are given by

$$u^\infty(\hat{x}) = \gamma_m \int_{\partial D} \left\{ \frac{\partial e^{i\kappa \hat{x} \cdot y}}{\partial \nu(y)} - i e^{i\kappa \hat{x} \cdot y} \right\} \varphi(y) \, ds(y), \quad \hat{x} \in \Omega \qquad (2.1.44)$$

with density $\varphi = -2(I + K - iS)^{-1} u^i$ in the sound-soft case and by

$$u^\infty(\hat{x}) := \gamma_m \int_{\partial D} \left\{ e^{i\kappa \hat{x} \cdot y} \varphi(y) + i \frac{\partial e^{i\kappa \hat{x} \cdot y}}{\partial \nu(y)} (S_0^2 \varphi)(y) \right\} ds(y), \quad \hat{x} \in \Omega$$
$$(2.1.45)$$

with density $\varphi = 2(I - K^* - iT S_0^2)^{-1} \frac{\partial u^i}{\partial \nu}$ for the sound-hard scatterer. The mapping $u^i \mapsto u^\infty$ can be split into the inversion of the corresponding boundary integral operator and the mapping (2.1.44) or (2.1.45), respectively. We will show that each of these mappings is bounded uniformly for $D \in \mathcal{C}_{obst}$.

The boundary integral operators $(I + K - iS)^{-1}$ and $(I - K^* - iT S_0^2)^{-1}$ are uniformly bounded according to Theorem 2.1.13. The functions (2.1.44) and (2.1.45) are considered as linear operators from $C(\partial D)$ into $C^1(\Omega)$. The first operator (2.1.44) has continuous kernel and thus depends continuously on $\partial D \in \mathcal{C}_{obst}$ with respect to the $C^{1,\alpha}$-norm. The second operator (2.1.45) is a sum and composition of integral operators with continuous kernels and the operator S_0, which due to Theorem 2.1.12 depends continuously on ∂D. Thus (2.1.45) depends continuously on ∂D.

As in the proof of Theorem 2.1.13, we may use the compactness of the imbedding $C^2(B_{r_i}(0))$ into $C^{1,\alpha}(B_{r_i}(0))$ and the boundedness of continuous functions on compact sets to obtain a uniform bound for scatterers $D \in \mathcal{C}_{obst}$. $\square$

The scattered field for incident point-sources and multipoles. Our main idea and the ongoing theme will be the use of point-sources, dipoles or multipoles for the investigation and solution of inverse scattering problems. As indicated by the mixed reciprocity relations above, the far field pattern of point-sources can be useful for reconstructions, since, up to a constant factor, it is equal to the scattered field of a plane wave. This observation will lead us to the point-source method in Chapter 5. For contributions to the questions of uniqueness, stability, and for the method of singular sources we will investigate the scattered field of incident point-sources, dipoles or multipoles.

We now investigate the behavior of the scattered field for incident point-sources. In the sound-soft case, for points x on the boundary ∂D from the boundary condition we have

$$\Phi^s(x,z) = -\Phi(x,z), \quad x \in \partial D,$$

and thus

$$\Phi^s(x,z) \to \infty, \quad z \to x.$$

This leads to the idea to use $\Phi^s(x,z)$ for reconstructions of the unknown boundary. But from the viewpoint of the inverse problem, where the boundary ∂D is not known, we would like to replace the point $x \in \partial D$ by something which is known. This new function should not assume knowledge about the boundary, but lead to the same singular behavior. In the following theorem a corresponding behavior is found for

$$\Phi^s(z,z)$$

and we estimate the nature of the singularity.

By $d(z,D)$ or $d(D_1,D_2)$ we denote the Hausdorff distance

$$d(z,D) := \inf\{|z-y| : \ y \in D\} \tag{2.1.46}$$

or

$$d(D_1,D_2) := \max\left\{\sup_{x \in D_1} d(x,D_2), \sup_{x \in D_2} d(x,D_1)\right\} \tag{2.1.47}$$

between z and the domain D or between the two domains D_1 and D_2, respectively.

THEOREM 2.1.15 *Consider the scattering of a point-source $\Phi(\cdot,z)$ by a sound-soft or sound-hard scatterer $D \in C_{obst}$. In $\mathbb{R}^2$ there exist constants $\tau, c > 0$, such that the scattered field Φ^s satisfies the lower estimate*

$$\left|\Phi^s(z,z)\right| \geq c \left|\ln d(z,D)\right| \tag{2.1.48}$$

in the strip $0 < d(z,D) < \tau$. With constants $C, E > 0$ we have the upper estimate

$$\left|\Phi^s(z,z)\right| \leq C \left|\ln d(z,D)\right| + E \tag{2.1.49}$$

for all $z \in B \setminus \overline{D}$. In $\mathbb{R}^3$ the corresponding estimates are

$$\left|\Phi^s(z, z)\right| \geq \frac{c}{|d(z, D)|} \tag{2.1.50}$$

and

$$\left|\Phi^s(z, z)\right| \leq \frac{C}{|d(z, D)|}. \tag{2.1.51}$$

All estimates hold uniformly for domains $D \in \mathcal{C}_{obst}$.

Proof. Consider the sound-hard boundary condition. We will investigate the behavior of Φ^s using the solution of the direct problem by means of boundary integral equations.

We abbreviate the modified acoustic single- and double-layer potential (2.1.15) by

$$(P\varphi)(x) := \int_{\partial D} \left\{ \Phi(x, y)\varphi(y) + i\frac{\partial \Phi(x, y)}{\partial \nu(y)}(S_0^2\varphi)(y) \right\} ds(y), \quad x \in \mathbb{R}^m \setminus \partial D. \tag{2.1.52}$$

Then from Theorem 2.1.2 we obtain a representation

$$\Phi^s(\cdot, z) = 2P(I - K' - iTS_0^2)^{-1}\frac{\partial \Phi(\cdot, z)}{\partial \nu}, \quad z \in \mathbb{R}^m \setminus \partial D, \tag{2.1.53}$$

for the scattered field of point-sources.

For a point $z \in \mathbb{R}^m \setminus D$ with $d(z, D)$ sufficiently small we have the unique representation $z = z_h$ with

$$z_h := z_0 + \nu(z_0)h, \quad z_0 \in \partial D$$

and $h \geq 0$.

We use z_h to decompose

$$2P(I - K^* - iTS_0^2)^{-1}\frac{\partial \Phi(\cdot, z_h)}{\partial \nu(\cdot)}$$

$$= -2P(I - K^* - iTS_0^2)^{-1}\frac{\partial \Phi(\cdot, z_{-h})}{\partial \nu(\cdot)}$$

$$\quad + 2P(I - K^* - iTS_0^2)^{-1}\left\{ \frac{\partial \Phi(\cdot, z_h)}{\partial \nu(\cdot)} + \frac{\partial \Phi(\cdot, z_{-h})}{\partial \nu(\cdot)} \right\}$$

$$= -\Phi^s(\cdot, z_{-h}) \tag{2.1.54}$$

$$\quad + 2P(I - K^* - iTS_0^2)^{-1}\left\{ \frac{\partial \Phi(\cdot, z_h)}{\partial \nu(\cdot)} + \frac{\partial \Phi(\cdot, z_{-h})}{\partial \nu(\cdot)} \right\}.$$

First, consider the term $-\Phi^s(\cdot, z_{-h})$ of (2.1.54). Since z_{-h} is in the interior of the scatterer, we have

$$-\Phi^s(x, z_{-h}) = \Phi(x, z_{-h}), \quad x \in \mathbb{R}^m \setminus D. \tag{2.1.55}$$

In two dimensions Φ has a logarithmic singularity; in three dimensions its singularity is of first order.

To obtain the estimates of the theorem we will show that the singularity of the second term of (2.1.54) is weaker than the singularity of the first term. To this end, in the three-dimensional case $m = 3$ we establish that

$$\left| 2P(I - K^* - iTS_0^2)^{-1} \left\{ \frac{\partial \Phi(\cdot, z_h)}{\partial \nu(\cdot)} + \frac{\partial \Phi(\cdot, z_{-h})}{\partial \nu(\cdot)} \right\} (z_h) \right| \le C \left| \ln h \right| \tag{2.1.56}$$

for all sufficiently small $h > 0$ with some constant C independent of ∂D. We decompose

$$P(I - K^* - iTS_0^2)^{-1} = P + P(I - K^* - iTS_0^2)^{-1}(K^* + iTS_0^2), \tag{2.1.57}$$

where P, given by (2.1.52), is the sum

$$P = P_1 + iP_2 S_0^2 \tag{2.1.58}$$

of a single-layer potential P_1 as defined in (1.2.53) and a term $P_2 S_0^2$ with a double-layer potential P_2 given by (1.2.54). Abbreviating

$$\Psi_h(y) := \left\{ \frac{\partial \Phi(y, z_h)}{\partial \nu(y)} + \frac{\partial \Phi(y, z_{-h})}{\partial \nu(y)} \right\}, \quad y \in \partial D, \ h \in [0, 1], \tag{2.1.59}$$

we obtain

$$\begin{aligned}
P(I - K^* - iTS_0^2)^{-1}\Psi_h &= \left(P_1 \Psi_h \right) + P_2 S_0 \left(S_0 \Phi_h \right) \\
&\quad + P(I - K^* - iTS_0^2)^{-1}\left(K^* \Psi_h \right) \\
&\quad + iP(I - K^* - iTS_0^2)^{-1}\left(TS_0 \right)\left(S_0 \Psi_h \right).
\end{aligned} \tag{2.1.60}$$

Below we will derive the logarithmic bounds

$$\begin{aligned}
\left| (P_1 \Psi_h)(z_h) \right| &\le c \, |\ln(h)|, \\
\left| (S_0 \Psi_h)(x) \right| &\le c \, |\ln(h)|, x \in \partial D, \\
\left| (K^* \Psi_h)(x) \right| &\le c \, |\ln(h)|, x \in \partial D,
\end{aligned} \tag{2.1.61}$$

with some constant c uniformly for $\mathcal{D} \in C_{obst}$. To keep the main line of reasoning, here we first use (2.1.61) and finish the proof. We note that by

$$\begin{aligned}
TS_0 &= (T - T_0)S_0 + T_0 S_0 \\
&= (T - T_0)S_0 + [(K_0^*)^2 - I],
\end{aligned}$$

the operator TS_0 can be extended from $C^{0,\alpha}(\partial D)$ to $C(\partial D)$ and is bounded in $C(\partial D)$ uniformly for all $\mathcal{D} \in \mathcal{C}_{obst}$. According to Theorems 2.1.8 and 2.1.13 the operators

$$(I - K^* - iTS_0^2)^{-1} : C(\partial D) \to C(\partial D),$$

$$TS_0 : C(\partial D) \to C(\partial D),$$

$$S_0 : C(\partial D) \to C^{0,\alpha}(\partial D)$$

and

$$P_2 : C^{0,\alpha}(\partial D) \to C(B \setminus \overline{D})$$

are bounded uniformly for $\mathcal{D} \in \mathcal{C}_{obst}$. The single-layer

$$P_1 : C(\partial D) \to C(B \setminus \overline{D})$$

is bounded uniformly for $\mathcal{D} \in \mathcal{C}_{obst}$. Using the properties of P_2 and S_0 the same result is obtained for the operator

$$P : C(\partial D) \to C(B \setminus \overline{D}).$$

Then from the decomposition (2.1.60) using the bounds (2.1.61) we derive the estimate (2.1.56). Finally, we combine (2.1.56), (2.1.55), and (2.1.54) to obtain the statements (2.1.50) and (2.1.51) of the Theorem.

We now reduce (2.1.61) to the potential theoretic case $\kappa = 0$ and estimate the corresponding integrals explicitly. The reduction is possible, since the difference $\Phi(x, y) - \Phi_0(x, y)$ and its normal derivative are continuous functions. Thus, to estimate the singularity in (2.1.61), we can replace Φ by Φ_0. We proceed in three steps.

1. For Ψ_h in the case $\kappa = 0$ we calculate

$$
\begin{aligned}
\Psi_h(y) &= \left\{ \frac{\partial \Phi_0(y, z_h)}{\partial \nu(y)} + \frac{\partial \Phi_0(y, z_{-h})}{\partial \nu(y)} \right\} \\
&= \frac{\nu(y) \cdot (y - z_0)}{|y - z_h|^3} + \frac{\nu(y) \cdot (y - z_0)}{|y - z_{-h}|^3} \qquad (2.1.62) \\
&\quad - \nu(y)\, \nu(z)\, h \left(\frac{1}{|y - z_h|^3} - \frac{1}{|y - z_{-h}|^3} \right).
\end{aligned}
$$

With the help of $|\nu(y) \cdot (y - z_0)| \leq L|y - z_0|^2$ given in Lemma 2.1.10 the first two termes of (2.1.62) can be estimated by

$$c\,|y - z_{\pm h}|^{-1} \qquad (2.1.63)$$

uniformly for scatterers $\mathcal{D} \in \mathcal{D}_{obst}$. The last term of (2.1.62) can be de-composed into

$$\nu(y)\,\nu(z)\,h\,\left(\frac{1}{|y-z_h|^3} - \frac{1}{|y-z_{-h}|^3}\right)$$
$$= \frac{\nu(y)\,\nu(z)\,h}{|y-z_h|^2}\left(\frac{1}{|y-z_h|} - \frac{1}{|y-z_{-h}|}\right)$$
$$+ \frac{\nu(y)\,\nu(z)\,h}{|y-z_h|\cdot|y-z_{-h}|}\left(\frac{1}{|y-z_h|} - \frac{1}{|y-z_{-h}|}\right)$$
$$+ \frac{\nu(y)\,\nu(z)\,h}{|y-z_{-h}|^2}\left(\frac{1}{|y-z_h|} - \frac{1}{|y-z_{-h}|}\right). \tag{2.1.64}$$

Then with the help of (2.1.29) and

$$h < |y - z_{\pm h}|$$

we derive the bound (2.1.63) for the third term of (2.1.62), also.

2. The kernel $\Phi(x_\eta, y)$ with $x_\eta = x + \eta\nu(x)$, $\eta \in [0, h]$, of the single-layer potential P_1 or the operator S_0 can be estimated by

$$\left|\Phi(x_\eta, y)\right| \le \frac{c}{|x_\eta - y|},$$

for $x, y \in \partial D$ and $h \ge 0$ sufficiently small. For the kernel $\frac{\partial\Phi(x,y)}{\partial\nu(x)}$ of K^* we have

$$\left|\frac{\partial\Phi(x,y)}{\partial\nu(x)}\right| \le \frac{c}{|x-y|}$$

for $x, y \in \partial D$. Thus to prove (2.1.61) we need to estimate

$$\left|\int_{\partial D} \frac{1}{|x_\eta - y|}\frac{1}{|y - z_{\pm h}|}\,ds(y)\right| \tag{2.1.65}$$

for $\eta \in [0, h]$ and $0 < h \le \tau$ with some sufficiently small parameter τ.

3. We decompose the domain of integration

$$\int_{\partial D} \ldots ds(y) = \int_{\partial D \cap B_R(x)} \ldots ds(y) + \int_{\partial D \setminus B_R(x)} \ldots ds(y) \tag{2.1.66}$$

with R chosen sufficiently small. Let us first consider the second integral. It is bounded by

$$\frac{1}{R}\left|\int_{\partial D \setminus B_R(x_0)} \frac{1}{|y - z_{\pm h}|}ds(y)\right|.$$

The weak singularity of $|y - z_{\pm h}|^{-1}$ is integrable with bounded integral. Thus the second integral of (2.1.66) is bounded uniformly for $\mathcal{D} \in \mathcal{C}_{obst}$ and $x \in \partial D$.

For the first integral of (2.1.66) we use the special coordinate system with origin x_0, third axis given by $\nu(x_0) = (0,0,1)$ and z_0 on the e_1-axis. The tangent plane T_{x_0} in x_0 coincides with the $e_1 - e_2$-plane. The tangent plane in z_0 is denoted by T_{z_0}. We consider only the case $|x_0 - z_0| \le 2R$, since otherwise we may proceed as above.

Let $\hat{y}$ denote the projection of y onto T_{z_0}. Using the bound on the C^2-norm of the surface by straightforward computation we estimate

$$|y - z_{\pm h}| \ge c\sqrt{|\hat{y} - z_0|^2 + h^2}, \quad |y - z_0| \le 4R, \qquad (2.1.67)$$

with some positive constant c uniformly for $\mathcal{D} \in \mathcal{C}_{obst}$. We now project $\hat{y}$ and z_0 onto the plane T_{x_0}. Let $\tilde{y}$ and $\tilde{z}_0$ be the projection of $\hat{y}$ and z_0, respectively. With the help of polar coordinates (r, ϕ) in T_{x_0} we obtain the estimate

$$
\begin{aligned}
|\hat{y} - z_0| &\ge |\tilde{y} - \tilde{z}_0| \\
&= \sqrt{r^2 - 2|x_0 - \tilde{z}_0|r\cos(\phi) + |x_0 - \tilde{z}_0|^2}
\end{aligned}
\qquad (2.1.68)
$$

and

$$|y - x_\eta| \ge |\tilde{y} - x_0| = r. \qquad (2.1.69)$$

Thus with $b := |x_0 - \tilde{z}_0| \in [0, 2R]$ an estimate for (2.1.65) is given by a constant C times the integral

$$
\begin{aligned}
\int_0^R \frac{1}{\sqrt{r^2 - 2br + b^2 + h^2}}\, dr &= -\ln(\sqrt{b^2 + h^2} - b) \\
&\quad + \ln\left(\sqrt{h^2 + (R - b)^2} + R - b\right).
\end{aligned}
$$

We derive

$$\left| \int_{\partial D} \frac{1}{|x_\eta - y|} \frac{1}{|y - z_h|}\, ds(y) \right| \le C|\ln h| \qquad (2.1.70)$$

for all $\eta \in [0, h]$ and sufficiently small $h > 0$ with a constant C uniformly for scatterers $\mathcal{D} \in \mathcal{C}_{obst}$. This completes the proof for the three-dimensional case.

For scattering in two dimensions $m = 2$ and in the sound-soft case the statements can be proven analogously. To avoid repetitions we leave this part to the reader. $\qquad \square$

In the following theorem we investigate the scattered fields $\Phi_{\mu,q}^s(z,z)$ for scattering of multipoles by a sound-soft or sound-hard impenetrable scatterer. We will use the results for the reconstruction of the shape of a scatterer, if the physical properties of the scatterer are unknown.

THEOREM 2.1.16 *We consider the scattering of a multipole $\Phi_{\mu,q}(\cdot, z)$ by a sound-soft or sound-hard scatterer $\mathcal{D} \in \mathcal{C}_{obst}$. In a sufficiently small neighbourhood $0 < d(z, D) < \tau$ of the boundary ∂D let z_0 be defined by the unique representation $z = z_0 + h\nu(z_0)$.*

There are constants $\tau, c > 0$, such that in the strip $0 < d(z, D) < \tau$ the scattered field $\Phi^s_{\mu, -\nu(z_0)}$ satisfies the lower estimate

$$\left| \Phi^s_{\mu, -\nu(z_0)}(z, z) \right| \geq c \left| d(z, D) \right|^{-\mu-m+2} \tag{2.1.71}$$

uniformly for $\mathcal{D} \in \mathcal{C}_{obst}$. For all $z \in B \setminus \overline{D}$ we have the upper estimate

$$\left| \Phi^s_{\mu, -\nu(z_0)}(z, z) \right| \leq C \left| d(z, D) \right|^{-\mu-m+2} \tag{2.1.72}$$

with a constant C uniformly for $\mathcal{D} \in \mathcal{C}_{obst}$.

Proof. For a proof we can go along the lines of the proof of Theorem 2.1.15. To avoid repetitions we will only sketch the differences for the case of the sound-hard scatterer. As in (2.1.54) we decompose into

$$\Phi^s_{\mu, -\nu(z_0)}(\cdot, z) = -\Phi^s_{\mu, \nu(z_0)}(\cdot, z_{-h}) \tag{2.1.73}$$
$$+ 2P(I - K^* - iTS_0^2)^{-1} \left\{ \frac{\partial \Phi_{\mu, -\nu(z_0)}(\cdot, z_h)}{\partial \nu(\cdot)} + \frac{\partial \Phi_{\mu, \nu(z_0)}(\cdot, z_{-h})}{\partial \nu(\cdot)} \right\}.$$

With the same arguments as in the proof of Theorem 2.1.15 the second term of (2.1.73) can be estimated by

$$\left| 2P(I - K^* - iTS_0^2)^{-1} \left\{ \frac{\partial \Phi_{\mu, -\nu(z_0)}(\cdot, z_h)}{\partial \nu(\cdot)} \right. \right. \tag{2.1.74}$$
$$\left. \left. + \frac{\partial \Phi_{\mu, \nu(z_0)}(\cdot, z_{-h})}{\partial \nu(\cdot)} \right\} \right| \leq C \frac{|\ln h|}{h^{\mu+m-3}}.$$

Then both estimates (2.1.71) and (2.1.72) are a consequence of (2.1.73) and (2.1.74).

□

2.2   The inhomogeneous acoustic medium

We now consider the scattering of time harmonic acoustic waves $u^i(x)e^{-i\omega t}$ by a penetrable inhomogeneous scatterer. The inhomogeneity of the scatterer is described by a refractive index

$$n(x) := \frac{c_0^2}{c^2(x)} + i\sigma(x),$$

where $c(x)$ is the sound speed at the point $x \in \mathbb{R}^m$, c_0 denotes the sound speed of a homogeneous background medium and $\sigma \geq 0$ is a function which models the influence of absorption. We assume the scatterer to be bounded, i.e., we have $n(x) = 1$ for x in the open exterior of a bounded domain D. The refractive index may have jumps on the boundary ∂D of D. Let the boundary ∂D of the inhomogeneity be of class C^2 and $n \in C^{0,\alpha}(\overline{D})$. As described in (1.2.91) we use the letter $\mathcal{D}$ for the full inhomogeneous scatterer and $\chi := 1 - n$.

DEFINITION 2.2.1 *Given an incident field u^i with wave number $\kappa = \omega/c_0$ and an inhomogeneous penetrable scatterer $\mathcal{D}$, the direct acoustic inhomogeneous medium scattering problem is to find a radiating scattered field*

$$u^s \in C^2(\mathbb{R}^m \setminus \partial D) \cap C^1(\mathbb{R}^m),$$

such that the total field $u = u^i + u^s$ satisfies

$$\Delta u + \kappa^2 n(x)u = 0 \qquad (2.2.1)$$

in $\mathbb{R}^m \setminus \partial D$.

Since $n(x) = 1$ for $x \in \mathbb{R}^m \setminus \overline{D}$, in the open exterior of D the scattered field is a radiating solution of the Helmholtz equation (1.1.1). Thus Green's formula (2.1.7) and the asymptotic behavior (2.1.8) remain valid for scattering from an inhomogeneous medium, i.e., the scattered field u^s has a far field pattern u^∞.

To solve the scattering problem by means of integral equations we introduce the volume potential

$$(V\varphi)(x) := \int_D \Phi(x,y)\varphi(y)\, dy, \quad x \in \mathbb{R}^m, \qquad (2.2.2)$$

defined on a bounded domain $D \subset \mathbb{R}^m$. For a proof of the following uniqueness and existence result for the inhomogeneous medium scattering problem we refer to Theorems 8.3 and 8.7 of [8].

THEOREM 2.2.2 *The inhomogeneous medium scattering problem has a unique solution and the solution u depends continuously on the incident field u^i with respect to the norm in $C(D)$. In particular, the scattered field u^s can be represented as a volume potential*

$$u^s(x) = -\kappa^2 \int_D \Phi(x,y)\chi(y)u(y) \, dy, \quad x \in \mathbb{R}^m, \qquad (2.2.3)$$

where the total field u is a solution of the Lippmann-Schwinger equation

$$(I + \kappa^2 V\chi)u = u^i. \qquad (2.2.4)$$

The integral operator $I + \kappa^2 V\chi$ is continuously invertible in $C(D)$. □

Reciprocity relations for an inhomogeneous medium. Recall that for incident plane waves $u^i(x,d) = e^{i\kappa x \cdot d}$ we denote the scattered field by $u^s(\cdot, d)$ and its far field pattern by $u^\infty(\cdot, d)$. If the incident field is given by a point-source $\Phi(\cdot, z)$ with source-point $z \in \mathbb{R}^m \setminus \overline{D}$, for the scattered field we write $\Phi^s(\cdot, z)$ and for its far field pattern $\Phi^\infty(\cdot, z)$. As for scattering by obstacles the symmetry between source and receiver is expressed by far field and mixed reciprocity relations.

THEOREM 2.2.3 (**Far field reciprocity relation.**) *The far field patterns for scattering of plane waves by an inhomogeneous medium $\mathcal{D}$ satisfy*

$$u^\infty(\hat{x}, d) = u^\infty(-d, -\hat{x}), \quad \hat{x}, d \in \Omega. \qquad (2.2.5)$$

Proof. We refer to [8], Theorem 8.8. □

THEOREM 2.2.4 (**Mixed reciprocity relation.**) *For acoustic scattering of plane waves $u^i(\cdot, d)$, $d \in \Omega$ and point-sources $\Phi(\cdot, z)$, $z \in \mathbb{R}^m \setminus \overline{D}$ from an inhomogeneous medium $\mathcal{D}$ we have*

$$\Phi^\infty(\hat{x}, z) = \gamma_m \, u^s(z, -\hat{x}), \quad z \in \mathbb{R}^m \setminus \overline{D}, \ \hat{x} \in \Omega, \qquad (2.2.6)$$

where the constant γ_m is defined in (2.1.18).

Proof. By Green's theorem (2.1.6) we have that

$$\int_{\partial D} \left(\Phi^s(y,z) \frac{\partial u^s(y,d)}{\partial\nu(y)} - \frac{\partial\Phi^s(y,z)}{\partial\nu(y)} u^s(y,d) \right) ds(y) = 0, \quad z \in \mathbb{R}^m \setminus \overline{D}, \ d \in \Omega. \qquad (2.2.7)$$

Green's formula (2.1.7) yields the representation

$$\Phi^{\infty}(\hat{x}, z) = \gamma_m \int_{\partial D} \left(\Phi^s(y, z) \frac{\partial e^{-i\kappa \hat{x} \cdot y}}{\partial \nu(y)} - \frac{\partial \Phi^s}{\partial \nu}(y, z) e^{-i\kappa \hat{x} \cdot y} \right) ds(y), \quad \hat{x} \in \Omega$$

(2.2.8)

with γ_m given by (2.1.18). Adding γ_m times (2.2.7) with d replaced by $-\hat{x}$ to equation (2.2.8) we obtain

$$\Phi^{\infty}(\hat{x}, z) = \gamma_m \int_{\partial D} \left(\Phi^s(y, z) \frac{\partial u(y, -\hat{x})}{\partial \nu(y)} - \frac{\partial \Phi^s}{\partial \nu}(y, z) u(y, -\hat{x}) \right) ds(y)$$

(2.2.9)

for $z \in \mathbb{R}^m \setminus \overline{D}$ and $\hat{x} \in \Omega$. We can now use Green's second theorem (2.1.6), the differential equation (2.2.1) and the representation (2.2.3) of the scattered field to derive (2.2.6). □

Classes of inhomogeneous media. We now define an appropriate class of media for further investigation. For the interior D of the support of the inhomogeneity we will have to demand the same restrictions as for obstacle scattering. To obtain stability estimates for the support of the function $\chi = 1 - n$ we will also need to uniformly specify the behavior of n at the boundary ∂D and to assume uniform smoothness of n or χ, respectively, on D.

DEFINITION 2.2.5 *Given positive constants R_e, r_0, a_0, C_0, β_e, C_n, c_{min}, c_{max}, $\mu_0 \in \mathbb{N}_0$, $l = 2$ and $\alpha \in [0, 1]$ we define the class C_m of inhomogeneous medium scatterers $\mathcal{D} = (D, n)$ by the following assumptions.*

1. *The domain D is of class $\mathcal{A}(R_e, r_0, a_0, l, \alpha, C_0, \beta_e)$.*

2. *The refractive index n is in $C^{\mu_0, \alpha}(\overline{D})$ with*

$$\|n\|_{C^{\mu_0, \alpha}(D)} \leq C_n.$$

(2.2.10)

   If $\mu_0 \geq 1$, the condition

$$n \in C^{\mu_0 - 1}(B_{R_e}(0))$$

   is satisfied.

3. *At the boundary ∂D the function $\chi = 1 - n$ has a jump in its μ_0-th derivatives uniformly for $\mathcal{D} \in C_m$ in the sense that*

$$0 < c_{min} \leq |\frac{\partial^{\mu_0} \chi}{\partial \nu^{\mu_0}}(x)| \leq c_{max}, \quad x \in \overline{D} \setminus D_{-r_i}$$

(2.2.11)

   with some constant $r_i > 0$, where

$$D_{-\rho} := \{y \in \mathbb{R}^m : d(y, \mathbb{R}^m \setminus D) > \rho\}.$$

(2.2.12)

Uniform bounds for the scattering map. We will now use the assumptions to derive uniform bounds for the solution of the scattering problem.

THEOREM 2.2.6 *For scatterers $\mathcal{D} \in C_m$ the norms*

$$\left\| (I + \kappa^2 V \chi)^{-1} \right\|_{C(B)}$$

of the integral operators $(I + \kappa^2 V \chi)^{-1}$ are uniformly bounded by some constant c.

Proof. We first note that due to [20], Section II, §13, the operator

$$V \chi : C(D) \to C^1(B)$$

is well defined. Thus $V\chi : C(B) \to C(B)$ is compact and by the Riesz theory the injectivity of $I + \kappa^2 V\chi$ yields its invertibility and the boundedness of the inverse operator. The injectivity of $I + \kappa^2 V\chi$ in $C(B)$ is proven in [8], Theorem 8.7. The operator $V\chi$ depends continuously on the function $\chi \in L^\infty(B)$, since the singularity of Φ is integrable and we can estimate

$$
\begin{aligned}
\|V\chi\varphi\|_{C(B)} &= \left\| \int_B \Phi(x,y)\chi(y)\varphi(y)\, dy \right\|_{C(B)} \\
&\leq c \, \|\chi\|_{L^\infty(B)} \|\varphi\|_{C(B)}
\end{aligned}
\tag{2.2.13}
$$

with some constant c. Then also the operator $(I + \kappa^2 V\chi)^{-1}$ depends continuously on χ for all $\chi \in L^\infty(B)$ for which $I + \kappa^2 V\chi$ is invertible. We will show that the set

$$\mathcal{N} := \{\chi : \; n = 1 - \chi \text{ is refractive index of } \mathcal{D} \in C_m\}$$

is relatively compact in $L^\infty(B)$. Let $(\mathcal{D}_j)_{j \in \mathbb{N}} \in C_m$ be a sequence of scatterers and n_j the refractive index of $\mathcal{D}_j$. Then $\chi_j = 1 - n_j \in \mathcal{N}$, $j \in \mathbb{N}$. With the parametrizations ψ_l constructed in part 5 of Lemma 1.2.2 we have

$$\left\| \chi_j \circ \psi_l \right\|_{C^{0,\alpha}(Z_{a_0,r_0})} \leq c \tag{2.2.14}$$

for all $j \in \mathbb{N}$ and $l = 1, ..., L_2$ with some constant c depending only on C_m. We use the compactness of the imbedding of $C^{0,\alpha}(Z_{a_0,r_0})$ into $C(Z_{a_0,r_0})$ to successively construct convergent subsequences of $(\chi_j \circ \psi_l)_{j \in \mathbb{N}}$ for $l = 1, ..., L_2$. We obtain a subsequence $(\mathcal{D}_{j_k})_{k \in \mathbb{N}}$ of $(\mathcal{D}_j)_{j \in \mathbb{N}}$, such that $(\chi_{j_k})_{k \in \mathbb{N}}$ is convergent towards a function $\chi \in L^\infty(B)$, i.e., we have shown that the set $\mathcal{N} \subset L^\infty(B)$ is relatively compact.

From the construction of the convergent subsequence above we derive that each element χ of $\overline{\mathcal{N}} \subset L^\infty(B)$ is in $C(D)$ with a domain $D \subset B$ of class C^1 and $\chi = 0$ in the exterior of $\overline{D}$. We have shown above that for such functions the operator $I + \kappa^2 V\chi$ is injective and thus $(I + \kappa^2 V\chi)^{-1}$ is well defined. Since on the compact set $\overline{N} \subset L^\infty(B)$ the operator

$$(I + \kappa^2 V\chi)^{-1} \in BL(C(B), C(B))$$

depends continuously on χ, it is bounded uniformly for $D \in C_m$ and the proof is complete. $\qquad\square$

THEOREM 2.2.7 *The mapping of the incident field $u^i \in C(D)$ onto the far field pattern $u^\infty \in C^1(\Omega)$ of the scattered field u^s for scattering from an inhomogeneous medium is bounded by a constant c_∞ uniformly for $D \in C_m$.*

Proof. We use the representation (2.2.3) for the scattered field. Passing to the limit $|x| \to \infty$ a representation of the far field pattern of u^s is given by

$$u^\infty(\hat{x}) = -\gamma_m \int_B e^{-i\kappa\hat{x}\cdot y} \chi(y) u(y)\, ds(y), \quad \hat{x} \in \Omega \qquad (2.2.15)$$

with $u = (I + \kappa^2 V\chi)^{-1} u^i$. By

$$\left\| \int_B e^{-i\kappa\hat{x}\cdot y} \chi(y)\varphi(y)\, ds(y) \right\|_{C^1(\Omega)} \qquad (2.2.16)$$

$$\leq \left\| \int_B (-i\kappa y) e^{-i\kappa\hat{x}\cdot y}\, ds(y) \right\|_{C^1(\Omega)} \|\varphi\|_{C(B)} \|\chi\|_{L^\infty(B)}$$

the integral operator (2.2.15) from $C(B)$ into $C^1(\Omega)$ depends continuously on $\chi \in L^\infty(B)$. We can use the compactness arguments of Theorem 2.2.6 to derive its boundedness uniformly for scatterers $D \in C_m$. From the bound for (2.2.15) and the uniform bound for the operators $(I + \kappa^2 V\chi)^{-1}$ given in Theorem 2.2.6 we obtain the statement of the theorem. $\qquad\square$

We will also need the mapping properties of $(I + \kappa^2 V\chi)^{-1}$ in $L^2(B)$ and of the potential $V\chi$ from $L^2(B)$ into $C(B)$.

THEOREM 2.2.8 *For scatterers $D \in C_m$ the operator $I + \kappa^2 V\chi$ is invertible in $L^2(B)$. The norms of*

$$(I + \kappa^2 V\chi)^{-1} : \ L^2(B) \to L^2(B) \qquad (2.2.17)$$

and

$$V\chi : \ L^2(B) \to C(B) \qquad (2.2.18)$$

are bounded uniformly for $D \in C_m$ by some constant c.

Proof. As shown in [8], Theorem 8.2, the operator V defines a bounded operator from $L^2(B)$ into $H^2(B)$. Thus also $V\chi$ is bounded from $L^2(B)$ into $H^2(B)$ and it is compact in $L^2(B)$. By the Riesz Theory for compact operators, injectivity of $I + \kappa^2 V\chi$ in $L^2(B)$ yields its invertibility. To show injectivity let us assume $u + \kappa^2 V\chi u = 0$ for $u \in L^2(B)$. Then we have $u = -\kappa^2 V\chi u \in H^2(B)$. Since $H^2(B)$ is a subset of $C(B)$ we obtain $u = 0$ from the injectivity of $I + \kappa^2 V\chi$ in $C(B)$.

To get uniform bounds for the norms with the help of the Cauchy Schwarz inequality we first derive the estimate

$$
\begin{aligned}
\|V\chi\varphi\|^2_{C(B)} &= \left\|\int_B \Phi(x,y)\chi(y)\varphi(y)\,dy\right\|^2_{C(B)} \\
&\leq \left(\sup_{x\in B}\int_B |\Phi(x,y)|^2\,dy\right)\int_B |\chi(y)\varphi(y)|^2\,dy \\
&\leq c\,\|\chi\|^2_{L^\infty(B)}\,\|\varphi\|^2_{L^2(B)}
\end{aligned}
\tag{2.2.19}
$$

with some constant c. Thus $V\chi : L^2(B) \to C(B)$ depends continuously on χ and we can proceed as in the proof of Theorem 2.2.6 to obtain the bounds uniformly for $\mathcal{D} \in \mathcal{C}_m$. $\qquad\square$

The scattered field of point-sources and multipoles. In the case of obstacle scattering two different ideas lead to the investigation of point-sources. First, with the help of the mixed reciprocity relation 2.1.4 we want to reconstruct the scattered field u^s in a constructive way from its far field pattern u^∞. Then we can use u^s to detect unknown scatterers using the boundary condition for the total field (see the point-source method in Chapter 5). Second, in Theorem 2.1.15 we obtained a characterization of the unknown scatterer by the behavior of the scattered field of point-sources $\Phi^s(z, z)$, which shall lead us to the method of singular sources for reconstructions in Chapter 6.

In the case of scattering by an inhomogeneous medium the situation is different. First, we will have to observe that the reconstruction of the scattered field u^s in the exterior of the unknown scatterer does not provide a straightforward possibility to detect the unknown scattering domain D or the size of the refractive index n. Second, we will show that the scattered field $\Phi^s(z, z)$ is bounded for $z \in \mathbb{R}^m \setminus D$. Thus the behavior of Φ^s does not characterize the unknown boundary.

LEMMA 2.2.9 *For scattering of a point-source $\Phi(\cdot, z)$ by an inhomogeneous medium we have*

$$
\left|\Phi^s(x,z)\right| \leq c, \quad x \in \mathbb{R}^m, z \in \mathbb{R}^m \setminus \overline{D}, \tag{2.2.20}
$$

with a constant c uniformly for scatterers $\mathcal{D} \in \mathcal{C}_m$.

Proof. We use the representation (2.2.3) of the solution to the scattering problem to derive the decomposition

$$
\begin{aligned}
\Phi^s(\cdot, z) &= -\kappa^2 V\chi(I + \kappa^2 V\chi)^{-1}\Phi(\cdot, z) & \quad (2.2.21)\\
&= -\kappa^2 V\chi\Phi(\cdot, z) + \kappa^4 V\chi(I + \kappa^2 V\chi)^{-1}V\chi\Phi(\cdot, z).
\end{aligned}
$$

Estimating the singularity of the kernel of the volume potential we derive that $\|V\chi\Phi(\cdot, z)\|_{C(B)}$ is bounded uniformly for $z \in \mathbb{R}^m \setminus D$ and $D \in \mathcal{C}_m$. Now the statement of the theorem follows with the help of Theorem 2.2.6.
□

To obtain a singular behavior for the scattered field for scattering by inhomogeneous media with an incident singular source when the source point tends to the boundary of the scatterer we have to work with multipoles of higher order. With γ_m given by (2.1.18) for the multipole of order $\mu \in \mathbb{N}_0$ we use the notation

$$
\Phi_{\mu,q}(x, y) := \begin{cases}
\gamma_2 H^{(1)}_\mu(\kappa|x - y|)e^{i\mu\theta}, & m = 2, \\
\gamma_3 h^{(1)}_\mu(\kappa|x - y|)P_\mu(\cos(\theta)), & m = 3,
\end{cases} \quad (2.2.22)
$$

where θ is the angle between $x - y$ and q. Point-sources Φ are the multipoles $\Phi_{0,q}$ of order zero. If the incident field is $\Phi_{\mu,q}(\cdot, z)$, we denote the scattered field by $\Phi^s_{\mu,q}(\cdot, z)$ and its far field pattern by $\Phi^\infty_{\mu,q}(\cdot, z)$.

We first calculate some integrals, which are needed for the proof of the following Lemma 2.2.11.

LEMMA 2.2.10 *For $\mu \in \mathbb{N}$ we have*

$$
\int_{-\pi/2}^{\pi/2} \cos^\mu(\phi)\cos((\mu + 2)\phi)\, d\phi = 0, \quad (2.2.23)
$$

$$
\int_{-\pi/2}^{\pi/2} \ln(\cos(\phi))\cos^\mu(\phi)\cos((\mu + 2)\phi)\, d\phi = \frac{\pi}{(\mu + 1)\, 2^{\mu+1}}, \quad (2.2.24)
$$

$$
\int_0^{\pi/2} \cos^\mu(\theta)P_{\mu+1}(\cos(\theta))\sin(\theta)\, d\theta = \frac{1}{(\mu + 1)\, 2^{\mu+1}}. \quad (2.2.25)
$$

Proof. We first treat the integral (2.2.23). By straightforward differentiation we verify

$$
\int_{-\pi/2}^{\pi/2} \cos^\mu(\theta)\cos((\mu + 2)\theta)\, d\theta = \left[\frac{\cos^{\mu+1}(\theta)\sin((\mu + 1)\theta)}{\mu + 1}\right]_{-\pi/2}^{\pi/2}.
$$

$$(2.2.26)$$

Since $\cos(\pm\pi/2) = 0$, we obtain (2.2.23). By partial integration we calculate

$$\int_{-\pi/2}^{\pi/2} \ln(\cos(\phi)) \, \cos^{\mu}(\phi) \, \cos((\mu+2)\phi) \, d\phi$$

$$= \left[\ln(\cos(\phi)) \frac{\cos^{\mu+1}(\phi)\sin((\mu+1)\phi)}{\mu+1} \right]_{-\pi/2}^{\pi/2}$$

$$+ \int_{-\pi/2}^{\pi/2} \frac{\sin(\phi)}{\cos(\phi)} \frac{\cos^{\mu+1}(\phi)\,\sin((\mu+1)\phi)}{\mu+1} \, d\phi$$

$$= \frac{1}{\mu+1} \int_{-\pi/2}^{\pi/2} \sin(\phi) \, \sin((\mu+1)\phi) \, \cos^{\mu}(\phi) \, d\phi.$$

$$= \frac{1}{2(\mu+1)} \int_0^{2\pi} \sin(\phi) \, \sin((\mu+1)\phi) \, \cos^{\mu}(\phi) \, d\phi. \quad (2.2.27)$$

With the help of the identity

$$\sin(\phi) \, \sin((\mu+1)\phi) = \frac{1}{2}\Big(\cos(\mu\phi) - \cos((\mu+2)\phi) \Big), \quad (2.2.28)$$

the expansion

$$\cos^{\mu} \phi = \frac{1}{2^{\mu}} \left(e^{i\phi} + e^{-i\phi} \right)^{\mu}$$

$$= \frac{1}{2^{\mu-1}} \sum_{k=0}^{\mu} \binom{\mu}{k} \cos(k\phi)$$

$$= \frac{1}{2^{\mu-1}} \cos(\mu\phi) + \text{ lower terms}, \quad (2.2.29)$$

the integral

$$\int_0^{2\pi} \cos^2(\mu\phi) d\phi = \pi, \quad \mu \in \mathbb{N}, \quad (2.2.30)$$

and the orthogonality of $\cos(k\phi)$ and $\cos(j\phi)$ for $k \neq j$ from (2.2.27) we derive (2.2.24). With the substitution $\cos(\theta) = x$ we get

$$\int_0^{\pi/2} \cos^{\mu}(\theta) P_{\mu+1}(\cos(\theta)) \sin(\theta) \, d\theta = \int_0^1 x^{\mu} P_{\mu+1}(x) \, dx, \quad (2.2.31)$$

where the Legendre polynomial $P_{\mu+1}(x)$ is given by formula (1.2.26). By induction we calculate

$$\int_0^1 x^{\mu} \frac{d^{\mu+1}}{dx^{\mu+1}} \Big\{ (x^2-1)^{\mu+1} \Big\} dx \quad (2.2.32)$$

$$= (-1)^j \frac{\mu!}{(\mu-j)!} \int_0^1 x^{\mu-j} \frac{d^{\mu+1-j}}{dx^{\mu+1-j}} \Big\{ (x^2-1)^{\mu+1} \Big\} \, dx,$$

$1 \leq j \leq \mu$. For $j = \mu$ this yields

$$
\begin{aligned}
\int_0^1 x^\mu P_{\mu+1}(x) \, dx &= \frac{(-1)^\mu}{(\mu+1)2^{\mu+1}} \int_0^1 \frac{d}{dx} \left\{ (x^2-1)^{\mu+1} \right\} \, dx \\
&= \frac{1}{(\mu+1)\,2^{\mu+1}}
\end{aligned}
\tag{2.2.33}
$$

and the proof is complete. $\qquad\square$

LEMMA 2.2.11 *With μ_0 given by the definition of C_m and*

$$
\mu := \begin{cases} \mu_0 + 2 & m = 2 \\ \mu_0 + 1 & m = 3, \end{cases}
\tag{2.2.34}
$$

we obtain the estimate

$$
\left| V\chi \Phi_{\mu,q}(\cdot,z)(x) \right| \leq C \begin{cases} \left| \ln|x-z| \right| + E & m = 2 \\ \frac{1}{|x-z|} & m = 3 \end{cases}
\tag{2.2.35}
$$

for all $x \in B$, $z \in B \setminus \overline{D}$, $x \neq z$, and $q \in \Omega$ with constants C, E uniformly for scatterers $D \in C_m$.

There are constants $\tau, c > 0$, such that in a strip $0 < d(z,D) \leq \tau$ for the special choice $q(z) := -\nu(z_0)$ with z_0 defined by the unique representation

$$
z = z_0 + \nu(z_0)h
$$

the lower estimate

$$
\left| V\chi \Phi_{\mu,-\nu(z_0)}(\cdot,z)(z) \right| \geq c \left| \ln d(z,D) \right|
\tag{2.2.36}
$$

holds uniformly for $D \in C_m$. For $z \in B \setminus \overline{D}$ we have the upper estimate

$$
\left| V\chi \Phi_{\mu,q}(\cdot,z)(z) \right| \leq C \left| \ln d(z,D) \right| + E
\tag{2.2.37}
$$

with constants C, E uniformly for scatterers $D \in C_m$.

Proof. We first investigate the behavior of χ near a point $z_0 \in \partial D$. From the continuity of the derivatives $\chi^{(\gamma)}$ for $|\gamma| < \mu_0$ on B we obtain $\chi^{(\gamma)}(z_0) = 0$ for $|\gamma| < \mu_0$. Thus the tangential derivatives of χ up to the order μ_0 vanish in z_0. Then in a neighbourhood $B_R(z_0) \cap \overline{D}$ of z_0 in $\overline{D}$, $0 < R < r_o < 1$, Taylor's expansion of $\chi \in C^{\mu_0,\alpha}(\overline{D})$ assumes the form

$$
\chi(y) = \frac{(-1)^{\mu_0}}{\mu_0!} \frac{\partial^{\mu_0}\chi}{\partial\nu^{\mu_0}}(z_0)(r\cos(\theta))^{\mu_0} + O\left(r^{\mu_0+\alpha}\right)
\tag{2.2.38}
$$

uniformly for $\mathcal{D} \in \mathcal{C}_m$ with

$$r := |y - z_0|, \quad \cos(\theta) := -\nu(z_0) \cdot (y - z_0)/r \qquad (2.2.39)$$

and

$$c_{min} \leq \left| \frac{\partial^{\mu_0} \chi}{\partial \nu^{\mu_0}} (z_0) \right| \leq c_{max}.$$

To prove (2.2.36) we split the domain of integration into $D \setminus B_R(z)$ and $D \cap B_R(z)$. The integral

$$\int_{D \setminus B_R(z)} \Phi(z, y) \chi(y) \Phi_{\mu,q}(y, z) \, dy \qquad (2.2.40)$$

has a bounded integrand and is bounded uniformly for $\mathcal{D} \in \mathcal{C}_m$. To treat the integral on the ball $B_R(z_0)$ we define

$$\chi_0(y) := \frac{(-1)^{\mu_0}}{\mu_0!} \frac{\partial^{\mu_0} \chi}{\partial \nu^{\mu_0}} (z_0)(r \cos(\theta))^{\mu_0}, \qquad (2.2.41)$$

where r and θ are given by (2.2.39), and decompose

$$\int_{D \cap B_R(z)} \Phi(z, y) \chi(y) \Phi_{\mu,q}(y, z) dy$$

$$= \int_{D \cap B_R(z)} \Phi(z, y) \chi_0(y) \Phi_{\mu,q}(y, z) dy \qquad (2.2.42)$$

$$+ \int_{D \cap B_R(z)} \Phi(z, y) \Big(\chi(y) - \chi_0(y) \Big) \Phi_{\mu,q}(y, z) dy.$$

We may use (2.2.38), the definition (2.2.34) of μ and the singularity of the multipoles $\Phi_{\mu,q}$ as given by (1.2.32) and (1.2.45) to estimate the second integral on the right-hand side of (2.2.42) by

$$\left| \int_{D \cap B_R(z)} \Phi(z, y) \Big(\chi(y) - \chi_0(y) \Big) \Phi_{\mu,q}(y, z) dy \right| \qquad (2.2.43)$$

$$\leq c \begin{cases} \displaystyle\int_h^R |\ln(r)| r^{\alpha-1} \, dr \; \leq \; R^\alpha \frac{|\ln(R)|}{\alpha} + \frac{R^\alpha}{\alpha^2}, & m = 2, \\[4mm] \displaystyle\int_h^R r^{\alpha-1} \, dr \; \leq \; \frac{1}{\alpha} R^\alpha, & m = 3, \end{cases}$$

with some constant c, i.e., for $z \in B \setminus \overline{D}$ and $q \in \Omega$ the integral (2.2.43) is bounded by a constant uniformly for $\mathcal{D} \in \mathcal{C}_m$.

We now investigate the first integral on the right-hand side of (2.2.42). In a neighbourhood of a point $z_0 \in \partial D$ a domain D with C^2-boundary is close to a half space

$$H(z_0) := \{ y \in \mathbb{R}^m : (y - z_0) \cdot \nu(z_0) < 0 \}.$$

To estimate the difference of the integral

$$\int_{D \cap B_R(z)} \Phi(z,y) \chi_0(y) \Phi_{\mu,q}(y,z) \, dy \qquad (2.2.44)$$

to the corresponding integral where D is replaced by the half-space $H(z_0)$, we define the set

$$\Delta := (D \cap B_R(z)) \setminus H(z_0) \cup (H(z_0) \cap B_R(z)) \setminus D.$$

Using the bound on the second derivatives of the parametrizations of ∂D by straightforward calculations we obtain the bound

$$\left| \int_\Delta \Phi(z,y) \chi_0(y) \Phi_{\mu,q}(y,z) \, dy \right| \leq c \qquad (2.2.45)$$

with some constant c uniformly for $z \in B \setminus \overline{D}$ and $\mathcal{D} \in \mathcal{C}_m$. Thus a lower bound for (2.2.44) is provided by

$$\left| \int_{D \cap B_R(z)} \Phi(z,y) \chi_0(y) \Phi_{\mu,q}(y,z) \, dy \right| \qquad (2.2.46)$$

$$\geq \left| \int_{H(z_0) \cap B_R(z)} \Phi(z,y) \chi_0(y) \Phi_{\mu,q}(y,z) \right| - c$$

with c given by (2.2.45). For $q = -\nu(z_0)$ we now explicitly calculate the leading term of (2.2.46). In two dimensions we derive

$$\left| \int_{H(z_0) \cap B_R(z)} \Phi(z,y) \chi_0(y) \Phi_{\mu,q}(y,z) \right|$$

$$= c \left| \int_{-\psi_0}^{\psi_0} \int_{\frac{d(z,D)}{\cos(\psi)}}^{R} \ln(r) \, (r \cos(\psi))^{\mu_0} \, r^{-\mu_0 - 2} \cos((\mu_0 + 2)\psi) \, r \, dr \, d\psi \right|$$

$$+ O(1)$$

$$= \frac{c}{2} \left| \int_{-\psi_0}^{\psi_0} \Big[\ln^2(R) - \ln^2 d(z,D) - \ln^2 \cos(\psi) \right. \qquad (2.2.47)$$

$$\left. + 2 \ln d(z,D) \ln \cos(\psi) \Big] \, \cos^{\mu_0}(\psi) \, \cos((\mu_0 + 2)\psi) \, d\psi \right|$$

$$+ O(1)$$

2.2 The inhomogeneous acoustic medium

with ψ_0 defined by $h = R\cos(\psi_0)$ and some constant c. We can now use Lemma 2.2.10 to calculate the integrals in (2.2.47) and derive

$$\left| \int_{H(z_0) \cap B_R(z)} \Phi(z,y)\chi_0(y)\Phi_{\mu,q}(y,z) \right| \geq c\,|\ln d(z,D)| \qquad (2.2.48)$$

for $d(z,D)$ sufficiently small with some constant c uniformly for $D \in C_m$. From (2.2.40), (2.2.43), (2.2.46) and (2.2.48) we obtain (2.2.36).

The estimate (2.2.37) can be proven analogously to (2.2.36), where now an upper estimate for (2.2.44) has to be calculated. Since basically all arguments are the same as above, we leave this part to the reader.

In three dimensions for (2.2.46) we calculate

$$\left| \int_{H(z_0) \cap B_R(z)} \Phi(z,y)\chi_0(y)\Phi_{\mu,q}(y,z) \right| \qquad (2.2.49)$$

$$= c\left| \int_0^{\theta_0} \int_{\frac{d(z,D)}{\cos(\theta)}}^R \int_0^{2\pi} \frac{1}{r}\,(r\cos(\theta))^{\mu_0}\,r^{-\mu_0-2} \right.$$

$$\left. P_{\mu_0+1}(\cos(\theta))\,\sin(\theta)\,r^2 d\varphi\,dr\,d\theta Big \right| + O(1)$$

$$= 2\pi c\left| \int_0^{\theta_0} \left[\ln(R) - \ln d(z,D) + \ln\cos(\theta) \right] \right.$$

$$\left. \cos^{\mu_0}(\theta)\,P_{\mu_0+1}(\cos(\theta))\,\sin(\theta)\,d\theta \right| + O(1)$$

with θ_0 defined by $h = R\cos(\theta_0)$ and some constant c. We use equation (2.2.25) of Lemma 2.2.10 and proceed as in the two-dimensional case.

We now prove the upper estimates (2.2.35). For the function

$$\Psi(y,z) := \chi(y)\Phi_{\mu,q}(y,z)$$

we have

$$|\Psi(y,z)| \leq c\,|y-z|^{-2}, \quad y \in B, z \in B \setminus \overline{D}, x \neq y \qquad (2.2.50)$$

with some constant c uniformly for $D \in C_m$. We split the domain of integration into three parts. For $x \in \mathbb{R}^m \setminus \{z\}$ we define $R := |x-z|/2$ and use the sets $D_1 := D \cap B_R(x)$, $D_2 := D \cap B_R(z)$ and $D_3 := D \setminus (D_1 \cup D_2)$. We first treat the three-dimensional case $m = 3$. For the following calculations we use C as a generic constant, i.e., C may vary from line to line. The integral over D_1 can be estimated by

$$\left| \int_{D_1} \Phi(x,y)\Psi(y,z)\,dy \right| \leq C\,R^{-2}\int_0^R r\,dr$$

$$\leq C \qquad (2.2.51)$$

uniformly for $\mathcal{D} \in \mathcal{C}_m$. To estimate the integral over D_2 we use the decomposition

$$\left| \int_{D_2} \Phi(x,y)\Psi(y,z)\,dy \right| \leq \left| \int_{D_2} \Big(\Phi(x,y) - \Phi(x,z)\Big)\Psi(y,z)\,dy \right|$$
$$+ \left| \int_{D_2} \Phi(x,z)\Psi(y,z)\,dy \right|. \qquad (2.2.52)$$

From the mean value theorem we derive

$$|\Phi(x,y) - \Phi(x,z)| \leq \frac{C}{|x-z|^2}|y-z|, \quad y \in D_2$$

and estimate (2.2.52) by

$$\left| \int_{D_2} \Phi(x,y)\Psi(y,z)\,dy \right| \leq C\left[\frac{1}{R^2}\int_0^R r\,dr + \frac{1}{R}\int_0^R dr \right]$$
$$\leq C. \qquad (2.2.53)$$

The integral over D_3 can be estimated by

$$\left| \int_{D_3} \Phi(x,y)\Psi(y,z)\,dy \right| \leq \frac{C}{R}\int_{D_3}\frac{1}{|y-z|^2}ds(y)$$
$$\leq \frac{C}{R} \qquad (2.2.54)$$

with some constant C. Note that the last estimate is not sharp, but sufficient for our purposes. Now, from (2.2.51), (2.2.53) and (2.2.54) we obtain the estimate (2.2.35) for the case $m = 3$.

In principle in two dimensions we can proceed analogously. A modification is necessary where $\Psi(\cdot, z)$ is considered, since for $m = 2$ the function $|y|^{-2}$ is not integrable. To obtain the estimate (2.2.52) we need to proceed as in (2.2.47) and estimate

$$\left| \int_{D_2} \chi(y)\Phi_{\mu,q}(y,z)\,dy \right| \leq C. \qquad (2.2.55)$$

Since all other parts of the proof are analogous to $m = 3$, to avoid repetitions we leave these parts to the reader. □

We are now prepared to prove estimates for the behavior of the scattered fields of multipoles.

THEOREM 2.2.12 *With μ given by (2.2.34) consider the scattering of a multipole $\Phi_{\mu,q}$ from an inhomogeneous medium scatterer $\mathcal{D} \in \mathcal{C}_m$. There are constants $\tau, c > 0$, such that in the strip $0 < d(z, D) < \tau$ the scattered field $\Phi_{\mu,q}^s$ satisfies the lower estimate*

$$\left| \Phi_{\mu,-\nu(z_0)}^s(z,z) \right| \geq c \left| \ln d(z,D) \right| \qquad (2.2.56)$$

uniformly for scatterers $\mathcal{D} \in \mathcal{C}_m$, where $z_0 \in \partial D$ is defined by the unique representation $z = z_0 + h\nu(z_0)$. With constants C, E for all $z \in B \setminus \overline{D}$ and $q \in \Omega$ we have the upper estimate

$$\left| \Phi_{\mu,q}^s(z,z) \right| \leq C \left| \ln d(z,D) \right| + E \qquad (2.2.57)$$

uniformly for $\mathcal{D} \in \mathcal{C}_m$.

Proof. As in (2.2.21) we decompose

$$\begin{aligned} \Phi_{\mu,q}^s(\cdot, z) &= -\kappa^2 V \chi \Phi_{\mu,q}(\cdot, z) \qquad (2.2.58)\\ &\quad + \kappa^4 V \chi (I + \kappa^2 V \chi)^{-1} V \chi \Phi_{\mu,q}(\cdot, z). \end{aligned}$$

We have shown in Lemma 2.2.11 that the function $V\chi\Phi_{\mu,q}(\cdot, z)$ is bounded in $L^2(B)$ uniformly for $\mathcal{D} \in \mathcal{C}_m$. From Theorem 2.2.8 we know that $(I + \kappa^2 V \chi)^{-1}$ is bounded in $L^2(B)$ and that $V\chi$ is bounded from $L^2(B)$ into $C(B)$ uniformly for $\mathcal{D} \in \mathcal{C}_m$. Thus

$$\kappa^4 V \chi (I + \kappa^2 V \chi)^{-1} V \chi \Phi_{\mu,q}(\cdot, z)$$

is bounded for $z \in B \setminus \overline{D}$ by some constant c uniformly for $\mathcal{D} \in \mathcal{C}_m$.

Since the second term of (2.2.58) is bounded, the statements of Theorem 2.2.12 are a consequence of the estimates (2.2.36) and (2.2.37) of Lemma 2.2.11 for the function $\left(V \chi \Phi_{n,q}(\cdot, z) \right)(z)$. $\qquad \square$

2.3   Electromagnetic scattering by a perfect conductor

We now switch from the Helmholtz equation to Maxwell's equations to investigate the scattering of electromagnetic waves. As in the case of acoustic scattering we study the cases of impenetrable and penetrable scatterers separately. Here we start with the problem of scattering by a perfectly conducting obstacle D in $\mathbb{R}^3$. We assume the boundary ∂D of the support of the scatterer $\mathcal{D}$ to be of class $C^{2,\alpha}$. A time-harmonic incident field E^i, H^i is a solution to the reduced Maxwell equations

$$\operatorname{curl} E - i\kappa H = 0, \qquad \operatorname{curl} H + i\kappa E = 0 \tag{2.3.1}$$

where the wave number κ is a constant given by

$$\kappa^2 = \left(\epsilon + \frac{i\sigma}{\omega} \right) \mu\omega^2 \tag{2.3.2}$$

with the electric permittivity ϵ, the magnetic permeability μ, the electric conductivity σ, the frequency ω of the time-harmonic wave and the sign of κ chosen such that $\operatorname{Im}\kappa \geq 0$.

DEFINITION 2.3.1 *Given an incident electromagnetic field E^i, H^i and a domain D, the direct electromagnetic scattering problem with perfect conductor boundary condition is to find a scattered electromagnetic field E^s, H^s which solves the reduced Maxwell equations (2.3.1) in $\mathbb{R}^3 \setminus \overline{D}$ and satisfies the Silver-Müller radiation condition*

$$\lim_{r\to\infty} (E \times x + rH) = 0, \quad r = |x|, \tag{2.3.3}$$

where the limit is assumed to hold uniformly in all directions $x/|x|$, such that the total field

$$E = E^i + E^s, \qquad H = H^i + H^s \tag{2.3.4}$$

satisfies the perfect conductor boundary condition

$$\nu \times E = 0 \text{ on } \partial D. \tag{2.3.5}$$

A solution of the reduced Maxwell equations in the exterior of some ball B which satisfies (2.3.3) is called radiating.

An important tool for the treatment of the direct and inverse scattering problems will again be Green's first and second theorem and a version of Green's formula for electromagnetic waves. For a radiating solution

$$E, H \in C^1(\mathbb{R}^3 \setminus \overline{D}) \cap C(\mathbb{R}^3 \setminus D)$$

to Maxwell's equations we have the Stratton-Chu formulas

$$\begin{aligned} E(x) &= \operatorname{curl} \int_{\partial D} \nu(y) \times E(y)\Phi(x,y)\, ds(y) \\ &\quad - \frac{1}{i\kappa}\operatorname{curl}\operatorname{curl} \int_{\partial D} \nu(y) \times H(y)\Phi(x,y)\, ds(y), \quad x \in \mathbb{R}^3 \setminus \overline{D}, \end{aligned} \tag{2.3.6}$$

and

$$H(x) = \operatorname{curl} \int_{\partial D} \nu(y) \times H(y) \Phi(x,y) \, ds(y)$$
$$+ \frac{1}{i\kappa} \operatorname{curl} \operatorname{curl} \int_{\partial D} \nu(y) \times E(y) \Phi(x,y) \, ds(y), \quad x \in \mathbb{R}^3 \setminus \overline{D}.$$

$$(2.3.7)$$

Every radiating solution E, H to the Maxwell equations has the asymptotic form

$$E(x) = \frac{e^{i\kappa|x|}}{|x|} \left\{ E^{\infty}(\hat{x}) + O\left(\frac{1}{|x|}\right) \right\}, \quad |x| \to \infty$$

$$(2.3.8)$$

$$H(x) = \frac{e^{i\kappa|x|}}{|x|} \left\{ H^{\infty}(\hat{x}) + O\left(\frac{1}{|x|}\right) \right\}, \quad |x| \to \infty$$

uniformly for all directions $\hat{x} = x/|x|$. The vector fields E^{∞} and H^{∞} are defined on the unit sphere Ω and known as the electric and magnetic far field pattern, respectively. They satisfy

$$H^{\infty} = \nu \times E^{\infty} \quad \text{and} \quad \nu \cdot E^{\infty} = \nu \cdot H^{\infty} = 0. \qquad (2.3.9)$$

Solutions E, H to Maxwell's equations are divergence free and satisfy the vector Helmholtz equation

$$\Delta E + \kappa^2 E = 0 \quad \text{and} \quad \Delta H + \kappa^2 H = 0. \qquad (2.3.10)$$

Passing to the far fields in (2.3.6) and (2.3.7) we obtain

$$E^{\infty}(x) = i\kappa \operatorname{curl} \int_{\partial D} \nu(y) \times E(y) e^{-i\kappa \hat{x} \cdot y} \, ds(y)$$
$$- \frac{\kappa}{i} \operatorname{curl} \operatorname{curl} \int_{\partial D} \nu(y) \times H(y) e^{-i\kappa \hat{x} \cdot y} \, ds(y), \quad x \in \mathbb{R}^3 \setminus \overline{D},$$

$$(2.3.11)$$

and

$$H^{\infty}(x) = i\kappa \operatorname{curl} \int_{\partial D} \nu(y) \times H(y) e^{-i\kappa \hat{x} \cdot y} \, ds(y)$$
$$+ \frac{\kappa}{i} \operatorname{curl} \operatorname{curl} \int_{\partial D} \nu(y) \times E(y) e^{-i\kappa \hat{x} \cdot y} \, ds(y), \quad x \in \mathbb{R}^3 \setminus \overline{D}.$$

$$(2.3.12)$$

As for acoustic scattering we will use integral equations to solve the direct scattering problem and study the properties of the scattered fields. With the fundamental solution $\Phi(x,y)$ of the Helmholtz equation we use the magnetic dipole operator

$$(Ma)(x) := 2 \int_{\partial D} \nu(x) \times \operatorname{curl}_x \{a(y) \Phi(x,y)\} \, ds(y), \quad x \in \partial D \quad (2.3.13)$$

and the electric dipole operator

$$(Nb)(x) := 2\nu(x) \times \operatorname{curl} \operatorname{curl} \int_{\partial D} \nu(y) \times b(y) \, \Phi(x,y) \, ds(y), \qquad (2.3.14)$$

$x \in \partial D$. Further, we define the projection operator P by

$$(Pb)(x) := (\nu(x) \times b(x)) \times \nu(x), \quad x \in \partial D. \qquad (2.3.15)$$

THEOREM 2.3.2 *The direct electromagnetic scattering problem with perfect conductor boundary condition has a unique solution and the solution depends continuously on the incident field in the sense that the mapping of the boundary data $\nu \times E^i$ onto the scattered fields is continuous from*

$$T_d^{0,\alpha}(\partial D) := \left\{ a \in C^{0,\alpha}(\partial D) : \operatorname{Div} a \in C^{0,\alpha}(\partial D), \nu \cdot a = 0 \right\} \qquad (2.3.16)$$

into $C^{0,\alpha}(\mathbb{R}^3 \setminus D) \times C^{0,\alpha}(\mathbb{R}^3 \setminus D)$.

In particular, the combined magnetic and electric dipole potential

$$(P_E a)(x) = \operatorname{curl} \int_{\partial D} a(y)\Phi(x,y)ds(y) \qquad (2.3.17)$$

$$+ i \operatorname{curl} \operatorname{curl} \int_{\partial D} \nu(y) \times (S_0^2 a)(y)\Phi(x,y)ds(y),$$

$$(P_H a)(x) = \frac{1}{i\kappa}\operatorname{curl} E^s(x), \quad x \in \partial D,$$

with density $a \in T_d^{0,\alpha}(\partial D)$ *and* S_0 *defined as in (2.1.15) solves the electromagnetic scattering problem provided the density a solves the integral equation*

$$a + Ma + i\,NPS_0^2 a = -2\nu \times E^i. \qquad (2.3.18)$$

The inverse operator $(I + M + iNPS_0^2)^{-1}$ *exists and is bounded in the space* $T_d^{0,\alpha}(\partial D)$.

Proof. We refer to Theorem 6.18 and 6.19 of [8].   □

Reciprocity relations for electromagnetic scatterer. For the electromagnetic scattering problems we will proceed along the lines of the acoustic problems. We first prove reciprocity relations; i.e., we show that for electromagnetic scattering the role of source and receiver can be exchanged.

In the following D is a perfect conductor. Let $q \in \Omega$ be a constant vector. The electric field of incident plane waves with polarization q and direction of incidence d is given by

$$E_{pl}^i(x,d,q) = i\kappa(d \times q) \times d \, e^{i\kappa x \cdot d}, \quad H_{pl}^i(x,d,q) = i\kappa d \times q \, e^{i\kappa x \times d}. \qquad (2.3.19)$$

The corresponding scattered fields and far field patterns are E_{pl}^s, H_{pl}^s and $E_{pl}^\infty, H_{pl}^\infty$, respectively.

First we formulate reciprocity for the far field patterns.

THEOREM 2.3.3 (**Far field reciprocity relation.**) *For scattering of electromagnetic plane waves by a perfect conductor we have the reciprocity relation*

$$q \cdot E_{pl}^\infty(\hat{x}, d, p) = p \cdot E_{pl}^\infty(-d, -\hat{x}, q) \qquad (2.3.20)$$

for all $\hat{x}, d, p, q \in \Omega$.

Proof. See [8], Theorem 6.28. □

To obtain a reciprocity relation if either the source or the receiver is in the near field, we consider the electromagnetic field of an electric dipole with polarization p, which is given by

$$
\begin{aligned}
E_{edp}^i(x, z, p) &:= \frac{-1}{i\kappa} \operatorname{curl}_y \operatorname{curl}_y \Big(p \Phi(x, z) \Big), \\[2mm]
H_{edp}^i(x, z, p) &:= \operatorname{curl}_y \Big(p \Phi(x, z) \Big)
\end{aligned}
\qquad (2.3.21)
$$

for $x \neq z$. We denote the corresponding scattered field by

$$E_{edp}^s(\cdot, z, p), \ H_{edp}^s(\cdot, z, p),$$

its far field pattern by

$$E_{edp}^\infty(\cdot, z, p), \ H_{edp}^\infty(\cdot, z, p).$$

For the total field, i.e., the sum of incident and scattered field, we use

$$E_{edp}(\cdot, z, p), \ H_{edp}(\cdot, z, p).$$

The total field for plane waves is denoted by

$$E_{pl}(\cdot, d, q), \ H_{pl}(\cdot, d, q).$$

THEOREM 2.3.4 (**Mixed electromagnetic reciprocity.**) *For scattering by a perfect conductor we have*

$$q \cdot E_{edp}^\infty(\hat{x}, z, p) = \gamma\, p \cdot E_{pl}^s(z, -\hat{x}, q) \qquad (2.3.22)$$

for $\hat{x} \in \Omega$, $z \in \mathbb{R}^M \setminus \overline{D}$ and $p, q \in \Omega$, where $\gamma = \frac{1}{4\pi}$.

Proof. We proceed in two steps. First, from Green's Vector Theorem for electromagnetic plane waves we derive the equation

$$0 = \text{curl} \int_{\partial D} \nu(y) \times E^i_{pl}(y, -\hat{x}, q)\Phi(z, y)ds(y) \tag{2.3.23}$$

$$-\frac{1}{i\kappa}\text{curl curl} \int_{\partial D} \nu(y) \times H^i_{pl}(y, -\hat{x}, q)\Phi(z, y)ds(y), \quad z \in \mathbb{R}^3 \setminus \overline{D}.$$

We add (2.3.23) to the representation formula (Stratton-Chu formula)

$$E^s_{pl}(z, -\hat{x}, q) = \text{curl} \int_{\partial D} \nu(y) \times E^s_{pl}(y, -\hat{x}, q)\Phi(z, y)ds(y) \tag{2.3.24}$$

$$-\frac{1}{i\kappa}\text{curl curl} \int_{\partial D} \nu(y) \times H^s_{pl}(y, -\hat{x}, q)\Phi(z, y)ds(y),$$

$z \in \mathbb{R}^3 \setminus \overline{D}$, and calculate

$$E^s_{pl}(z, -\hat{x}, q) = \text{curl} \int_{\partial D} \nu(y) \times E_{pl}(y, -\hat{x}, q)\Phi(z, y)ds(y) \tag{2.3.25}$$

$$-\frac{1}{i\kappa}\text{curl curl} \int_{\partial D} \nu(y) \times H_{pl}(y, -\hat{x}, q)\Phi(z, y)ds(y)$$

$$= -\frac{1}{i\kappa}\text{curl curl} \int_{\partial D} \nu(y) \times H_{pl}(y, -\hat{x}, q)\Phi(z, y)ds(y),$$

$z \in \mathbb{R}^3 \setminus \overline{D}$, where for the last equality we used the boundary condition for a perfect conductor. By elementary calculations we may verify the equality

$$p \cdot \text{curl}_z\text{curl}_z(a(y)\Phi(y, z)) = a(y) \cdot \text{curl}_z\text{curl}_z(p\Phi(y, z)). \tag{2.3.26}$$

Using (2.3.26) we derive from (2.3.25)

$$p \cdot E^s_{pl}(z, -\hat{x}, q)$$

$$= \int_{\partial D} \nu(y) \times H_{pl}(y, -\hat{x}, q) \cdot E^i_{edp}(y, z, p) \, ds(y),$$

$$= -\int_{\partial D} \nu(y) \times E^i_{edp}(y, z, p) \cdot H_{pl}(y, -\hat{x}, q) \, ds(y) \tag{2.3.27}$$

for $z \in \mathbb{R}^3 \setminus \overline{D}$ and $p, q \in \Omega$.

Second, from the Stratton-Chu formula (2.3.11) and (2.3.12) for far field patterns and the definition of the electromagnetic plane waves we get

$$q \cdot E^\infty_{edp}(\hat{x}, z, p)$$

$$= \gamma \int_{\partial D} \{\nu(y) \times E^s_{edp}(y, z, p) \cdot H^i_{pl}(y, -\hat{x}, q) \tag{2.3.28}$$

$$+ \nu(y) \times H^s_{edp}(y, z, p) \cdot E^i_{pl}(y, -\hat{x}, q)\} \, ds(y)$$

for $\hat{x} \in \Omega$ and $z \in \mathbb{R}^3 \setminus \overline{D}$. Analogously to (2.3.23) we derive the formula

$$
0 = i\kappa \int_{\partial D} \{ \nu(y) \times E^s_{edp}(y, z, p) \cdot H^s_{pl}(y, -\hat{x}, q) \qquad (2.3.29)
$$
$$
+ \nu(y) \times H^s_{edp}(y, z, p) \cdot E^s_{pl}(y, -\hat{x}, q) \} ds(y)
$$

from Green's vector formula and the Maxwell equations applied to the scattered electromagnetic fields. We multiply (2.3.29) by $\gamma/i\kappa$ and add it to (2.3.28) to obtain

$$
q \cdot E^\infty_{edp}(\hat{x}, z, p)
$$
$$
= \gamma \int_{\partial D} \{ \nu(y) \times E^s_{edp}(y, z, p) \cdot H_{pl}(y, -\hat{x}, q)
$$
$$
+ \nu(y) \times H^s_{edp}(y, z, p) \cdot E_{pl}(y, -\hat{x}, q) \} ds(y)
$$
$$
= \gamma \int_{\partial D} \nu(y) \times E^s_{edp}(y, z, p) \cdot H_{pl}(y, -\hat{x}, q) ds(y) \qquad (2.3.30)
$$

for $z \in \mathbb{R}^3 \setminus \overline{D}$ and $p, q \in \Omega$, where we have used the boundary condition for the total field $E_{pl}(., -\hat{x}, q)$.

Now, (2.3.27) and (2.3.30) and the boundary condition for $E_{edp}(., z, p)$ yield the statement of the theorem. □

Classes of scatterers. We now discuss appropriate assumptions on the boundary of a perfectly conducting scatterer D. For the electromagnetic scattering problems we will need slightly more regularity than for the acoustic problems, since we need to work with the mapping properties of M and N and S_0 in appropriate Hölder spaces on the boundary ∂D of the domain D.

DEFINITION 2.3.5 *Given positive constants $R_e, r_0, a_0, C_0, \beta_e, l = 2$ and $\alpha \in (0, 1]$ we define the class*

$$
\mathcal{C}_{pc} = \mathcal{C}_{pc}(R_e, r_0, a_0, l, \alpha, C_0, \beta_e)
$$

of impenetrable perfectly conducting scatterers $\mathcal{D}$ as the set of perfect conductors with scattering domain $D \in \mathcal{A}(R_e, r_0, a_0, l, \alpha, C_0, \beta_e)$.

Clearly, for electromagnetic scatterers of class $\mathcal{C}_{pc}$ the statements of Lemmas 1.2.3 and 1.2.2 remain true. Instead of Lemma 2.1.7 we need the following property.

LEMMA 2.3.6 *Given a sequence $(\mathcal{D}_j)_{j \in \mathbb{N}}$ of scatterers $\mathcal{D}_j \in \mathcal{C}_{pc}$ and $0 < \alpha' < \alpha$, there is a subsequence $(\mathcal{D}_{j_k})_{k \in \mathbb{N}}$ of $(\mathcal{D}_j)_{j \in \mathbb{N}}$, for which the sequence of scattering domains $(D_{j_k})_{k \in \mathbb{N}}$ converges in the $C^{2,\alpha'}$-norm to a domain $D \subset B$.*

Proof. The proof is analogous to the proof of Lemma 2.1.7, where we now use the compactness of the imbedding of the space $C^{2,\alpha}(B_{r_0}(0))$ into the space $C^{2,\alpha'}(B_{r_0}(0))$.   □

Properties of electromagnetic boundary integral operators. We need to collect some of the well-known mapping properties of the operators M and N. Let $CT(\partial D)$ be the space of continuous tangential vector fields on the boundary ∂D of the scatterer $\mathcal{D}$. $CT^{n,\alpha}(\partial D)$ are the tangential vector fields of $C^{n,\alpha}(\partial D)$ for $n \in \mathbb{N}_0$. The space $T_d^{0,\alpha}(\partial D)$ has been defined in (2.3.16). We also need the space

$$T_r^{0,\alpha}(\partial D) := \left\{ b \in CT^{0,\alpha}(\partial D) : \nu \times b \in T_d^{0,\alpha}(\partial D) \right\}. \qquad (2.3.31)$$

For a more detailed study we refer to [8]

THEOREM 2.3.7 *The operator M is bounded as a mapping from $CT(\partial D)$ into $CT^{0,\alpha}(\partial D)$, $CT^{0,\alpha}(\partial D)$ into $CT^{1,\alpha}(\partial D)$ and $T_d(\partial D)$ into $T_d^{0,\alpha}(\partial D)$. The operator N is bounded from $T_r^{0,\alpha}(\partial D)$ into $T_d^{0,\alpha}(\partial D)$.*

Proof. The first statement for M is given by Theorem 3.32 of [7], the second by Theorem 3.3 of [46], the third by Theorem 6.16 of [8]. The mapping properties of N are proven in Theorem 6.17 of [8].   □

Since the kernel of N has a strongly singular part, for $M + iNPS_0^2$ it is more difficult to obtain continuity statements with respect to the boundary of the domain than for the weakly singular acoustic potential operators S, K, K^* or $T - T_0$. In the following we use the results of [79] on the Fréchet differentiability of the operators with respect to the boundary of the domain, which imply the continuous dependence.

THEOREM 2.3.8 *For each $0 < \alpha < 1$ the operators*

$$S_0 : CT(\partial D) \to C^{0,\alpha}(\partial D),$$
$$S_0 : C^{0,\alpha}(\partial D) \to C^{1,\alpha}(\partial D)$$
$$M : CT(\partial D) \to CT^{0,\alpha}(\partial D)$$

depend continuously on the boundary ∂D of the scatterer $\mathcal{D}$ with respect to the C^2-norm of ∂D. The operator

$$N : C^{1,\alpha}(\partial D) \to CT^{0,\alpha}(\partial D)$$

depends continuously on the boundary ∂D of the scatterer $\mathcal{D}$ with respect to the $C^{2,\alpha}$-norm of ∂D.

Proof. The proof for the first statement for S_0 and M is a consequence of Theorem 3.14 of [79]. The second statement for S_0 is given by Theorem 3.18 of [79]. The statement for N can be found in Corollary 3.15 of [79]. □

THEOREM 2.3.9 *For $0 < \alpha' < \alpha$ the operators*

$$S_0 : CT(\partial D) \to C^{0,\alpha'}(\partial D),$$
$$S_0 : C^{0,\alpha'}(\partial D) \to C^{1,\alpha'}(\partial D)$$
$$M : CT(\partial D) \to CT^{0,\alpha'}(\partial D)$$
$$N : C^{1,\alpha'}(\partial D) \to CT^{0,\alpha'}(\partial D)$$

are bounded uniformly for $\mathcal{D} \in \mathcal{C}_{pc}$.

Proof. By the compactness of the imbedding of the space $C^{2,\alpha}(B_{r_i}(0))$ into the space $C^{2,\alpha'}(B_{r_i}(0))$ for $\alpha' < \alpha$ the uniform bounds are a consequence of the continuous dependence given by Theorem 2.3.8.

□

THEOREM 2.3.10 *For $0 < \alpha' < \alpha$ the boundary integral operators*

$$(I + M + iNPS_0^2)^{-1} : CT(\partial D) \to CT(\partial D)$$

and

$$(I + M + iNPS_0^2)^{-1} : CT^{0,\alpha'}(\partial D) \to CT^{0,\alpha'}(\partial D)$$

are bounded by a constant c uniformly for scatterers $\mathcal{D} \in \mathcal{C}_{pc}$.

Proof. We first prove the invertibility of the integral operator $I + M + iNPS_0^2$ in $CT(\partial D)$ and $CT^{0,\alpha'}(\partial D)$. Since M and NPS_0^2 are compact in $CT(\partial D)$ and in $CT^{0,\alpha'}(\partial D)$ we have to prove injectivity of the integral operator to obtain invertibility by the Riesz-Theory for compact operators. Injectivity in $T_d^{0,\alpha'}(\partial D)$ is stated in Theorem 2.3.2. We assume that a is a density in $CT(\partial D)$ such that

$$(I + M + iNPS_0^2)a = 0.$$

Then we have $a = -Ma - iNPS_0^2 a$. By the mapping properties of M, N and S_0^2 as given by Theorem 2.3.7 we first obtain $a \in CT^{0,\alpha'}(\partial D)$ and in a second step $a \in T_d^{0,\alpha'}(\partial D)$. We obtain $a = 0$ from Theorem 2.3.2, i.e., the integral operator is injective in $CT(\partial D)$ and $CT^{0,\alpha'}(\partial D)$ and thus continuously invertible.

In $CT(\partial D)$ and $CT^{0,\alpha'}(\partial D)$ for $0 < \alpha' < \alpha$ the integral operator depends continuously on the boundary ∂D with respect to the $C^{2,\alpha'}$-norm for the domain. We use the compactness of the imbedding of the space $C^{2,\alpha}(U_0)$ into the space $C^{2,\alpha'}(U_0)$ and the fact that continuous functions on compact sets are bounded to derive the statement of the theorem. □

The preceding theorems enable us to derive bounds for the mapping of the incident field onto the far field pattern of the scattered electromagnetic wave uniformly for scatterers $\mathcal{D} \in \mathcal{C}_{pc}$.

THEOREM 2.3.11 *For scattering by a perfect conductor the mapping of the incident electromagnetic field* E^i, H^i *in* $T_d^{0,\alpha}(\partial D)$ *onto the far field pattern* E^∞ *in* $C^1(\Omega, \mathbb{R}^3)$ *of the scattered electric field* E^s *is bounded by a constant* c_∞ *uniformly for scatterers* $\mathcal{D} \in \mathcal{C}_{pc}$.

Proof. We use the combined magnetic and electric dipole potential (2.3.17) for the solution of the scattering problem. Then the far field pattern of the electric field is given by

$$(P_E^\infty a)(\hat{x}) := \frac{i\kappa}{4\pi} \; \hat{x} \times \int_{\partial D} a(y)e^{-i\kappa\hat{x}\cdot y} \, ds(y) \tag{2.3.32}$$

$$+ i \frac{\kappa^2}{4\pi} \; \hat{x} \times \int_{\partial D} \Big(\nu(y) \times (S_0^2 a)(y) \Big) \times \hat{x} \, e^{-i\kappa\hat{x}\cdot y} \, ds(y), \hat{x} \in \Omega,$$

with a density a given by (2.3.18). By the now well-known compactness arguments, the mapping (2.3.32) is bounded from $T_d^{0,\alpha}(\partial D)$ into $C^1(\Omega, \mathbb{R}^3)$ uniformly for $\mathcal{D} \in \mathcal{C}_{pc}$. Together with the uniform bound for the integral operator $(I + M + iNPS_0^2)^{-1}$ we obtain the statement of the theorem. $\square$

The scattered fields of incident electric dipoles. We now study the behavior of the scattered field for incident electric dipoles. The following theorem is the basis and main ingredient for the proof of stability in Chapter 4 and the convergence properties of the method of singular sources for the reconstruction of electromagnetic scatterers in Chapter 6.

THEOREM 2.3.12 *Consider scattering of an incident field given by an electric dipole* E_{edp}^i, H_{edp}^i *by a perfect conductor* $\mathcal{D} \in \mathcal{C}_{pc}$. *There are constants* $\tau, c > 0$, *such that in the strip* $0 < d(z, D) < \tau$ *the scattered electric field* $E_{edp}^s(z, z, \nu(z_0))$, *with* $z_0 \in \partial D$ *defined by the unique representation* $z = z_0 + h\nu(z_0)$, *satisfies the lower estimate*

$$\Big| E_{edp}^s(z, z, \nu(z_0)) \Big| \geq \frac{c}{|d(z, D)|^3}. \tag{2.3.33}$$

With a constant C *we have for all* $p \in \Omega$ *and* $z \in B \setminus \overline{D}$ *the upper estimate*

$$\Big| E_{edp}^s(z, z, p) \Big| \leq \frac{C}{|d(z, D)|^3}. \tag{2.3.34}$$

The estimates are satisfied uniformly for scatterers $\mathcal{D} \in \mathcal{C}_{pc}$.

Proof. With $z_h := z_0 + h\nu(z_0)$ and a representation of the scattered electric field by means of the combined potential P_E we decompose

$$E^{s_{edp}}(\cdot, z_h, p)$$
$$= -2\, P_E\, (I + M + iNPS_0^2)^{-1} \left(\nu \times E_{edp}^i(\cdot, z_h, p)\right)$$
$$= 2\, P_E\, (I + M + iNPS_0^2)^{-1} \left(\nu \times E_{edp}^i(\cdot, z_{-h}, p)\right)$$
$$\quad - 2\, P_E\, (I + M + iNPS_0^2)^{-1} \left(\nu \times \{E_{edp}^i(\cdot, z_h, p)\right.$$
$$\left. +E_{edp}^i(\cdot, z_{-h}, p)\}\right)$$
$$= -E_{edp}^s(\cdot, z_{-h}, p) \qquad (2.3.35)$$
$$\quad - 2\, P_E\, \left(\nu \times \{E_{edp}^i(\cdot, z_h, p) + E_{edp}^i(\cdot, z_{-h}, p)\}\right)$$
$$\quad +2\, P_E\, (I + M + iNPS_0^2)^{-1}(M + iNPS_0^2)$$
$$\quad \left(\nu \times \{E_{edp}^i(\cdot, z_h, p) + E_{edp}^i(\cdot, z_{-h}, p)\}\right).$$

For the first term of (2.3.35) we have $-E_{edp}^s(\cdot, z_{-h}, p) = E_{edp}^i(\cdot, z_{-h}, p)$. By straightforward differentiation with the help of

$$\text{curl curl} = -\Delta + \text{grad div} \qquad (2.3.36)$$

we calculate

$$E_{edp}^i(x, z, p) = -\frac{c\, p}{|x - z|^3} + \frac{3c\, p \cdot (x - z)\, (x - z)}{|x - z|^5} \qquad (2.3.37)$$
$$+ O(\frac{1}{|x - z|^2})$$

with some constant c. We estimate

$$E_{edp}^i(z_h, z_{-h}, \nu(z_0)) = O(\frac{1}{h^2}) + \frac{c}{4\, h^3}\nu(z_0), \qquad (2.3.38)$$

i.e., the first term of the right-hand side of (2.3.35) has a singularity of order three in h. To obtain the statement of the theorem we will show that all other terms can be estimated by a constant times h^{-2} uniformly for scatterers $\mathcal{D} \in \mathcal{C}_{pc}$. We will proceed in four steps.

1. As a first step from (2.3.37) by straightforward computation as in (2.1.64) we derive the estimate

$$\left\|\nu \times \{E_{edp}^i(\cdot, z_h, \nu(z_0))\right. \qquad (2.3.39)$$
$$\left. +E_{edp}^i(\cdot, z_{-h}, \nu(z_0))\}\right\|_{C(\partial D)} = O(\frac{1}{h^2})$$

uniformly for scatterers $\mathcal{D} \in \mathcal{C}_{pc}$. Using the bounds of Theorem 2.3.9 we obtain

$$\left\| (M + iNPS_0^2)\left(\nu \times \left\{ E_{edp}^i(\cdot, z_h, \nu(z_0)) \right. \right. \right. \tag{2.3.40}$$
$$\left. \left. \left. + E_{edp}^i(\cdot, z_{-h}, \nu(z_0)) \right\} \right) \right\|_{CT^{0,\alpha'}(\partial D)} = O(\frac{1}{h^2})$$

and from Theorem 2.3.10 we calculate the estimate

$$\left\| (I + M + iNPS_0^2)^{-1}(M + iNPS_0^2) \right. \tag{2.3.41}$$
$$\left. \left(\nu \times \left\{ E_{edp}^i(\cdot, z_h, \nu(z_0)) + E_{edp}^i(\cdot, z_{-h}, \nu(z_0)) \right\} \right) \right\|_{CT^{0,\alpha'}(\partial D)}$$
$$= O(\frac{1}{h^2}).$$

2. We now investigate the potential P_E defined by (2.3.17). We first note that using

$$\text{grad }_x \Phi(x,y) = -\text{grad }_y \Phi(x,y), \tag{2.3.42}$$

the relation

$$\text{grad } v = \text{Grad } v + \frac{\partial v}{\partial \nu}\nu \tag{2.3.43}$$

in a neighbourhood of ∂D and the partial integration

$$\int_{\partial D} v\text{Div } a \, ds = -\int_{\partial D} \text{Grad } v \cdot a \, ds \tag{2.3.44}$$

for continuously differentiable functions v and tangential fields a and for points $x \in \mathbb{R}^3 \setminus \partial D$ we obtain

$$\text{div} \int_{\partial D} a(y)\Phi(x,y) \, ds(y) = \int_{\partial D} \text{Div } a(y)\Phi(x,y) \, ds(y). \tag{2.3.45}$$

By an application of (2.3.36) with the help of (2.3.45) we derive

$$(P_E a)(x) = \text{curl} \int_{\partial D} a(y)\Phi(x,y) \, ds(y)$$
$$+ \kappa^2 \int_{\partial D} a(y)\Phi(x,y) \, ds(y) \tag{2.3.46}$$
$$+ \text{grad} \int_{\partial D} \text{Div } (\nu(y) \times S_0^2 a(y))\Phi(x,y) \, ds(y),$$

$x \in \mathbb{R}^3 \setminus \partial D$.

3. We now investigate the behavior of the potential P_E for a Hölder continuous density a with $\|a\|_{CT^{0,\alpha}(\partial D)} = O(h^{-2})$. Since the gradient of

the single-layer potential with Hölder continuous density is bounded (see Theorem 2.17 of [7] for the boundedness and [79] for the uniformity of the bounds), we obtain

$$\left|(P_E a)(\cdot)\right| = O(\frac{1}{h^2})$$

uniformly for scatterers $\mathcal{D} \in \mathcal{C}_{pc}$, i.e., we obtain the desired estimate for the last term of (2.3.35).

4. Finally, we investigate the term

$$P_E \left(\nu \times \left\{ E^i_{edp}(\cdot, z_h, p) + E^i_{edp}(\cdot, z_{-h}, p) \right\} \right), \qquad (2.3.47)$$

where no smoothing operators are involved. For the second and third summand of (2.3.46) appropriate bounds can be found by the same arguments as above. We will have to explicitly calculate the leading term of the first summand of P_E, i.e.,

$$\int_{\partial D} \left(\nu(y) \times \left\{ E^i_{edp}(y, z_h, p) + E^i_{edp}(y, z_{-h}, p) \right\} \right) \times \nabla_x \Phi(x, y) \Big|_{x=z_h} ds(y),$$
$$(2.3.48)$$

since for the gradient of the single-layer potential with mere continuous density we have no general estimates available. Proceeding as in (2.1.64) and (2.3.39) we derive a bound

$$\left| \nu(y) \times \left\{ E^i_{edp}(y, z_h, p) + E^i_{edp}(y, z_{-h}, p) \right\} \right| \leq \frac{c}{|y - z_{\pm h}||y - z_{\pm h}|}. \quad (2.3.49)$$

The leading term of (2.3.48) is thus bounded by

$$\int_{\partial D} \frac{C}{|y - z_h|^2 |y - z_{\pm h}||y - z_{\pm h}|} ds(y) \qquad (2.3.50)$$

with some constant C. We proceed analogously to part 3 of the proof of Theorem 2.1.15. With the help of the integral

$$\int_0^R \int_0^{2\pi} \frac{r d\varphi dr}{(r^2 + h^2)^2} = \frac{1}{2h^2} \frac{1}{2(R^2 + h^2)} \qquad (2.3.51)$$

we obtain $O(h^{-2})$ for (2.3.50) and the proof is complete.

$\square$

2.4   The electromagnetic inhomogeneous medium

The inhomogeneity of an inhomogeneous medium for scattering of an electromagnetic wave is described by the refractive index

$$n(x) := \frac{1}{\epsilon_0}\left(\epsilon(x) + i\frac{\sigma(x)}{\omega}\right), \tag{2.4.1}$$

where $\epsilon = \epsilon(x) > 0$ denotes the electric permittivity, $\sigma = \sigma(x)$ the electric conductivity of the medium, and where ω is the frequency of the wave. The magnetic permeability is considered to be a constant $\mu = \mu_0 > 0$. We assume the medium to be bounded, i.e., $\epsilon(x) = \epsilon_0$ and $\sigma(x) = 0$ for $x \notin \overline{D}$ for some domain D contained in a fixed ball $B = B_{R_e}(0)$. Let the domain D have the boundary ∂D of class $C^{2,\alpha}$ and $n \in C^{1,\alpha}(\mathbb{R}^3)$ for some $0 < \alpha < 1$. The inhomogeneous electromagnetic scatterer is denoted by $\mathcal{D} = (D, n)$ and we use $\chi := 1 - n$.

DEFINITION 2.4.1 *Given an incident electromagnetic field* E^i, H^i,

$$\text{curl } E^i - i\kappa H^i = 0, \qquad \text{curl } H^i + i\kappa E^i = 0 \tag{2.4.2}$$

with wave number $\kappa = \epsilon_0\mu_0\omega^2$ *and an inhomogeneous penetrable scatterer* $\mathcal{D}$, *the electromagnetic inhomogeneous medium scattering problem is to find a radiating scattered field* $E^s, H^s \in C^1(\mathbb{R}^3)$, *such that the total field*

$$E = E^i + E^s, \qquad H = H^i + H^s \tag{2.4.3}$$

satisfies the time-harmonic Maxwell equations

$$\text{curl } E - i\kappa H = 0, \qquad \text{curl } H + i\kappa n(x)E = 0 \tag{2.4.4}$$

in $\mathbb{R}^3 \setminus \partial D$.

Outside of the support of the scatterer $\mathcal{D}$ the scattered field E^s, H^s solves the Maxwell equations (2.3.1). Thus the Stratton-Chu formulas (2.3.6) and (2.3.7) are valid and the scattered field has the asymptotic behavior (2.3.8) uniformly for all directions, i.e., the scattered fields have a far field pattern E^∞, H^∞.

To study the properties of the scattered field E^s, H^s we use a solution of the direct scattering problem by means of volume integral equations. For a continuous vector field a we define the potential

$$\begin{aligned}(T_e a)(x) &:= -\kappa^2 \int_D \Phi(x,y)\chi(y)a(y)\,dy \\ &\quad + \text{grad} \int_D \frac{1}{n(y)}\text{grad}\,n(y) \cdot a(y)\Phi(x,y)\,dy, \quad x \in \mathbb{R}^3.\end{aligned} \tag{2.4.5}$$

We summarize uniqueness and existence results in the following theorem.

THEOREM 2.4.2 *The problem of scattering by an electromagnetic inhomogeneous medium has a unique solution and the solution depends continuously on the incident field with respect to the maximum norm on D. In particular, the scattered electric field E^s can be represented as a potential $T_e E$, where the total electric field E satisfies the integral equation*

$$(I - T_e)E = E^i \qquad (2.4.6)$$

on D. The integral operator $I - T_e$ is continuously invertible in $C(D)$.

Proof. We refer to Theorems 9.1, 9.2, 9.4 and 9.5 of [8].   □

Reciprocity relations and classes of inhomogeneous media. As for obstacle scattering we denote an incident plane wave with polarization $q \in \Omega$ and direction of incidence $d \in \Omega$ by $E_{pl}^i(\cdot, d, q), H_{pl}^i(\cdot, d, q)$. The corresponding scattered fields and far field patterns are E_{pl}^s, H_{pl}^s and $E_{pl}^\infty, H_{pl}^\infty$, respectively. An incident electric dipole E_{edp}^i, H_{edp}^i produces the scattered field E_{edp}^s, H_{edp}^s with far field pattern $E_{edp}^\infty, H_{edp}^\infty$. We obtain electromagnetic reciprocity relations as follows.

THEOREM 2.4.3 (**Far field reciprocity relation.**) *The far field patterns for scattering of plane waves by an inhomogeneous medium D satisfy*

$$q \cdot E_{pl}^\infty(\hat{x}, d, p) \ = \ p \cdot E_{pl}^\infty(-d, -\hat{x}, q) \qquad (2.4.7)$$

for $\hat{x}, d, p, q \in \Omega$.

Proof. See [8], Theorem 9.6.   □

THEOREM 2.4.4 (**Mixed reciprocity relation.**) *The far field patterns for scattering of plane waves by an inhomogeneous medium D satisfy*

$$q \cdot E_{edp}^\infty(\hat{x}, z, p) \ = \ \gamma \, p \cdot E_{pl}^s(z, -\hat{x}, q) \qquad (2.4.8)$$

for $\hat{x}, p, q \in \Omega$ and $z \in \mathbb{R}^3 \setminus \overline{D}$, where $\gamma = \frac{1}{4\pi}$.

Proof. The proof is literally the same as for Theorem 2.3.4.   □

To detect the boundary of an inhomogeneous medium using incident singular sources, in acoustic scattering we needed a jump in one of the derivatives of n. For the treatment of electromagnetic inhomogeneous medium scattering by means of integral equations, to avoid the use of boundary integral terms we will have to restrict our presentation to a refractive index

$n \in C^{1,\alpha}(\mathbb{R}^3)$. Thus the order of the derivatives of n, where jumps can occur, must be larger or equal to two. Here we will restrict ourselves to a jump in the second derivative of the refractive index at the boundary of the inhomogeneous medium. In addition, we need some smoothness conditions on n to obtain stability estimates.

DEFINITION 2.4.5 *Given positive constants* R_e, r_0, a_0, C_0, β_e, C_n, c_{min}, c_{max}, $l = 2$ *and* $\alpha \in (0,1]$ *we define the class* C_{elm} *of electromagnetic inhomogeneous medium scatterers* D *by the following assumptions.*

1. *The scattering domain* D *is of class* $A(R_e, r_0, a_0, l, \alpha, C_0, \beta_e)$.

2. *The refractive index* n *is in* $C^{1,\alpha}(\mathbb{R}^3)$ *and in* $C^{2,\alpha}(\overline{D})$ *with*

$$\|n\|_{C^{2,\alpha}(D)} \leq C_n. \tag{2.4.9}$$

3. *At the boundary* ∂D *the function* $\chi = 1 - n$ *has a jump in its second derivatives uniformly for* $D \in C_m$ *in the sense that*

$$0 < c_{min} \leq |\frac{\partial^2 \chi}{\partial \nu^2}(x)| \leq c_{max}, \quad x \in D \setminus D_{-r_i} \tag{2.4.10}$$

with some constant $r_i > 0$ *and* $D_{-\rho}$ *defined by (2.2.12).*

We need to study the mapping properties of the operators T_e and $(I - T_e)^{-1}$ in the spaces of continuous and of L^2-integrable functions.

THEOREM 2.4.6 *For the integral operator* $(I - T_e)^{-1}$ *the norms*

$$\|(I - T_e)^{-1}\|_{C(B)} \tag{2.4.11}$$

are bounded by some constant c *uniformly for scatterers* $D \in C_{elm}$.

Proof. The operator T_e as a mapping from $C(B)$ into $C(B)$ depends continuously on the refractive index $n \in C^1(B)$ and the same is true for the inverse of the operator $I - T_e$. Thus the uniform bound is obtained by a compactness argument using the imbedding from $C^{1,\alpha}(B)$ into $C^1(B)$ as in Theorem 2.2.6.   □

With the same compactness arguments we derive the following uniform bound for the scattering map. We leave the straightforward proof to the reader.

THEOREM 2.4.7 *For scattering of electromagnetic waves from an inhomogeneous medium the mapping of the incident electric* $E^i \in C(D)$ *onto the electric far field patterns* $E^\infty \in C^1(\Omega)$ *is bounded uniformly for* $D \in C_{elm}$ *by a constant* c_∞.

The behaviour of the scattered fields of multipoles. As for acoustic scattering we would like to estimate the singularity of the scattered field of multipoles. But since the singularity of the operator T_e is one order stronger than the singularity of the acoustic volume potential V, the proofs of upper and lower estimates for E^s_{edp} will be more complicated than in the acoustic case.

The following theorem investigates the behavior of $E^s_{edp}(z, z, p)$, if z tends to the boundary of an electromagnetic inhomogeneous medium scatterer $\mathcal{D} \in \mathcal{C}_{elm}$. As a preparation we prove a lemma.

LEMMA 2.4.8 *For the kernel $\Psi(y, z)$ we assume*

$$|\Psi(y, z)| \leq c|y - z|^{-2}, \quad y \in D, \ z \in B, \ y \neq z \qquad (2.4.12)$$

with some constant c. Then we have

$$\left| \int_D \mathrm{grad}_z \Phi(x, y) \Psi(y, z) \, dy \right| \leq \frac{C}{|x - z|}, \quad x, z \in B, \ x \neq z, \qquad (2.4.13)$$

with a constant C uniformly for $\mathcal{D} \in \mathcal{C}_{elm}$.

Proof. We need to work out the proof only for the potential theoretic case $\kappa = 0$. We split the domain of integration into three parts. For $x \in \mathbb{R}^3 \setminus \{z\}$ we define $R := |x - z|/2$ and use $D_1 := D \cap B_R(x)$, $D_2 := D \cap B_R(z)$ and $D_3 := D \setminus (D_1 \cup D_2)$.

1. The integral over D_1 can be estimated by

$$\left| \int_{D_1} \nabla_x \Phi(x, y) \Psi(y, z) \, dy \right| \leq C \, R^{-2} \int_0^R dr \leq \frac{C}{R} \qquad (2.4.14)$$

with some constant C.

2. To estimate the integral over D_2 we use the decomposition

$$\left| \int_{D_2} \nabla_x \Phi(x, y) \Psi(y, z) \, dy \right| \leq \left| \int_{D_2} \left(\nabla_x \Phi(x, y) - \nabla_x \Phi(x, z) \right) \Psi(y, z) \, dy \right|$$

$$+ \left| \int_{D_2} \nabla_x \Phi(x, z) \Psi(y, z) \, dy \right|. \qquad (2.4.15)$$

With

$$|\nabla_x \Phi(x, y) - \nabla_x \Phi(x, z)| \leq \frac{c}{|x - z|^3} |y - z|, \quad |y - z| \leq R$$

for some constant c we estimate (2.4.15) by

$$\left| \int_{D_2} \nabla_x \Phi(x, y) \Psi(y, z) \, dy \right| \leq C \left[\frac{1}{R^3} \int_0^R r \, dr + \frac{1}{R^2} \int_0^R dr \right]$$

$$\leq \frac{C}{R}, \qquad (2.4.16)$$

where C is a generic constant; i.e., C may change from line to line.

3. The domain D_3 can be decomposed again into the subdomains $D_4 :=$ $D \setminus B_{3R}(z)$ and $D_5 := D \cap (B_{3R}(z) \setminus (B_R(x) \cup B_R(z)))$. We use polar coordinates with origin z and third axis given by $x - z$ to calculate

$$|y - x|^2 = 4R^2 + r^2 - 4rR \cos(\theta).$$

Now the integral over D_4 is estimated by a constant times

$$\int_{3R}^{R_*} \frac{1}{4R^2 + r^2 - 4rR}\, dr = \left[-\frac{1}{r - 2R} \right]_{3R}^{R_*}. \qquad (2.4.17)$$

The integral over D_5 can be estimated by a constant times

$$\frac{1}{R^2} \int_R^{3R} dr = \frac{2}{R}. \qquad (2.4.18)$$

The estimates (2.4.17) and (2.4.18) yield

$$\left| \int_{D_3} \nabla_z \Phi(x, y) \Psi(y, z)\, dy \right| \le \frac{C}{R} \qquad (2.4.19)$$

with some constant C. Now from (2.4.14), (2.4.16) and (2.4.19) we obtain the estimate (2.4.13). $\qquad\qquad\qquad\qquad\qquad\qquad\qquad\qquad\qquad\qquad\square$

THEOREM 2.4.9 *Consider scattering of an electric dipole by an inhomogeneous medium scatterer $D \in C_{elm}$. There are constants $\tau, c > 0$ such that in the strip $0 < d(z, D) < \tau$ the scattered field $E_{edp}^s(z, z, \nu(z_0))$ satisfies the lower estimate*

$$\left| E_{edp}^s(z, z, \nu(z_0)) \right| \ge \frac{c}{|d(z, D)|}, \qquad (2.4.20)$$

where $z_0 \in \partial D$ is defined by the unique representation $z = z_0 + h\nu(z_0)$. With a constant C we have for all $z \in B \setminus \overline{D}$ the upper estimate

$$\left| E_{edp}^s(z, z, \nu(z_0)) \right| \le \frac{C}{|d(z, D)|}. \qquad (2.4.21)$$

Proof. By Theorem 2.4.2 we have a representation of the scattered field of an incident electric dipole by

$$
\begin{aligned}
E_{edp}^s(\cdot, z, p) &= T_e(I - T_e)^{-1} E_{edp}^i(\cdot, z, p) \\
&= T_e E_{edp}^i(\cdot, z, p) + T_e(I - T_e)^{-1} T_e E_{edp}^i(\cdot, z, p) \\
&= T_e E_{edp}^i(\cdot, z, p) + T_e T_e E_{edp}^i(\cdot, z, p) \qquad (2.4.22) \\
&\quad + T_e(I - T_e)^{-1} T_e T_e E_{edp}^i(\cdot, z, p),
\end{aligned}
$$

where we twice inserted the identity operator $I = (I - T_e) + T_e$. We first give upper and lower estimates for the singularity of $T_e E^i_{edp}(\cdot, z, p)$, in a second step prove the boundedness of $T_e T_e E^i_{edp}(\cdot, z, p)$ and in a third step derive bounds for $T_e(I - T_e)^{-1} T_e T_e E^i_{edp}(\cdot, z, p)$ in $\mathbb{R}^3$.

1. We need to investigate the refractive index near the boundary. From Definition 2.4.5 as in (2.2.38) we obtain

$$\chi(y) = \frac{1}{2} \frac{\partial \chi}{\partial \nu}(z_0) \Big(r \cos(\theta)\Big)^2 + O(r^{2+\alpha}) \qquad (2.4.23)$$

with polar coordinates $r = |y - z_0|$, $\cos(\theta) = -\nu(z_0) \cdot (y - z_0)/r$ and

$$\mathrm{grad}\, n(y) = \frac{\partial \chi}{\partial \nu}(z_0)\, r \left\{ \cos(\theta)^2 \cdot \begin{pmatrix} \sin(\theta)\cos(\varphi) \\ \sin(\theta)\sin(\varphi) \\ \cos(\theta) \end{pmatrix} \right. \qquad (2.4.24)$$

$$\left. -\cos(\theta)\sin(\theta) \cdot \begin{pmatrix} \cos(\theta)\cos(\varphi) \\ \cos(\theta)\sin(\varphi) \\ -\sin(\theta) \end{pmatrix} \right\} + O(r^{1+\alpha})$$

uniformly for scatterers $\mathcal{D} \in \mathcal{C}_{elm}$.

Consider the operator T_e as defined by (2.4.5). It consists of two terms, the first one of which is a single-layer potential and the second the gradient of a single-layer potential. From (2.4.23) we derive

$$\Big|\chi(y) E^i_{edp}(y, p, z)\Big| \leq \frac{c}{|y - z|}, \quad y \in D, \qquad (2.4.25)$$

with some constant c. Thus by standard arguments the potential

$$\int_D \Phi(x, y)\chi(y) E^i_{edp}(z, p, z)\, dy$$

is bounded uniformly for $z \in B \setminus \overline{D}$ and for all scatterers $\mathcal{D} \in \mathcal{C}_{elm}$. We now investigate the second term of T_e. We use Lemma 2.4.8 to estimate the integral

$$\int_D \mathrm{grad}\,_z\Phi(x, y) \underbrace{\frac{1}{n(y)} \mathrm{grad}\, n(y) \cdot E^i_{edp}(y, p, z)}_{=:\Psi(y,z)}\, dy \qquad (2.4.26)$$

and derive the upper estimate

$$\Big|\Big(T_e E^i_{edp}(\cdot, z, p)\Big)(x)\Big| \leq \frac{C}{|x - z|}, \quad p \in \Omega,\ x, z \in B,\ x \neq z, \qquad (2.4.27)$$

with some constant C uniformly for scatterers $\mathcal{D} \in \mathcal{C}_{elm}$.

We now calculate a lower bound for $T_e E^i_{edp}(z, z, \nu(z_0))$. The leading term of the potential is given by

$$\int_D \frac{z-y}{|z-y|^3} \frac{1}{n(y)} \operatorname{grad} n(y) \tag{2.4.28}$$

$$\cdot \left\{ \frac{3c\, \nu(z_0) \cdot (y-z)\,(y-z)}{|y-z|^5} - \frac{c\, \nu(z_0)}{|y-z|^3} \right\} dy, \quad z \in \mathbb{R}^3 \setminus \overline{D}.$$

We insert (2.4.24) into (2.4.28) and use an argumentation analogous to the derivation of (2.2.47) and (2.2.49) to derive

$$T_e E^i_{edp}(z, z, \nu(z_0)) = C \int_0^{2\pi} \int_0^{\theta_0} \int_{d(z,D)/\cos(\theta)}^{R_0}$$

$$\left\{ \cos(\theta)^2 \begin{pmatrix} \sin(\theta)\cos(\varphi) \\ \sin(\theta)\sin(\varphi) \\ \cos(\theta) \end{pmatrix} - \cos(\theta)\sin(\theta) \begin{pmatrix} \cos(\theta)\cos(\varphi) \\ \cos(\theta)\sin(\varphi) \\ -\sin(\theta) \end{pmatrix} \right\}$$

$$\cdot \left\{ 3\cos(\theta) \begin{pmatrix} \sin(\theta)\cos(\varphi) \\ \sin(\theta)\sin(\varphi) \\ \cos(\theta) \end{pmatrix} - \begin{pmatrix} 0 \\ 0 \\ 1 \end{pmatrix} \right\} \cdot \begin{pmatrix} \sin(\theta)\cos(\varphi) \\ \sin(\theta)\sin(\varphi) \\ \cos(\theta) \end{pmatrix} \frac{1}{r^2}\, dr\, d\theta\, d\varphi$$

$$+ O(1)$$

with a sufficiently small fixed constant R_0 and θ_0 defined by $R_0 \cos(\theta_0) = d(z, D)$. We expand the products and first integrate over φ to obtain zero for the e_1 and e_2 components of the vector. Evaluating the integral over r we derive

$$T_e E^i_{edp}(z, z, \nu(z_0))$$

$$= \frac{C}{d(z,D)} \begin{pmatrix} 0 \\ 0 \\ 1 \end{pmatrix} \int_0^{\theta_0} \cos^3(\theta) \left\{ 3\cos^2(\theta) - 1 \right\} d\theta + O(1)$$

$$= \frac{C}{d(z,D)} \begin{pmatrix} 0 \\ 0 \\ 1 \end{pmatrix} + O(1), \quad z \in \mathbb{R}^3 \setminus \overline{D},$$

with a generic constant C. This yields

$$\left| T_e E^i_{edp}(z, z, \nu(z_0)) \right| \geq \frac{C}{|d(z,D)|}, \quad z \in \mathbb{R}^3 \setminus \overline{D}, \tag{2.4.29}$$

with some constant C uniformly for scatterers $\mathcal{D} \in \mathcal{C}_{elm}$.

2. We need to investigate $T_e T_e E^i_{edp}(\cdot, z, p)$. From (2.4.27) we derive

$$\left| \left(T_e T_e E^i_{edp}(\cdot, z, p) \right)(x) \right| \leq C \int_D \frac{1}{|x-y|^2} \frac{\operatorname{grad} n(y)}{|y-z|}\, dy$$

$$+ O(1). \tag{2.4.30}$$

Since the integral in (2.4.30) can be estimated by a constant, the right-hand side is bounded. These bounds hold uniformly for $\mathcal{D} \in \mathcal{C}_{elm}$.

3. To complete the proof we collect the estimates (2.4.27), (2.4.29) and (2.4.30), and use the decomposition (2.4.22) and the uniform bounds for $(I - T_e)^{-1}$ in $C(B)$ as given by Theorem 2.4.6 to derive the upper and lower bounds (2.4.20) and (2.4.21). $\qquad\qquad\qquad\qquad\qquad\square$

2.5 Scattering by orthotropic media

To introduce the model of orthotropic media consider time-harmonic electromagnetic wave propagation in an inhomogeneous anisotropic medium in $\mathbb{R}^3$. The electric field E and the magnetic field H satisfy the reduced Maxwell equations

$$
\begin{aligned}
\operatorname{curl} E - i\kappa H &= 0 \\
\operatorname{curl} H + i\kappa \mathcal{N}(x)E &= 0,
\end{aligned}
\qquad (2.5.1)
$$

where the wave number $\kappa > 0$ is defined by $\kappa^2 = \epsilon_0 \mu_0 \omega^2$ and where we assume the refractive index $\mathcal{N} = \mathcal{N}(x)$ to be a matrix which is given by

$$
\mathcal{N}(x) = \frac{1}{\epsilon_0}\left(\epsilon(x) + i\frac{\sigma(x)}{w}\right).
\qquad (2.5.2)
$$

Here μ_0 denotes the magnetic permeability which is assumed to be constant, $\epsilon = \epsilon(x)$ denotes the tensor of the electric permittivity, $\sigma = \sigma(x)$ denotes the tensor of the electric conductivity and ω is the frequency of the electromagnetic wave. For physical reasons we can assume that ϵ is positive definite and σ is positive semi-definite. In the case of a diagonal tensor this can be reduced to the usual positivity-conditions which are imposed on ϵ and σ in the scalar case. Further we assume that $\mathcal{N}(x) - I$ has compact support $D \subset \mathbb{R}^2$, i.e., the inhomogeneity is bounded.

Let E^i, H^i be a solution of Maxwell's equations for a homogeneous medium $\mathcal{N}(x) \equiv I$ representing an incident field. We want to find a sufficiently smooth solution E, H of (2.5.1) in $\mathbb{R}^3$ such that the scattered field E^s, H^s defined by

$$
\begin{aligned}
E &= E^i + E^s, \\
H &= H^i + H^S,
\end{aligned}
\qquad (2.5.3)
$$

satisfies the Silver-Müller radiation condition

$$
\lim_{r \to \infty} (H^s \times x - rE^s) = 0
\qquad (2.5.4)
$$

uniformly for all directions $\hat{x} = x/|x|$ where $r = |x|$. We restrict our investigations to continuously differentiable matrix functions $\mathcal{N}$ and an orthotropic medium, i.e., we assume for $\mathcal{N}$ the form

$$\mathcal{N}(x) = \begin{pmatrix} n_{11}(x) & n_{12}(x) & 0 \\ n_{21}(x) & n_{22}(x) & 0 \\ 0 & 0 & n_{33}(x) \end{pmatrix} \qquad (2.5.5)$$

where the matrix is independent of the z-coordinate. We consider electromagnetic fields which are also independent of the z-coordinate. Then we get the following two groups of scalar equations which are called TM-mode and TE-mode, respectively:

$$\frac{\partial E_3}{\partial y} = i\kappa H_1$$

$$\frac{\partial E_3}{\partial x} = -i\kappa H_2$$

$$\frac{\partial H_2}{\partial x} - \frac{\partial H_1}{\partial y} = -i\kappa n_{33} E_3 \qquad (2.5.6)$$

and

$$\frac{\partial H_3}{\partial y} = -i\kappa(n_{11}E_1 + n_{12}E_2)$$

$$\frac{\partial H_3}{\partial x} = i\kappa(n_{21}E_1 + n_{22}E_2)$$

$$\frac{\partial E_2}{\partial x} - \frac{\partial E_1}{\partial y} = i\kappa H_3, \qquad (2.5.7)$$

where the first group involves only H_1, H_2 and E_3 and the second group involves only E_1, E_2 and H_3. The first group of equations describes the scattering problem for an electromagnetic wave polarized perpendicular to the z-axis. The solution to this problem is well known (cf. [8]). From the second group of equations we obtain for $u = H_3$ the equation

$$(\nabla \cdot N(x)\nabla + \kappa^2)u = 0, \qquad (2.5.8)$$

or in a weak sense

$$\int_D \nabla v \cdot N\nabla u \, dx - \int_{\partial D} v \cdot (\nu N \nabla u) \, ds = \kappa^2 \int_D vu \, dx \qquad (2.5.9)$$

for all domains D with C^1-boundary having unit outward normal vector ν and all $v \in H^1(D)$, where the matrix $N(x)$ is given by

$$N(x) = \frac{1}{n_{11}n_{22} - n_{12}n_{21}} \begin{pmatrix} n_{11} & n_{21} \\ n_{12} & n_{22} \end{pmatrix}. \qquad (2.5.10)$$

In the case of the TM or TE-mode we replace the Silver-Müller radiation condition for E^s, H^s by the Sommerfeld radiation condition for the field $u^s := H_3^s$

$$\lim_{r \to \infty} \sqrt{r} \left(\frac{\partial u^s}{\partial r} - iku^s \right) = 0, \tag{2.5.11}$$

uniformly in all directions where $r = |x|$, $x \in \mathbb{R}^2$.

We will work with matrices $N(x)$ which can be pointwise diagonalized with an unitary complex matrix $U(x)$, i.e., we have

$$N(x) = U^*(x)N_D(x)U(x)$$

with a diagonal matrix $N_D(x)$ and $U^*(x)U(x) = I$ for every $x \in \mathbb{R}^2$. Further we will assume that N_D has a positive definite real part and a negative semi-definite imaginary part (for diagonal matrices this can be obtained from the assumptions on the matrix $\mathcal{N}$). For a wide range of practical applications these conditions are fulfilled. We call the corresponding class of orthotropic medium scatterers $\mathcal{C}_{orth}$ and will also consider the matrices N as elements $\mathcal{C}_{orth}$ as abbreviation of $\mathcal{D} = (D, N) \in \mathcal{C}_{orth}$ with the notation of (1.2.91).

Uniqueness and Existence. In this section we will establish the uniqueness and the existence of solutions to the scattering problem (2.5.8) - (2.5.11) following [72]. For the uniqueness proof we use the unique continuation principle. The existence will be shown by means of integral equations. A complex matrix $N(x)$ is called coercive (semi-coercive) if

$$\text{Im} \left(a \cdot \overline{Na} \right) \geq \gamma(x)|a|^2 \tag{2.5.12}$$

for every $a \in \mathbb{C}^2$ where $\gamma(x) > 0$ ($\gamma \geq 0$), compare [4]. Matrices $N \in \mathcal{C}_{orth}$ are semi-coercive. A complex matrix $N(x)$ is called elliptic, if $a \cdot Na \neq 0$ for all $a \neq 0$, $a \in \mathbb{R}^2$. Matrices $N \in \mathcal{C}_{orth}$ are elliptic.

THEOREM 2.5.1 *Assume that N is semi-coercive and elliptic. Then the scattering problem (2.5.8) has at most one solution.*

Proof. Let u be a solution of (2.5.8) with $u^i = 0$. Apply equation (2.5.9) to $v = \bar{u}$ for $D \supset supp(M)$ and use the coercivity of N to obtain

$$\text{Im} \left(\int_{\partial D} u \frac{\overline{\partial u}}{\partial \nu} ds \right) = \int_D \text{Im} \left(\nabla u \cdot (\overline{N\nabla u}) \right) dy \geq 0. \tag{2.5.13}$$

From Theorem 2.12 of [8] we obtain $u \equiv 0$ in the exterior of D (note that $u \in H^1_{loc}(\mathbb{R}^2)$ solves the Helmholtz equation in a weak sense in the exterior of $supp(M)$ and is therefore analytic) and from the unique continuation principle (see [32], Theorem 17.2.1) we get $u = 0$ (in Hörmander's theorem it is shown that the set of gereralized normal vectors to the subset $supp(u) \subset \mathbb{R}^2$ is empty and therefore $u = 0$ in all of $\mathbb{R}^2$). This ends the proof. $\square$

We now transform problem (2.5.8) - (2.5.11) into an integral equation. Let $D \subset \mathbb{R}^2$ be a bounded domain which contains the support of M. Recall that we denote the ball with center x and radius ϵ by $B(\epsilon, x)$. From (2.5.9) for $v(y) = \Phi(x, y)$ we obtain for $D(\epsilon, x) := D \setminus \overline{B(\epsilon, x)}$ the equation

$$
\int_{D(\epsilon,x)} \nabla_y \Phi(x,y) \cdot \nabla u(y) \, dy - \int_{D(\epsilon,x)} \nabla_y \Phi(x,y) \cdot M(y) \nabla u(y) \, dy
$$

$$
- \int_{\partial D(\epsilon,x)} \nu(y) \cdot \Phi(x,y) \nabla u(y) \, ds(y)
$$

$$
+ \int_{\partial B(\epsilon,x)} \nu(y) \cdot \Phi(x,y) M(y) \nabla u(y) \, ds(y)
$$

$$
= \kappa^2 \int_{D(\epsilon,x)} \Phi(x,y) u(y) dy. \tag{2.5.14}
$$

The function $\Phi(x,.)$ solves (2.5.9) in $D(\epsilon, x)$ for $N \equiv I$. Therefore with $v = u$ we obtain

$$
\int_{D(\epsilon,x)} \nabla u(y) \cdot \nabla \Phi(x,y) \, dy - \int_{\partial D(\epsilon,x)} \nu(y) \cdot u(y) \nabla \Phi(x,y) \, ds(y)
$$

$$
= \kappa^2 \int_{D(\epsilon,x)} u(y) \Phi(x,y) dy. \tag{2.5.15}
$$

We subtract (2.5.15) from (2.5.14) and get

$$
- \int_{D(\epsilon,x)} \nabla \Phi(x,y) \cdot M(y) \nabla u(y) \, dy
$$

$$
- \int_{\partial D(\epsilon,x)} \left\{ \frac{\partial u(y)}{\partial \nu(y)} \Phi(x,y) - u(y) \frac{\partial \Phi(x,y)}{\partial \nu(y)} \right\} ds(y)
$$

$$
= - \int_{\partial B(\epsilon,x)} \nu(y) \cdot \Phi(x,y) M(y) \nabla u(y) \, ds(y). \tag{2.5.16}
$$

Using Green's formula (2.1.7) applied to u^i and Green's theorem (2.1.6) and the radiation condition (2.5.11) applied to u^s we derive

$$
\int_{\partial D} \left\{ \Phi(x,y) \frac{\partial u^i(y)}{\partial \nu(y)} - u^i(y) \frac{\partial \Phi(x,y)}{\partial \nu(y)} \right\} ds(y) = u^i(x),
$$

$$
\int_{\partial D} \left\{ \Phi(x,y) \frac{\partial u^s(y)}{\partial \nu(y)} - u^s(y) \frac{\partial \Phi(x,y)}{\partial \nu(y)} \right\} ds(y) = 0. \tag{2.5.17}
$$

Finally we estimate the integrals over $\partial B(\epsilon, x)$ in (2.5.16). We estimate (note that ν is directed into the interior of $B(\epsilon, x)$)

$$\int_{\partial B(\epsilon,x)} u(y)\frac{\partial \Phi(x,y)}{\partial \nu(y)} ds(y) = \frac{1}{2\pi}\int_{\partial B(\epsilon,x)} u(y)\frac{ds(y)}{\epsilon} + O(\epsilon) \qquad (2.5.18)$$

and

$$\int_{\partial B(\epsilon,x)} \frac{\partial u(y)}{\partial \nu(y)}\Phi(x,y)ds(y) = O(\epsilon \ln(\epsilon))$$

uniformly for x in compact subsets of $\mathbb{R}^2$. Using the convergence

$$\lim_{\epsilon \to 0} \| \int_{\partial B(\epsilon,x)} \left(u(y) - u(x) \right)\frac{ds(y)}{\epsilon} \|_{L^2(D)} = 0,$$

which is valid for functions in $H^1(D)$, we get the L^2-convergence

$$-\int_{\partial B(\epsilon,x)} \left\{ \frac{\partial u(y)}{\partial \nu(y)}\Phi(x,y) - u(y)\frac{\partial \Phi(x,y)}{\partial \nu(y)} \right\} ds(y) \;\to\; u(x). \qquad (2.5.19)$$

Now collecting the equations (2.5.14) - (2.5.19) and passing to the limit $\epsilon \to 0$ we obtain the integral equation

$$u(x) - \int_D \nabla\Phi(x,y) \cdot M\nabla u(y)dy = u^i(x), \quad x \in D, \qquad (2.5.20)$$

which is an equation of Lippmann-Schwinger type. Note that the integral exists in the sense of an improper integral.

We will prove existence of a solution of (2.5.20) in $H^1(D)$. For a domain $D \subset \mathbb{R}^2$ define the space $X(D) := L^2(D) \times L^2(D)$ equipped with the norm

$$\|v\|_X := (\|v_1\|_{L^2(D)}^2 + \|v_2\|_{L^2(D)}^2)^{\frac{1}{2}}$$

for $v = (v_1, v_2) \in X(D)$. We consider matrices $M(x), x \in \mathbb{R}^2$, to be multiplication operators on $X(D)$ and we abbreviate $X := X(\mathbb{R}^2)$. Take the gradient of equation (2.5.20), i.e.,

$$\nabla u(x) + \nabla(\nabla \cdot \int_D \Phi(x,y)(M\nabla u)(y)dy) = \nabla u^{in}(x). \qquad (2.5.21)$$

Here we work with the weak derivative of a function in $H^1(D)$; the integrals exist in the sense of Cauchy's principal value. We use the formula

$$\frac{\partial}{\partial x_k}\frac{\partial}{\partial x_i}\int_D \Phi(x,y)\varphi(y)dy = \int_D \frac{\partial}{\partial x_k}\frac{\partial}{\partial x_i}\Phi(x,y)\varphi(y)dy - \frac{1}{2}\delta_{ki}\varphi(x) \qquad (2.5.22)$$

(see [60], page 310) for L^2-densities to obtain the equation

$$\nabla u(x) - \frac{1}{2}(M\nabla u)(x) + \int_D \nabla_x(\nabla_x \cdot \{\Phi(x,y)(M\nabla u)(y)\})dy = \nabla u^{in}(x). \tag{2.5.23}$$

Using the diagonal matrix

$$I - N_D =: M_D = \begin{pmatrix} d_1(x) & 0 \\ 0 & d_2(x) \end{pmatrix},$$

we have

$$(I - \frac{1}{2}M)(x) = U^*(x) \begin{pmatrix} 1 - \frac{1}{2}d_1(x) & 0 \\ 0 & 1 - \frac{1}{2}d_2(x) \end{pmatrix} U(x) \tag{2.5.24}$$

where from the conditions on $\mathcal{N}$ we obtain $1 - \text{Re}(d_1) > 0$ and $1 - \text{Re}(d_2) > 0$. We use this to derive that the matrix $I - \frac{1}{2}M$ is invertible. We multiply the vector equation (2.5.23) by $M(I - \frac{1}{2}M)^{-1}$. Using the new function $v := M\nabla u$ we get the equation

$$v(x) + M(I - \frac{1}{2}M)^{-1}(x) \int_D \nabla_x(\nabla_x \cdot \{\Phi(x,y)v(y)\})dy \tag{2.5.25}$$

$$= M(I - \frac{1}{2}M)^{-1}\nabla u^{in}(x), \quad x \in D$$

which is a system of strongly singular integral equations for the components v_i, $i = 1, 2$ of v.

We first show the equivalence of (2.5.25) to (2.5.20). Assume that the L^2-vector field v solves (2.5.25). We note that the integral in (2.5.25) maps $X(D)$ into $X(D)$ (see [60], p.257). We define

$$\hat{v}(x) \quad := \quad -(I - \frac{1}{2}M)^{-1}(x) \int_D \nabla_x(\nabla_x \cdot \{\Phi(x,y)v(y)\})dy$$

$$+(I - \frac{1}{2}M)^{-1}\nabla u^{in}(x)$$

and obtain $M\hat{v} = v$ from (2.5.25). From the definition of $\hat{v}$ using (2.5.22) we derive that $\hat{v}$ satisfies

$$\hat{v}(x) + \nabla\left(\nabla \cdot \int_D \Phi(x,y)(M\hat{v})(y)dy\right) = \nabla u^{in}(x). \tag{2.5.26}$$

From (2.5.26) we see that $\hat{v}$ is the gradient of a vector field

$$u(x) := -\nabla \cdot \int_D \Phi(x,y)(M\hat{v})(y)dy + u^{in}(x) \tag{2.5.27}$$

and that u satisfies

$$u(x) + \nabla \cdot \int_D \Phi(x,y)(M\nabla u)(y)dy = u^{in}(x),$$

which is equation (2.5.20).

Note that if $u \in H^2(D)$, we also obtain easily (2.5.8) from (2.5.20) by partial integration, application of $\Delta + \kappa^2$ and (2.5.22). Let us consider the integral

$$(Tv)(x) := \int_{R^2} \nabla_x(\nabla_x \cdot \Phi(x,y)v(y))dy, \quad x \in R^2. \tag{2.5.28}$$

T defines a strongly singular bounded integral operator in the space $X(R^2)$ (cf. [60], p. 257) and therefore also on the space $X(D) \subset X(R^2)$. Let T_0 denote the operator T in the case $\kappa = 0$, i.e., where Φ is replaced by the fundamental solution of Laplace's equation. Using the operators T_0 and $T - T_0$ we can rewrite equation (2.5.25) in the form

$$v + M(I - \frac{1}{2}M)^{-1}T_0 v + M(I - \frac{1}{2}M)^{-1}(T - T_0)v = M(I - \frac{1}{2}M)^{-1}\nabla u^{in}. \tag{2.5.29}$$

We estimate the norm of the operator T_0 in the following

LEMMA 2.5.2 *On the space* $X = L^2(R^2) \times L^2(R^2)$ *equipped with the norm*

$$\|v\|_X = (\|v_1\|_{L^2}^2 + \|v_2\|_{L^2}^2)^{\frac{1}{2}}, \quad v = (v_1, v_2) \in X$$

the operator norm of T_0 *is given by*

$$\|T_0\|_{BL(X,X)} = 1/2$$

where $BL(X,X)$ *denotes the space of bounded linear operators on* X *which is equipped with the canonical operator norm.*

Proof. For the kernel k_0 of the operator T_0 we compute

$$k_0(x,y) = \frac{-1}{2\pi} \frac{1}{|x-y|^2} \begin{pmatrix} 1 - 2\frac{(x_1-y_1)^2}{|x-y|^2} & \frac{(x_1-y_1)(x_2-y_2)}{|x-y|^2} \\ \frac{(x_1-y_1)(x_2-y_2)}{|x-y|^2} & 1 - 2\frac{(x_2-y_2)^2}{|x-y|^2} \end{pmatrix}. \tag{2.5.30}$$

We use $r = |x-y|$, $\phi = \arg[(y-x)/r]$ and the trigonometric formulas $1 - 2\cos^2\phi = -\cos(2\phi)$ and $2\sin\phi\cos\phi = \sin(2\phi)$ to obtain

$$k_0(x,y) = \frac{-1}{2\pi} \frac{1}{r^2} \begin{pmatrix} -\cos(2\phi) & \sin(2\phi) \\ \sin(2\phi) & \cos(2\phi) \end{pmatrix}. \tag{2.5.31}$$

Since $\sin(2\phi)$ and $\cos(2\phi)$ are the spherical harmonics of order 2 in two dimensions we can use [60] Chapter X, § 2 (6) to compute the Fourier transform $\hat{k}_0 = \mathcal{FT}k_0$ of k_0. We obtain

$$\hat{k}_0(\xi) = \frac{1}{2} \left(\begin{array}{cc} -\cos(2\theta) & \sin(2\theta) \\ \sin(2\theta) & \cos(2\theta) \end{array} \right) \qquad (2.5.32)$$

with $\xi = |\xi|(\cos\theta, \sin\theta)$. In the Fourier-space the operator T_0 becomes a multiplication operator $\hat{k}_0(\xi)$. For every point $\xi \in I\!\!R^2$ the matrix

$$\left(\begin{array}{cc} -\cos(2\theta) & \sin(2\theta) \\ \sin(2\theta) & \cos(2\theta) \end{array} \right) \qquad (2.5.33)$$

simply rotates and reflects the vector $(\mathcal{FT}v_1(\xi), \mathcal{FT}v_2(\xi))$ which does not affect its L^2-norm $\|.\|_X$. Since v and $\mathcal{FT}v$ have the same norms we obtain the statement of the lemma. $\qquad\Box$

For $N \in \mathcal{C}_{orth}$ the norm of the matrix operator $M(I - \frac{1}{2}M)^{-1}$ in $X(D)$ can be estimated by

$$\|M(I - \frac{1}{2}M)^{-1}v\|_{X(D)} \leq c\|v\|_{X(D)} \qquad (2.5.34)$$

with some constant $c < 2$. This can be done by again using the diagonal matrix $M_D := I - N_D$. We have

$$M(1 - \frac{1}{2}M)^{-1}(x) = U^*(x) \left(\begin{array}{cc} \frac{d_1(x)}{1-\frac{1}{2}d_1(x)} & 0 \\ 0 & \frac{d_2(x)}{1-\frac{1}{2}d_2(x)} \end{array} \right) U(x) \qquad (2.5.35)$$

with $1 - \mathrm{Re}(d_1) > 0$ and $1 - \mathrm{Re}(d_2) > 0$ for the continuous and compactly supported functions d_1 and d_2. For the diagonal matrix we have

$$\left\| \left(\begin{array}{cc} \frac{d_1(x)}{1-\frac{1}{2}d_1(x)} & 0 \\ 0 & \frac{d_2(x)}{1-\frac{1}{2}d_2(x)} \end{array} \right) \tilde{v} \right\|_{X(D)}$$

$$= \left(\left\| \frac{d_1(x)}{1-\frac{1}{2}d_1(x)}\tilde{v}_1(x) \right\|_{L^2(D)} + \left\| \frac{d_2(x)}{1-\frac{1}{2}d_2(x)}\tilde{v}_2(x) \right\|_{L^2(D)} \right)^{\frac{1}{2}}$$

$$\leq \max_{i=1,2} \sup_{x \in D} \left| \frac{d_i(x)}{1-\frac{1}{2}d_i(x)} \right| \|\tilde{v}\|_{X(D)}$$

$$= \max_{i=1,2} \sup_{x \in D} 2\left(\frac{(\mathrm{Re}d_i(x))^2 + (\mathrm{Im}d_i(x))^2}{(2 - \mathrm{Re}d_i(x))^2 + (\mathrm{Im}d_i(x))^2} \right)^{\frac{1}{2}} \|\tilde{v}\|_{X(D)},$$

from which we get the estimate. Therefore by restricting T_0 to $X(D)$ we obtain that the operator norm of $M(I - \frac{1}{2}M)^{-1}T_0$ in $X(D)$ is less than one. Looking at the power series (1.2.44) and (1.2.45) of $\Phi(x,y)$ we see that the kernel of $T - T_0$ is weakly singular. Hence the operator is compact in $X(D)$ for the bounded domain D. We have thus shown that in $X(D)$ the left-hand side of (2.5.29) consists of the sum of the identity operator, an operator with norm lower than one and a compact operator.

LEMMA 2.5.3 *Assume that the matrix $N = I - M$ is semi-coercive and elliptic. Then the integral operator on the left hand side of (2.5.25) is injective and the integral equation (2.5.25) has at most one solution in $X(D)$.*

Proof. Consider a solution $v \in X(D)$ of (2.5.25) with zero right hand side. We first obtain $v \equiv 0$ in $\mathbb{R}^2 \setminus supp(M)$. From the regularity results below we obtain $v \in H^1(D) \times H^1(D)$. Then we get as shown above a solution $u \in H^2(\mathbb{R}^2)$ to the scattering problem with a vanishing incident field. The uniqueness of the scattering problem yields $u \equiv 0$. From this we obtain $v = M\nabla u \equiv 0$, which proves the injectivity of the integral equation in $X(D)$. □

We now come to the existence theorem.

THEOREM 2.5.4 *For $N = I - M \in C_{orth}$ the operator given by the left hand side of (2.5.25) is invertible in $X(D)$ and the inverse operator is bounded.*

Proof. Since the integral operator is injective we obtain the invertibility and the boundedness of the inverse by the Neumann series Theorem 1.2.7 applied to $I + M(x)(I - \frac{1}{2}M(x))^{-1}T_0$ and the Riesz-Fredholm theory for compact operators as described in Theorem 1.2.5. □

Regularity properties of the solution. We need to prove regularity properties of the solution of (2.5.25) which were already used for the proof of the existence Theorem 2.5.4. Here, we will present a proof based on the theory of singular integral equations.

For a domain $D \subset \mathbb{R}^2$ introduce the Sobolev space

$$X^{(l)}(D) := H^{(l)}(D) \times H^{(l)}(D)$$

equipped with the norm

$$\|v\|_{X^{(l)}(D)} := \sum_{|\alpha| \leq l} \left(\|D^\alpha v_1\|_{L^2(D)}^2 + \|D^\alpha v_2\|_{L^2(D)}^2 \right)^{\frac{1}{2}} \qquad (2.5.36)$$

for $v = (v_1, v_2) \in X^{(l)}(D)$. $X_0^{(l)}(D)$ denotes the subspace of functions in $X^{(l)}(D)$ with compact support in D. Note that the term in brackets is the $X(D)$-norm of the function $D^\alpha v$. We use the abbreviation $X^{(l)} := X^{(l)}(\mathbb{R}^2)$ and $\mathcal{M}(x) := M(x)(I - \frac{1}{2}M(x))^{-1}$.

Let us first consider the solution of the equation

$$(I + \mathcal{M}(x)T_0)v = f. \tag{2.5.37}$$

We will show that for $M \in C_0^l(D)$ and $f \in X_0^{(l)}(D)$ the solution of (2.5.37) is in $X_0^{(l)}(D)$. From the Neuman series Theorem 1.2.7 we obtain the invertibility of the operator $I + \mathcal{M}(x)T_0$ in $X(D)$ and the representation

$$\begin{aligned} v &= (I + \mathcal{M}(x)T_0)^{-1}f \\ &= \sum_{\nu=0}^{\infty}(-\mathcal{M}(x)T_0)^\nu f \end{aligned} \tag{2.5.38}$$

where we have $q := \|\mathcal{M}(x)T_0\|_{X(D)} < 1$. From

$$\|D^\alpha T_0 v\|_X = \|\xi^\alpha \mathcal{FT}(k_0)\mathcal{FT}(v)\|_X = \frac{1}{2}\|\xi^\alpha \mathcal{FT}(v)\|_x = \frac{1}{2}\|D^\alpha v\|_X$$

we see that T_0 maps $X_0^{(l)}(D)$ into $X^{(l)}(D)$. Thus $\mathcal{M}(x)T_0$ maps $X_0^{(l)}(D)$ into itself. We conclude that every term in the expansion (2.5.38) is l times differentiable with respect to x. We want to show that the sum of the derivatives converges in X. For the derivative of T_0 we have

$$D^\alpha T_0 f = T_0(D^\alpha f) \tag{2.5.39}$$

(see for example [60], page 303). Therefore we obtain

$$\nabla_j\left(\mathcal{M}(x)T_0 f\right) = (\nabla_j\mathcal{M}(x))T_0 f + \mathcal{M}(x)T_0(\nabla_j f),$$

$$\begin{aligned} \nabla_j\left((\mathcal{M}(x)T_0)^2 f\right) &= (\nabla_j\mathcal{M}(x))T_0\left(\mathcal{M}(x)T_0\right)f \\ &+ \left(\mathcal{M}(x)T_0\right)(\nabla_j\mathcal{M}(x))T_0\, f + \left(\mathcal{M}(x)T_0\right)^2(\nabla_j f) \end{aligned} \tag{2.5.40}$$

and an analogous expression for the terms $(-\mathcal{M}(x)T_0)^\nu f$. Let us estimate the norm of $\nabla_j\left((-\mathcal{M}(x)T_0)^\nu f\right)$ in $X(D)$. Inductively we obtain

$$\|\nabla_j\left((\mathcal{M}(x)T_0)^\nu f\right)\|_{X(D)} \le \nu\frac{1}{2}Cq^{\nu-1}\|f\|_{X(D)} + q^\nu\|f\|_{X^1(D)} \tag{2.5.41}$$

where C is a bound for the derivatives of $\mathcal{M}$. From (2.5.41) we clearly see that the sum of the derivatives of (2.5.38) is convergent in $X(D)$, i.e., the function v is in $X_0^{(1)}(D)$. An analogous estimate holds true for the higher derivatives up to the order l. We have proven that $v \in X_0^{(l)}(D)$.

Next, we investigate the term $M(x)(I - \frac{1}{2}M(x))^{-1}(T - T_0)$ of equation (2.5.29).

THEOREM 2.5.5 *Assume that $M \in C_0^l(D)$. Then the operator*

$$M\left(I - \frac{1}{2}M\right)^{-1}(T - T_0)$$

maps the space $X_0^{(j)}(D)$ continuously into the space $X_0^{(j+1)}(D)$ for all integers $j + 1 \leq l$.

Proof. We first examine the operator $T - T_0$. From the expansions of the Hankel function we obtain for the difference $\Phi - \Phi_0$ the form

$$\Phi(x,y) - \Phi_0(x,y) = c_1|x-y|^2 \ln|x-y| + c_2|x-y|^2 + O\left(|x-y|^3\right) \quad (2.5.42)$$

as $|x - y| \to 0$ with some constants c_1 and c_2. Since the expansions are absolutely convergent we can differentiate each term and we obtain for the kernel k of $T - T_0$ the form

$$k_{ij}(x,y) = 2c_1\delta_{ij}\ln|x-y| + 2c_1\frac{(x-y)_i(x-y)_j}{|x-y|^2} + 2c_2\delta_{ij} + R_{ij}(x,y) \quad (2.5.43)$$

where $R_{ij}(x,y) = O(|x-y|)$ for $|x-y| \to 0$. According to [8], Theorem 8.2, the leading term in (2.5.43) defines a bounded integral operator $V : L^2(D) \to H^2(D)$. Denote the second term of the right hand side of (2.5.43) by t_2. The term t_2 is continuously differentiable for $x \neq y$ with weakly singular partial derivatives

$$\frac{\delta_{i\mu}(x-y)_j}{|x-y|^2} + \frac{\delta_{j\mu}(x-y)_i}{|x-y|^2} - 2\frac{(x-y)_i(x-y)_j(x-y)_\mu}{|x-y|^4}, \quad \mu = 1, 2. \quad (2.5.44)$$

Assume first that $u \in C_0(D)$. Then we get

$$\frac{\partial}{\partial x_\mu}\int_{D\setminus B(\epsilon,x)} t_2(x,y)u(y)dy = \int_{D\setminus B(\epsilon,x)} \frac{\partial t_2(x,y)}{\partial x_\mu}u(y)dy \quad (2.5.45)$$

$$- \int_{\partial B(\epsilon,x)} t_2(x,y)u(y)\cos(y-x,x_\mu)ds(y).$$

In the limit $\epsilon \to 0$ the last term vanishes. We obtain that the integral

$$w(x) := \int_D \frac{(x-y)_i(x-y)_j}{|x-y|^2} u(y) dy, \quad x \in D$$

with the second term of the right hand side of (2.5.43) as its kernel is a differentiable function w, that differentiation can be carried out by differentiating the kernel and that the first derivatives are bounded with respect to the norm on $L^2(D)$, i.e., the operator maps the space $C_0(D) \subset L^2(D)$ continuously into $H^1(D)$. Since $C_0(D)$ is dense in $L^2(D)$ there exists a unique continuous extension of the operator from $L^2(D)$ into $H^1(D)$.

For the function $2c_2\delta_{ij} + R_{ij}$ in (2.5.43) we now can proceed in the same way as for the second term. Together we obtain that the operator $T - T_0$ maps $X(D) = X^{(0)}(D)$ continuously into $X^{(1)}(D)$. Therefore the operator $M(I - \frac{1}{2}M)^{-1}(T - T_0)$ maps $X_0^{(0)}(D)$ continuously into $X_0^{(1)}(D)$.

To prove the statement for higher derivatives we will use induction. Assume that we have $m \in C_0^l(D)$. For $u \in C_0^l(D)$, $l \geq 1$ we get, using Gauss' divergence theorem,

$$\frac{\partial}{\partial x_\mu}\left(m(x)\int_D k_{ij}(x,y)u(y)dy\right) = \frac{\partial m(x)}{\partial x_\mu}\int_D k_{ij}(x,y)u(y)dy \quad (2.5.46)$$

$$+ \quad m(x)\int_D k_{ij}(x,y)\frac{\partial u}{\partial y_\mu}(y)dy, \quad x \in D.$$

Assume that the operators on the right-hand side of (2.5.46) define bounded operators from $H_0^{l-1}(D)$ into $H_0^l(D)$. For $l = 1$ this was proven in the first part of our proof. Then from (2.5.46) we obtain that m times the integral over k_{ij} is bounded from $C_0^l(D) \cap H_0^l(D)$ into $H_0^{l+1}(D)$ with respect to the Sobolev norms. Since $C_0^l(D)$ is dense in $H_0^l(D)$ we now obtain the statement by induction. □

Again by induction we now obtain the following theorem.

THEOREM 2.5.6 *Assume that $M \in C_0^l(\mathbb{R}^2)$ for some $l \in \mathbb{N}$ and that $N = I - M \in C_{orth}$ and let D be a domain with $\mathrm{supp}(M) \subset D$. Then the solution of (2.5.25) is in $X_0^{(l)}(D)$, i.e., the solution u to the scattering problem (2.5.8) is in $H_{loc}^{(l+1)}(\mathbb{R}^2)$. We have the estimate*

$$\|u\|_{H^{l+1}(D)} \leq C\|u^i\|_{H^l(D)} \quad (2.5.47)$$

with some constant C depending on M.

REMARK. Note that we have avoided using both the classical regularity results for elliptic partial differential equations of second order with coefficients in C^l (cf.[18]) as well as the use of the symbol of singular integral operators (cf. [60]).

Fréchet differentiability with respect to M. Finally, we study the differentiability properties of the mapping $M \mapsto u^s(M)$. Let

$$\mathcal{C}_{orthM} := \{M : I - M \in \mathcal{C}_{orth}\}. \tag{2.5.48}$$

For a summary of the basic properties of Fréchet differentiable and analytic functions in complex Banach spaces we refer to Berger (cf. [1]). As for the one-dimensional counterpart of holomorphic functions to obtain analyticity in a complex space we only have to prove simple Fréchet differentiability.

Fréchet differentiability and analyticity of the scattered field and its far field pattern with respect to the refractive index is of basic interest in scattering theory. They are used for example to apply expansion methods for the computation of the scattered fields (see [2]) or to use Newton type methods for the corresponding inverse scattering problems.

THEOREM 2.5.7 *The mapping of the matrix M onto the solution u of (2.5.20) is analytic from $\mathcal{C}_{orthM} \subset C_0^1(\mathbb{R}^2)$ into $H^1(D)$. The n-th derivative is given by*

$$\frac{\partial^n u}{\partial M^n}(dM) = n! \left((I - L(M))^{-1} L(dM) \right)^n (I - L(M))^{-1} u^{in}. \tag{2.5.49}$$

Proof. Define $L(M) : H^1(D) \to H^1(D)$ by

$$L(M)u := \int_D \nabla \Phi(.,y) \cdot M \nabla u(y) dy. \tag{2.5.50}$$

From its linearity in M we see that the mapping

$$\mathcal{C}_{orthM} \to BL(H^1(D), H^1(D)), \quad M \mapsto L(M) \tag{2.5.51}$$

is Fréchet differentiable on $\mathcal{C}_{orthM}$. Therefore the operator

$$I - L(M)$$

is Fréchet differentiable $\mathcal{C}_{orthM} \to BL(H^1(D), H^1(D))$. Since the operator is invertible for $M \in \mathcal{C}_{orthM}$ from Theorem 2 of [78] we obtain that its inverse is Fréchet differentiable as a mapping $\mathcal{C}_{orthM} \to BL(H^1(D), H^1(D))$ and we obtain the given form of the inverse. The statement for higher derivatives can be obtained by induction.

$$\square$$

Analogously to [44], [77] or [78] we can obtain a characterization of the Fréchet derivative as the solution to a special scattering problem.

THEOREM 2.5.8 *The n-th Fréchet derivative $u^{(n)}$ of the scattering problem (2.5.8) with respect to the $M = I - N$ at the point M and direction δ satisfies the Sommerfeld radiation condition and solves the PDE*

$$(\nabla \cdot N\nabla)u^{(n)} + \kappa^2 u^{(n)} = n\,(\nabla \cdot \delta\nabla)u^{(n-1)}.$$

Proof. The statement is an immediate consequence of equation (2.5.49).
□

2.6 Anisotropic electromagnetic media

The aim of this section is to obtain an integral equation formulation for the time harmonic electromagnetic scattering problem from a bounded anisotropic medium following [71]. It will be a basis for the development of the linear sampling method for the reconstruction of the support of a scatterer in Chapter 7.

Different approaches have been developed by other authors. In the monograph of Leis [58] one can find a Hilbert space approach to the problem which is valid for real refractive matrices. Kirsch and Monk [48] more recently gave a variational proof of the existence under the assumption of uniqueness and describe a finite element/spectral method for the numerical solution of the problem. Strongly singular integral equations have been used extensively to solve problems in elasticity.

Again, we start with the time-harmonic Maxwell equations. The electric field E and the magnetic field H are assumed to satisfy the reduced Maxwell equations

$$\begin{aligned}
\nabla \times E - i\kappa H &= 0 \\
\nabla \times H + i\kappa N(x)E &= 0.
\end{aligned} \qquad (2.6.1)$$

Here the wave number $\kappa > 0$ is defined by $\kappa^2 = \epsilon_0 \mu_0 \omega^2$ and we assume the refractive index $N = N(x)$ to be a complex 3×3-matrix which is given by

$$N(x) = \frac{1}{\epsilon_0}\left(\epsilon(x) + i\frac{\sigma(x)}{\omega}\right), \qquad (2.6.2)$$

where $\epsilon = \epsilon(x)$ is the tensor of the electric permittivity, $\sigma = \sigma(x)$ the tensor of the electric conductivity and ω is the frequency of the electromagnetic wave. We consider scattering from bounded obstacles, i.e., we assume that the support of $M := I - N$ is bounded in $\mathbb{R}^3$.

Let $E^i, H^i \in H^{(1)}_{loc}(\mathbb{R}^3)$ be a solution of Maxwell's equations for the background medium, i.e., E^i, H^i solves (2.6.1) with $N(x) \equiv I$. We want to find a solution $E, H \in H_{loc}^{(1)}(\mathbb{R}^3)$ of (2.6.1) such that the scattered field E^s, H^s defined by

$$
\begin{aligned}
E &= E^i + E^s, \\
H &= H^i + H^S
\end{aligned}
\tag{2.6.3}
$$

satisfies the Silver-Müller radiation condition

$$
\lim_{r \to \infty} (H^s \times x - rE^s) = 0
\tag{2.6.4}
$$

uniformly for all directions $\hat{x} = x/|x|$ where $r = |x|$.

Here, we consider at least twice continuously differentiable matrix functions N. We will work with matrices N which can be diagonalized with a unitary complex matrix $U(x)$, i.e., we have $N(x) = U(x)N_D(x)U^*(x)$ with a diagonal matrix N_D and $U^*(x)U(x) = I$ for every $x \in \mathbb{R}^3$. Motivated by the isotropic case we assume that the real part of the diagonal elements of N_D is greater or equal to one. We further assume that N_D has a positive semi-definite imaginary part (i.e., the conductivity is positive or zero). For a wide range of practical applications these conditions are fulfilled. We call the corresponding class of anisotropic medium scatterers $\mathcal{C}_{aniso}$ and will usually consider matrices N as elements of $\mathcal{C}_{aniso}$ as abbreviations of $\mathcal{D} = (D, N) \in \mathcal{C}_{aniso}$ with the notation of (1.2.91).

An integral equation formulation. First we transform our scattering problem (2.6.1) - (2.6.4) into an integral equation. We use

$$
\Phi(x,y) := \frac{1}{4\pi} \frac{e^{i\kappa|x-y|}}{|x-y|}, \quad x \neq y,
$$

for the fundamental solution to the Helmholtz equation and $M := I - N$.

THEOREM 2.6.1 *Let $E, H \in H_{loc}^{(1)}(\mathbb{R}^3)$ be a solution of the scattering problem (2.6.1) - (2.6.4). Then E satisfies the integral equation*

$$
E(x) + \nabla \int_{\mathbb{R}^3} \nabla_x \Phi(x,y) \cdot (ME)(y) \, dy
\tag{2.6.5}
$$

$$
+ \kappa^2 \int_{\mathbb{R}^3} \Phi(x,y) M(y)E(y) \, dy = E^i(x), \quad x \in \mathbb{R}^3.
$$

Proof. Let us first assume that we have $E, H \in C^1(\mathbb{R}^3)$. Let $B \subset \mathbb{R}^3$ be a ball which contains the support of M and the point $x \in \mathbb{R}^3$ in its interior. Using the Stratton-Chu formula (2.3.6) we obtain

$$
E(x) = -\nabla \times \int_{\partial B} \Phi(x,y) \, \nu \times E(y) \, dy
$$

$$+\nabla \int_{\partial B} \Phi(x,y)\, \nu(y) \cdot E(y)\, ds(y)$$

$$-i\kappa \int_{\partial B} \Phi(x,y)\, \nu(y) \times H(y)\, ds(y) \qquad (2.6.6)$$

$$-\kappa^2 \int_B \Phi(x,y)\, M(y)E(y)\, dy$$

$$-\nabla \int_B \Phi(x,y)\, \nabla \cdot (ME)(y)\, ds(y)$$

where we have used $\nabla \times H + i\kappa N(x)E = 0$ and $\nabla \cdot (NE) = \nabla \cdot ((I-M)E) = 0$. The Stratton-Chu formula applied to E^i, H^i yields

$$\begin{aligned} E^i(x) &= -\nabla \times \int_{\partial B} \Phi(x,y)\, \nu(y) \times E^i(y)\, ds(y) \\ &\quad +\nabla \int_{\partial B} \Phi(x,y)\, \nu(y) \cdot E^i(y)\, ds(y) \qquad (2.6.7) \\ &\quad -i\kappa \int_{\partial B} \Phi(x,y)\, \nu(y) \times H^i(y)\, ds(y). \end{aligned}$$

From the Stratton-Chu formula for radiating solutions to the Maxwell equations we obtain

$$\begin{aligned} 0 &= -\nabla \times \int_{\partial B} \Phi(x,y)\, \nu(y) \times E^s(y)\, ds(y) \\ &\quad +\nabla \int_{\partial B} \Phi(x,y)\, \nu(y) \cdot E^s(y)\, ds(y) \qquad (2.6.8) \\ &\quad -i\kappa \int_{\partial B} \Phi(x,y)\, \nu(y) \times H^s(y)\, ds(y). \end{aligned}$$

From Gauss' theorem we have

$$\int_B \Phi(x,y)\, \nabla_y \cdot (ME)(y)\, ds(y) = \int_B \nabla_x \Phi(x,y) \cdot (ME)(y)\, ds(y). \quad (2.6.9)$$

We combine (2.6.6) - (2.6.9) to obtain equation (2.6.5). We now remark that the equations (2.6.6) to (2.6.9) are also valid for functions $E, H \in H^{(1)}_{loc}(\mathbb{R}^3)$ to obtain the theorem. $\qquad\qquad\qquad\qquad\qquad\qquad\qquad\qquad\qquad\qquad\quad \square$

We want to derive a slightly different form for equation (2.6.5). For the second term in (2.6.5) we obtain for $E \in L^2(\mathbb{R}^3)$ (see [60], page 310)

$$\nabla \int_{\mathbb{R}^3} \nabla_x \Phi(x,y) \cdot (ME)(y)\, dy \qquad\qquad\qquad (2.6.10)$$

$$= \int_{\mathbb{R}^3} \nabla\Big(\nabla_x \cdot \{\Phi(x,y)(ME)(y)\}\Big)\, dy - \frac{1}{3}(ME)(x)$$

for $x \in \mathbb{R}^3$. For the case $\kappa = 0$ we denote $\Phi(x, y)$ by $\Phi_0(x, y)$. We define T by

$$(Tv)(x) := \int_{\mathbb{R}^3} \nabla \left(\nabla_x \cdot \{\Phi(x, y)v(y)\} \right) dy, \quad x \in \mathbb{R}^3, v \in L^2(\mathbb{R}^2, \mathbb{C}^3)$$

(2.6.11)

and T_0 as T with Φ replaced by Φ_0. The kernel of the operator T is a matrix of strongly singular functions; the integral is defined in the sense of a Cauchy principal value for Hölder-continuous vector fields and for L^2-vector fields by extending the operator. With the volume potential operator V given by (2.2.2) the weakly singular integral in (2.6.5) is given by $\kappa^2 V(ME)$. If we use $T = T_0 + (T - T_0)$ and (2.6.10) we obtain from (2.6.5) the equivalent equation

$$(I - \frac{1}{3}M)E + T_0(ME) + (T - T_0)(ME) + \kappa^2 V(ME) = E^i. \quad (2.6.12)$$

For $N = I - M \in \mathcal{C}_{aniso}$ the matrix $I - \frac{1}{3}M$ is invertible. To see this observe the representation

$$I - \frac{1}{3}M(x) = U^*(x) \begin{pmatrix} 1 - \frac{1}{3}d_1(x) & 0 & 0 \\ 0 & 1 - \frac{1}{3}d_2(x) & 0 \\ 0 & 0 & 1 - \frac{1}{3}d_3(x) \end{pmatrix} U(x)$$

where d_1, d_2 and d_3 are the diagonal elements of $M_D := I - N_D$. Using $\mathrm{Re}(d_i(x)) \leq 0$ for $x \in \mathbb{R}^3$ and $i = 1, 2, 3$ we obtain the invertibility of $1 - \frac{1}{3}d_i$, and thus that also $I - \frac{1}{3}M$ is invertible. We can now multiply equation (2.6.12) by $M(I - \frac{1}{3}M(x))^{-1}$ to obtain for $v(x) := (ME)(x), x \in \mathbb{R}^3$ and $\mathcal{M}(x) := M(x)(I - \frac{1}{3}M(x))^{-1}, x \in \mathbb{R}^3$ the equation

$$v(x) + \mathcal{M}(x)(T_0 v)(x)$$
$$+\mathcal{M}(x)((T - T_0)v)(x) + \mathcal{M}(x)(\kappa^2 V M v)(x) \quad (2.6.13)$$
$$= \mathcal{M}(x)E^i(x), \quad x \in \mathbb{R}^3.$$

This is a system of strongly singular integral equations.

The symbol matrix of singular integral operators. In this part we collect some definitions and theorems on the symbol matrix of a system of singular integral operators. The concept of a symbol of a singular integral operator was introduced by Michlin (1936). We refer to the monograph of Michlin and Prößdorf [60] for a more detailed treatment.

By Ω we denote the unit sphere in $\mathbb{R}^m, m \in \mathbb{N}$. A singular integral operator of the form

$$(Au)(x) := a(x)u(x) + \int_{\mathbb{R}^m} k(x, x - y)u(y)dy, \quad k(x, x - y) = r^{-m}f(x, \theta)$$

(2.6.14)

with matrix-valued functions a and k with $r = |x - y|$ and $\theta := (y - x)/r \in \Omega$ is called simple singular operator. The symbol matrix of A is defined formally by

$$\mathrm{Smb}A = a + \hat{k}$$

where $\hat{\ }$ denotes the Fourier transform with respect to the second variable y of $k(x, y)$. It is possible to give a rather simple representation of the symbol of A by means of the characteristic f(see [60]). Since the symbol is a homogeneous function of degree zero in ξ we may speak of the symbol as a function on $I\!\!R^3 \times \Omega$, i.e., for $\Psi := \mathrm{Smb}A$ we have $\Psi = \Psi(x, \theta), x \in I\!\!R^m, \theta \in \Omega$. The determinant of the symbol matrix is called the symbol determinant. We quote the following theorem which is proven in [60], page 271.

THEOREM 2.6.2 *Let A be an $m \times m$ matrix of singular operators, $m \in I\!\!N$, $\Psi(x, \theta)$ the corresponding symbol matrix and $\Psi(x, .)$ belong to $H^{(l)}(\Omega)$ for $l > 1$ uniformly with respect to $x \in \overline{I\!\!R^m}$ where $\overline{I\!\!R^m}$ denotes that compact which is obtained from $I\!\!R^m$ by adding infinity. The symbol matrices corresponding to all such singular matrix operators form a non-commutative ring with respect to the usual operations of addition and multiplication of matrices. Let $\mathcal{R}$ denote the restriction of the ring mentioned which consists of all symbol matrices that are continuous on $\overline{I\!\!R}^m \times \Omega$. A matrix $\Psi \in \mathcal{R}$ has an inverse in that ring if and only if*

$$\inf |\det \Psi(x, \theta)| > 0 \qquad\qquad (2.6.15)$$

for all $x \in \overline{I\!\!R^m}$. □

The preceding theorem opens the possibility to regularize the integral equation $Au = g$ using the singular integral operator B with symbol matrix Ψ^{-1}. For scalar multidimensional integral equations (i.e., m=1) Michlin and Prößdorf (cf. [60], page 319) prove the existence of an equivalent regularization in $L^p(I\!\!R^m)$ under the condition (2.6.15). However, for systems of singular integral equations in general there is not an equivalent regularizer (cf. [60], page 369).

For our study of an inhomogeneous anisotropic medium we will extend arguments used by Michlin (cf. [60], page 380) to construct an equivalent regularizer under an additional condition. This is worked out in the following theorem.

THEOREM 2.6.3 *I. Let A be a singular operator with symbol matrix $\Psi(x, \theta)$ for $x \in I\!\!R^3$ and $\theta \in \Omega$ which is continuous on $\overline{I\!\!R^m} \times \Omega$ and which satisfies $\Psi(x, .) \in H^{(3/2)}(\Omega)$ uniformly for $x \in \overline{I\!\!R^3}$. Assume that we have*

$$|\det[\zeta I - \Psi(x, \theta)]| \geq \gamma > 0 \qquad\qquad (2.6.16)$$

for $\zeta \in (-\infty, 0]$ *with a constant* γ. *Then there exists an equivalent right regularizer* R *of* A *in* $L^2(\mathbb{R}^3)$.

II. Let the assumptions of I be true and let $B \subset \mathbb{R}^3$ *be a bounded domain such that for the operator* A *given by (2.6.14) the functions* $a(.) - I$ *and* $f(., \theta)$ *are elements of* $C_0(B)$. *Then there exists an equivalent regularizer also in* $L_0^2(B)$.

Proof. I. We first show that the determinant of the symbol matrix $I - \lambda(I - \Psi)$ of the system

$$u - \lambda(I - A)u = g \qquad (2.6.17)$$

is non-zero for $\lambda \in [0, 1]$. For small $|\lambda|$ this can be seen using continuity arguments. We find a number $\eta > 0$ such that $|\det[(1 - \lambda)I + \lambda\Psi]| \geq q > 0$ for all $|\lambda| < \eta$. For $|\lambda| \geq \eta$ we estimate

$$|\det[(1 - \lambda)I + \lambda\Psi]| \geq \eta^3 \ |\det[\frac{(\lambda - 1)}{\lambda}I - \Psi]| \geq \eta^3\gamma > 0 \quad (2.6.18)$$

where we have used that the mapping $\lambda \mapsto \frac{(\lambda-1)}{\lambda}$ maps $[0, 1]$ onto $[-\infty, 0]$ and the assumption (2.6.16). Now from Theorem 2.6.2 we obtain that there is a singular operator $\tilde{P}_\lambda$ with symbol matrix $[(1 - \lambda)I + \lambda\Psi]^{-1}[I - \Psi]$. By Theorem 3.2 of [60], page 281, it is bounded in $L^2(\mathbb{R}^3)$ independently from λ, i.e., $\|\tilde{P}_\lambda\| \leq C$.

Consider the operators $P_j := \tilde{P}_{\lambda_j}$ with $\lambda_j := \frac{2[C]-j}{2[C]}$, $j = 0, ..., 2[C]$ where $[C]$ is the smallest integer greater or equal to C. Clearly we have the estimate

$$\|(\lambda_{j-1} - \lambda_j)P_j\| \leq \frac{1}{2}, \ j = 1, ..., 2[C]. \qquad (2.6.19)$$

Thus the operators $I - (\lambda_{j-1} - \lambda_j)P_j$ are continuously invertible. Let us for a moment abbreviate the symbol of the operator P_j by σ_j. We now show that

$$R_j := [I - (\lambda_{j-1} - \lambda_j)P_j]^{-1}, \ j = 0, ..., 2[C] - 1$$

has the symbol matrix

$$r_j := [1 - (\lambda_{j-1} - \lambda_j)\sigma_j]^{-1} = [(1 - \lambda_{j-1})I + \lambda_{j-1}\Psi]^{-1}[(1 - \lambda_j)I + \lambda_j\Psi].$$

The operator R_j can be written in terms of the Neumann series

$$R_j = \sum_{\nu=0}^{\infty} \left((\lambda_{j-1} - \lambda_j)P_j\right)^\nu. \qquad (2.6.20)$$

For C sufficiently large we also have

$$r_j = [1 - (\lambda_{j-1} - \lambda_j)\sigma_j]^{-1} = \sum_{\nu=1}^{\infty} (\lambda_{j-1} - \lambda_j)^\nu \sigma_j^\nu. \qquad (2.6.21)$$

Using the absolute convergence of the sum in (2.6.21) and Theorem 3.2, page 281 of [60] we can estimate the remainder in the series (2.6.20) and thus obtain that the infinite sum in (2.6.20) converges to the operator $\tilde{R}_j$ with symbol matrix r_j. Since it also converges to R_j we obtain $\tilde{R}_j = R_j$ and $\mathrm{Smb}(R_j) = r_j$.

It is now easy to compute the symbol matrix of the Operator $AR_1 \cdots R_k$. We get

$$[(1 - \lambda_k)I + \lambda_k \Psi]. \qquad (2.6.22)$$

For $k = 2[C]$ we obtain $\lambda_k = 0$. Then the operator A is equivalently transformed into an operator AR with symbol matrix I. The symbol of $K := AR - I$ vanishes. By §10, Chapter XI of [60] K must be compact, i.e., we have $AR = I + K$ with a compact operator K. Thus the operator $R := R_1 R_2 \cdots R_k$ is an equivalent right regularizer for A, which ends the proof of part I.

II. The main step to prove part II is the statement that the operator P_j can be chosen to map $L_0^2(B)$ into $L_0^2(B)$. Then the same is true for all R_j, therefore also for $K = AR - I$ and the proof is complete. First we note that the inverse of $[(1 - \lambda)I + \lambda\Psi]$ is given by the product of $(\det[(1 - \lambda)I + \lambda\Psi])^{-1}$ with a matrix $\tilde{\Psi}$ which consists of sums of products of elements of $[(1 - \lambda)I + \lambda\Psi]$. Since $(1 - \lambda)I + \lambda A$ maps $L_0^2(B)$ into $L_0^2(B)$ the same is true for the sums of products of its elements and we can construct an operator $\tilde{A}$ with symbol matrix $\tilde{\Psi}$ which maps $L_0^2(B)$ into itself. We have that

$$\Psi_{scal}(x, \theta) := (\det[(1 - \lambda)I + \lambda\Psi(x, \theta)])^{-1}$$

is equal to one for $x \notin B$. Thus it remains to show the existence of a singular integral operator P_{scal} with symbol Ψ_{scal}

$$(P_{scal}u)(x) = a(x)u(x) + \int_{R^3} \frac{f(x, \theta)}{r^3} u(y) dy$$

such that $a(.) - 1 \in C_0(B)$ and $f(., \theta) \in C_0(B)$. The existence of a singular operator with symbol Ψ_{scal} is a consequence (2.6.18) and we expand its characteristic $f(x, \theta)$ with respect to spherical functions (compare [60], page 251)

$$f(x, \theta) := \sum_{n=1}^{\infty} \sum_{k=1}^{k_{n,3}} a_n^{(k)}(x) Y_{n,3}^{(k)}(\theta), \quad x \in R^3, \theta \in \Omega.$$

This yields a corresponding expansion for the symbol

$$\Psi_{scal}(x,\theta) = \sum_{n=0}^{\infty}\sum_{k=1}^{k_{n,3}} \gamma_{3,n} a_n^{(k)}(x) Y_{n,3}^{(k)}(\theta), \quad x \in I\!\!R^3, \theta \in \Omega \qquad (2.6.23)$$

with

$$\gamma_{3,0} := 1, \quad \gamma_{3,n} := \frac{i^n \pi^{3/2}\Gamma(n/2)}{\Gamma((3+n)/2)}, n \geq 1,$$

and $a_0^{(1)}(x) = a(x)$. From the last expansion we obtain $a_n^{(k)}(x) = 0$ for $n \geq 1, x \notin B$ and $a_0^{(1)}(x) = 1$ for $x \notin B$. This ends the proof.

□

Regularity properties of solutions to the integral equation. Now, we prove regularity results for solutions of the integral equation (2.6.5). We will need to verify the condition

$$\inf |\det(\Psi)| > 0 \qquad (2.6.24)$$

for the symbol matrix Ψ of the singular integral operator $I + \mathcal{M}T_0$. Since for the existence proof below we need the stronger estimate

$$\inf |\det(\zeta I - \Psi)| \geq \gamma > 0 \qquad (2.6.25)$$

for $\zeta \in (-\infty, 0]$ with some constant γ, we will verify (2.6.25).

LEMMA 2.6.4 *For the symbol Ψ of the operator $I + \mathcal{M}T_0$ there exists $\gamma > 0$ such that (2.6.25) holds.*

Proof. The kernel k_0 of the operator T_0 is

$$k_0(x,y) = \qquad (2.6.26)$$

$$\frac{1}{4\pi r^3} \begin{pmatrix} 3\cos^2\varphi\sin^2\vartheta - 1 & 3\cos\varphi\sin\varphi\sin^2\vartheta & 3\cos\varphi\sin\vartheta\cos\vartheta \\ 3\cos\varphi\sin\varphi\sin^2\vartheta & 3\sin^2\varphi\sin^2\vartheta - 1 & 3\sin\varphi\sin\vartheta\cos\vartheta \\ 3\cos\varphi\sin\vartheta\cos\vartheta & 3\sin\varphi\sin\vartheta\cos\vartheta & 3\cos^2\vartheta - 1 \end{pmatrix}$$

with $r := |x - y|$ and $y - x = r(\cos\varphi\sin\vartheta, \sin\varphi\sin\vartheta, \cos\vartheta)$. The components of k_0 are spherical harmonics of order 2 in three dimensions and eigenvalues of the Beltrami operator (compare [60], page 214). We can use [60], page 251, to compute the Fourier transform

$$\hat{k}_0(\xi) = \qquad (2.6.27)$$

$$\frac{-1}{3} \begin{pmatrix} 3\cos^2\varphi\sin^2\vartheta - 1 & 3\cos\varphi\sin\varphi\sin^2\vartheta & 3\cos\varphi\sin\vartheta\cos\vartheta \\ 3\cos\varphi\sin\varphi\sin^2\vartheta & 3\sin^2\varphi\sin^2\vartheta - 1 & 3\sin\varphi\sin\vartheta\cos\vartheta \\ 3\cos\varphi\sin\vartheta\cos\vartheta & 3\sin\varphi\sin\vartheta\cos\vartheta & 3\cos^2\vartheta - 1 \end{pmatrix}$$

of the kernel k_0. The symbol $\Psi_\zeta(x, \varphi, \vartheta)$ of the operator $(1 - \zeta)I + MT_0$ is given by the matrix $(1 - \zeta)I + \mathcal{M}(x)\hat{k}_0$. Let us in a first step assume that $\mathcal{M}(x)$ is diagonal, i.e., $\mathcal{M}(x) = diag\{\tilde{m}_1(x), \tilde{m}_2(x), \tilde{m}_3(x)\}$. Then we get

$$\tilde{m}_j(x) = \frac{d_j(x)}{1 - \frac{1}{3}d_j(x)}, \quad j = 1, 2, 3, \ z \in \mathbb{R}^3 \tag{2.6.28}$$

with the diagonal elements d_j of the matrix $M_D := I - N_D$. For $N \in \mathcal{C}_{aniso}$ we know that we have $\mathrm{Re}(d_j) \leq 0$ and $\mathrm{Im}(d_j) \leq 0$. Observe that the matrix in (2.6.27) has the structure $((v_i v_j - \delta_{ij})_{i,j=1,...,3}) = v \cdot v^T - I$ with $v^T = \sqrt{3}(\cos\varphi \sin\vartheta, \sin\varphi \sin\vartheta, \cos\vartheta)$ where $\cdot$ denotes the usual matrix multiplication. This is very helpful to compute the determinant of $(1 - \zeta)I + \mathcal{M}(x)\hat{k}_0$. By lengthy but straightforward calculation we obtain

$$\det\left((1 - \zeta)I + \mathcal{M}(x)\hat{k}_0(\varphi, \vartheta) \right)$$

$$= \left((1 - \zeta) + \frac{\tilde{m}_1(x)}{3} \right)\left((1 - \zeta) + \frac{\tilde{m}_2(x)}{3} \right)\left((1 - \zeta) + \frac{\tilde{m}_3(x)}{3} \right)$$

$$+ \left((1 - \zeta) + \frac{\tilde{m}_1(x)}{3} \right)\left((1 - \zeta) + \frac{\tilde{m}_2(x)}{3} \right)(-\frac{\tilde{m}_3}{3})v_3^2 \tag{2.6.29}$$

$$+ \left((1 - \zeta) + \frac{\tilde{m}_1(x)}{3} \right)\left((1 - \zeta) + \frac{\tilde{m}_3(x)}{3} \right)(-\frac{\tilde{m}_2}{3})v_2^2$$

$$+ \left((1 - \zeta) + \frac{\tilde{m}_2(x)}{3} \right)\left((1 - \zeta) + \frac{\tilde{m}_3(x)}{3} \right)(-\frac{\tilde{m}_3}{1})v_1^2.$$

For real matrices $\mathcal{M}(x)$ (i.e., $\sigma \equiv 0$) we have $0 \geq \frac{\tilde{m}_j}{3} > -1$ from (2.6.28) and therefore $\det\left((1 - \zeta)I + \mathcal{M}(x)\hat{k}_0(\varphi, \vartheta) \right) > 0$ for all $\varphi \in [0, 2\pi]$, $\vartheta \in [0, \pi], \zeta \in (-\infty, 0]$ and for all $x \in \mathbb{R}^3$. For complex values $\tilde{m}_j$ let us multiply the expression (2.6.29) by

$$m_H(x) := \left((1 - \zeta) + \frac{\tilde{m}_1(x)}{3} \right)^{-1} \cdot$$

$$\left((1 - \zeta) + \frac{\tilde{m}_2(x)}{3} \right)^{-1}\left((1 - \zeta) + \frac{\tilde{m}_3(x)}{3} \right)^{-1}.$$

Then we obtain after some computation

$$m_H(x)\det\left((1 - \zeta)I + \mathcal{M}(x)\hat{k}_0(\varphi, \vartheta) \right) \tag{2.6.30}$$

$$= 1 - \frac{d_3(x)}{3(1 - \zeta) + \zeta d_3(x)}v_3^2$$

$$- \frac{d_2(x)}{3(1 - \zeta) + \zeta d_2(x)}v_2^2$$

$$- \frac{d_1(x)}{3(1 - \zeta) + \zeta d_1(x)}v_1^2.$$

The real part of $d_j, j = 1, 2, 3$ is negative or zero. Therefore taking the real part of (2.6.30) by straightforward calculation we obtain $\det\left((1 - \zeta)I + \mathcal{M}(x)\hat{k}_0(\varphi, \vartheta)\right) \neq 0$ for all $\varphi \in [0, 2\pi]$, $\vartheta \in [0, \pi]$, $\zeta \in (-\infty, 0]$ and for all $x \in \mathbb{R}^3$. Using that $\mathcal{M}$ is compactly supported we get (2.6.25) for the diagonal case.

We now have to consider the non-diagonal case. For $N = I - M \in \mathcal{C}_{aniso}$ we know that $M(x)$ and thus also $\mathcal{M}(x)$ can be diagonalized, i.e., we can find a matrix $U(x)$ with $U^*(x)U(x) = I$ such that $\mathcal{M}_D := U^*(x)\mathcal{M}(x)U(x)$ is diagonal. Furthermore, its diagonal elements are of the form (2.6.28). We get

$$U^*(x)\Big[(1 - \zeta)I + \mathcal{M}(x)\hat{k}_0(\phi, \vartheta)\Big]U(x) \qquad (2.6.31)$$
$$= (1 - \zeta)I + \mathcal{M}_D\Big[U^*(x)\hat{k}_0(\phi, \vartheta)U(x)\Big].$$

Let us define $w := U^*v$. Then we obtain

$$(-3)U^*(x)\hat{k}_0(\phi, \vartheta)U(x) = U^*(x) \cdot v \cdot v^T \cdot U - I = w \cdot \overline{w}^T - I.$$

We now compute the determinant of (2.6.31) and obtain again the expression (2.6.29) with v_j^2 replaced by $|w_j|^2$. In the same way as above we conclude that

$$\det\left((1 - \zeta)I + \mathcal{M}(x)\hat{k}_0(\varphi, \vartheta)\right) \geq \gamma > 0$$

for all $\varphi \in [0, 2\pi]$, $\vartheta \in [0, \pi]$, $\zeta \in (-\infty, 0]$ and for all $x \in \mathbb{R}^3$. $\qquad\square$

We now come to the proof of the regularity properties. We extended Theorem 5.1, page 325 of [60] to a system of strongly singular integral equations.

THEOREM 2.6.5 *Let* $v \in L^2(\mathbb{R}^3)$ *be a solution to (2.6.13) with* $\mathcal{M} \in C_0^{l+1}(\mathbb{R}^3)$ *and* $E^i \in H_{loc}^{(l)}(\mathbb{R}^3)$. *Then* v *belongs to the space* $H_0^{(l)}(\mathbb{R}^3)$.

Proof. First we note that the operators MVM and $\mathcal{M}(T - T_0)$ are smoothing of order l, i.e., they map $H_0^{(j)}(\mathbb{R}^3)$ into $H_0^{(j+1)}(\mathbb{R}^3)$ for all $j = 0, ..., l-1$. For MVM this can be seen by a simple inductive proof using the mapping properties of the volume potential and the fact that the volume potential has a convolution kernel. To show that $\mathcal{M}(T - T_0)$ is smoothing of order l we refer to the appendix of [77], where the proof is worked out for the two-dimensional case. All steps are basically the same. Next we observe that the symbol $\Psi(x, \varphi, \vartheta)$ of $I + \mathcal{M}(x)T_0$ is $l+1$ times continuously differentiable with respect to x and that $D_x^{\alpha}\Psi(x, ., .)$ is in $H^{(\lambda)}(\Omega)$ for all $|\alpha| \leq l + 1$ and all $\lambda \in I\!N$.

We now show that part 5.2 of the proof of Theorem 5.1, Chapter XII of [60] can be transferred to the case where we have the matrix valued singular integral operator $I + \mathcal{M}T_0 + \mathcal{M}(T - T_0) + \kappa^2 \mathcal{M}VM$ instead of a scalar operator. By B we denote the simple singular operator with symbol $\Psi(x, \theta)^{-1}$. The existence of B is a consequence of the estimate (2.6.24) due to Theorem 2.6.2. The operator B is of the form

$$(Bv)(x) = b(x)v(x) + \int_{I\!R^3} \frac{h(x,\theta)}{r^3} v(y) dy, \quad r = |x - y|, \theta = \frac{y - x}{r}, x \in I\!R^3$$

where b and h are now 3×3-matrices of functions. We now proceed in the same way as Michlin and Prößdorf. The only difference to the proof in [60] is that the multiplication is no longer commutative. It turns out that commutativity is not an essential point of the derivations (3) and (4) of part 5.2, [60], page 326. The parts 5.3 and 5.4 of [60], page 327, can then be applied to each component of the resulting operators and we obtain the statement of the theorem. $\Box$

Uniqueness of the scattering problem. In this part we will establish the uniqueness of solutions to the scattering problem (2.6.1) to (2.6.4) for complex refractive matrices N. Our proof is an extension of the proof given in [58], page 168, which is valid for real-valued symmetric N. We use the unique continuation principle as given in [32]. A complex matrix $M(x)$ is called coercive (semi-coercive) if

$$\text{Im}(a \cdot \overline{M a}) \geq \gamma(x)|a|^2 \qquad\qquad (2.6.32)$$

for every $a \in \mathbb{C}^3$ where $\gamma(x) > 0$ ($\gamma \geq 0$). For a matrix $N \in \mathcal{C}_{aniso}$ the matrix $M := I - N$ is semi-coercive. A complex matrix is called elliptic, if $a \cdot Na \neq 0$ for all $a \neq 0$, $a \in I\!R^3$. Matrices $N \in \mathcal{C}_{aniso}$ are elliptic.

THEOREM 2.6.6 (**Unique continuation principle.**) *Let the fields* $E, H \in H_{loc}^{(1)}(I\!R^3)$ *solve the Maxwell equations (2.6.1) with elliptic* $N \in C^3(I\!R^3)$. *Assume that* E *vanishes in the neighbourhood of some* $x_0 \in I\!R^3$. *Then* E, H *is identically zero in* $I\!R^3$.

Proof. First we note that we have $H, E \in H_{loc}^{(2)}(I\!R^3)$ due to Theorem 2.6.5. From the Maxwell equations we obtain

$$\nabla \cdot (NE) = 0, \quad \nabla \cdot H = 0$$

and thus (using the summation convention and curl curl $H = -\Delta H + \nabla(\nabla \cdot H)$)

$$N_{\xi j}\partial_\xi \partial_k E_j = -(\partial_k N_{\xi j})(\partial_\xi E_j) - \partial_k[(\partial_\xi N_{\xi j})E_j] \quad (2.6.33)$$
$$\Delta H = i\nabla \times (NE).$$

Again using the Maxwell equation $\nabla \times E = i\kappa H$, i.e.,

$$\partial_k E_j = \partial_j E_k + f_{jk}(H) \quad (2.6.34)$$

with some function f_{jk}, we transform (2.6.33) into

$$N_{\xi j}\partial_\xi \partial_j E_k = F_k(E, H, \nabla E, \nabla H) \quad (2.6.35)$$
$$\Delta H = G_k(E, H, \nabla E, \nabla H) \quad (2.6.36)$$

with functions F_k, G_k, which is a weakly coupled second order system. We can apply Proposition 17.2.3 of [32] to every component of E and H. If we proceed as in the proof of Theorem 17.2.1 of [32] we obtain that the statement of Theorem 17.2.1 of [32] holds for every component of E and H, i.e., the set of generalized normal vectors to $supp(E_j)$ and $supp(H_j)$ is empty. This yields $E = H = 0$ in $\mathbb{R}^3$ and the proof is complete. □

THEOREM 2.6.7 (**Uniqueness.**) *Assume that $M = I - N \in C_0^3(\mathbb{R}^3)$ is semi-coercive and that N is elliptic. Then the scattering problem (2.6.1) - (2.6.4) has at most one solution.*

Proof. Let E, H be the difference between two solutions, i.e., a solution with vanishing incident field. From Gauss' divergence theorem applied to a ball B with $supp(M) \subset B$ and the Maxwell equations (2.6.1) we have that

$$\int_{\partial B} \nu \times E \cdot \overline{H} ds = \int_B (\nabla \times E \cdot \overline{H} - E \cdot \nabla \times \overline{H}) dx$$
$$= i\kappa \int_B \left(|H|^2 - E \cdot \overline{NE}\right) dx \quad (2.6.37)$$

and therefore using $M = I - N$ and the coercivity of M

$$\operatorname{Re} \int_{\partial B} \nu \times E \cdot \overline{H} ds = -\kappa \int_B \operatorname{Im}\left(E \cdot \overline{ME}\right) dx \leq 0.$$

Since E, H are analytic in the exterior of $supp(M)$ we can apply Theorem 6.10 of [8]. We conclude that $E(x) = H(x) = 0$ for $x \in \mathbb{R}^3 \setminus \overline{B}$. Now from the unique continuation principle as given by Theorem 2.6.6 we obtain the statement of the theorem. □

Existence of solutions to the anisotropic scattering problem.
We now prove the uniqueness and invertibility of $I + \mathcal{M}T + \kappa^2 \mathcal{M}V\mathcal{M}$ and
use it to obtain the existence of solutions to the scattering problem (2.6.1)
to (2.6.4). By B denote a bounded domain containing the support of $\mathcal{M}$
in its interior.

Let us first assume that we have a solution $v \in L^2(\mathbb{R}^3)$ of equation
(2.6.13), that $\mathcal{M}$ is in $C_0^2(\mathbb{R}^3)$ and that E^i, H^i is an incident field. Note
that v has compact support. From Theorem 2.6.5 we obtain $v \in H^{(1)}(\mathbb{R}^3)$.
We define $E \in L^2(\mathbb{R}^3)$ by

$$E(x) := (I - \frac{1}{3}\mathcal{M}(x))^{-1}\Big(E^i(x) - Tv - \kappa^2 T\mathcal{M}v\Big). \qquad (2.6.38)$$

Then from (2.6.13) we get $\mathcal{M}E = v$ and from (2.6.12) that E solves the
integral equation (2.6.5). The field E^i is real analytic as a solution to the
free space Maxwell equations. The volume potential $\kappa^2 V\mathcal{M}v$ in (2.6.38)
is in $H_{loc}^{(2)}(\mathbb{R}^3)$. With the help of (2.6.9) and the mapping properties of
the volume potential we obtain $Tv \in H_{loc}^{(1)}(\mathbb{R}^3)$. Together this yields $E \in
H_{loc}^{(1)}(\mathbb{R}^3)$. Thus, for $H := \nabla \times E/i\kappa$, differentiating equation (2.6.5) we
obtain the equation

$$H(x) = H^i(x) + i\kappa\nabla \times \int_{\mathbb{R}^3} \Phi(x,y)v(y)dy, \quad x \in \mathbb{R}^3. \qquad (2.6.39)$$

This implies $H \in H_{loc}^{(2)}(\mathbb{R}^3)$. Thus we obtain with the help of $\nabla \times \nabla \times F =
-\Delta F + \nabla\nabla \cdot F$ for vector fields $F \in H_{loc}^{(2)}(\mathbb{R}^3)$, the Maxwell equations for
the incident fields and (2.6.9) the equalities

$$
\begin{aligned}
\nabla \times H(x) + i\kappa E(x) &= i\kappa\Big(\nabla \times \nabla \times -\kappa^2\Big)\int_{\mathbb{R}^3} \Phi(x,y)\mathcal{M}E(y)dy \\
&\quad -i\kappa\nabla\int_{\mathbb{R}^3} \nabla_x\Phi(x,y)(\mathcal{M}E)(y)dy \\
&= -i\kappa\Big(\Delta + \kappa^2\Big)\int_{\mathbb{R}^3} \Phi(x,y)\mathcal{M}E(y)dy \\
&= i\kappa\mathcal{M}(x)E(x). \qquad (2.6.40)
\end{aligned}
$$

Since the radiation condition is a consequence of the integral representations the fields E, H satisfy (2.6.1) to (2.6.4). We have proven

THEOREM 2.6.8 *Assume that we have $\mathcal{M} \in C_0^2(\mathbb{R}^3)$. Let $v \in L^2(\mathbb{R}^3)$ be a
solution of the integral equation (2.6.13). Then E defined by (2.6.38) and
$H := \nabla \times E/i\kappa$ solve the scattering problem (2.6.1) to (2.6.4).*

In Lemma 2.6.4 we verified the assumption (2.6.25) of Theorem 2.6.3 for the operator $I + \mathcal{M}T_0$. All regularity assumptions are also fulfilled. Thus we obtain that there exists an equivalent right regularizer for $I + \mathcal{M}T_0$ in the space $L_0^2(B)$. We further observe that $\mathcal{M}(T - T_0)$ has a weakly singular kernel and is therefore a compact operator in $L_0^2(B)$. The same is true for $\mathcal{M}$ times the volume potential VM. Thus by the standard arguments of the Riesz-Fredholm theory (see for example [51], Theorem 3.4) we obtain existence and the boundedness of the inverse of $I + \mathcal{M}T + \kappa^2 \mathcal{M}VM$ in $L_0^2(B)$ if we can prove the uniqueness of the homogeneous equation

$$(I + \mathcal{M}T + \kappa^2 \mathcal{M}VM)v = 0. \tag{2.6.41}$$

Let $v \in L_0^2(B)$ be a solution to equation (2.6.41). Clearly v can be considered as an element of $L^2(\mathbb{R}^3)$. From Theorem 2.6.5 we obtain $v \in H^1(\mathbb{R}^3)$ and E, H defined by (2.6.38), and $H := \nabla \times E/i\kappa$ satisfy the scattering problem (2.6.1) to (2.6.4) with vanishing incident field. To apply the uniqueness theorem we have to assume $\mathcal{M} \in C_0^3(\mathbb{R}^3)$. Then the uniqueness theorem yields $H = E = 0$; thus $v = ME = 0$ and the uniqueness proof is complete.

We can summarize our results in the following

THEOREM 2.6.9 (Existence.) *Assume that we have $\mathcal{M} \in C_0^3(B)$. Then the integral equation (2.6.13) has one and only one solution in $L_0^2(B)$ and the solution depends continuously on the right-hand side.*

Now from Theorem 2.6.9, Theorem 2.6.8 and Theorem 2.6.5 we obtain

THEOREM 2.6.10 *Assume that we have $\mathcal{M} \in C_0^{l+1}(B)$, $l \geq 2$. The scattering problem (2.6.1) - (2.6.4) for an anisotropic inhomogeneous medium has one and only one solution E, H in $H_{loc}^{(l)}(\mathbb{R}^3) \times H_{loc}^{(l)}(\mathbb{R}^3)$ and the field E, H depends continuously on the incident field E^i, H^i.*

Proof. It remains to verify the regularity statement. It is a consequence of Theorem 2.6.5, the mapping properties of the volume potential which is bounded $H_0^{(j)}(B) \to H^{(j+2)}(B)$ and the equations (2.6.38) and (2.6.39). □

Fréchet differentiability with respect to M. We conclude the section with an investigation of the differentiability properties of the mapping $M \mapsto u^s(M)$. Note that analyticity in a complex Banach space is a consequence of simple Fréchet differentiability.

Consider the operator $L(M) : H^{(1)}(B) \to H^{(1)}(B)$ where B is a ball containing the support of M defined by

$$(L(M)E) := \nabla \int_{\mathbf{R}^3} \nabla_x \Phi(x,y) \cdot (ME)(y)\, dy \qquad (2.6.42)$$

$$+ \kappa^2 \int_{\mathbf{R}^3} \Phi(x,y)\, M(y)E(y)\, dy. \qquad (2.6.43)$$

Using $\mathcal{C}_{anisoM} := \{M : I - M \in \mathcal{C}_{aniso}\}$ we are now able to formulate the differentiability statements.

THEOREM 2.6.11 *The mapping of the matrix M onto the solution E of (2.6.5) is analytic from $\mathcal{C}_{anisoM}$ into $H^1(B)$. The n-th derivative at the point M and direction δ is given by*

$$E^{(n)}(M,\delta) = (-1)^n n! \left((I + L(M))^{-1}\, L(\delta) \right)^n (I + L(M))^{-1} E^i. \qquad (2.6.44)$$

Proof. The mapping

$$\mathcal{C}_{anisoM} \to BL(H^{(1)}(B), H^{(1)}(B)), \quad M \mapsto L(M) \qquad (2.6.45)$$

is a linear mapping and thus Fréchet differentiable on $\mathcal{C}_{anisoM}$. The same holds for $I + L(M)$. Since for $M \in \mathcal{C}_{anisoM}$ the operator $I + L(M)$ is continuously invertible in $BL(H^{(1)}(B), H^{(1)}(B))$, from Theorem 2 of [78] we obtain that its inverse is Fréchet differentiable as a mapping $\mathcal{C}_{anisoM} \to BL(H^{(1)}(B), H^{(1)}(B))$ and we obtain the given form of the inverse. The statement for higher derivatives can be obtained by induction. $\qquad \square$

Equation (2.6.44) yields a characterization of the n-th derivative as a solution to an anisotropic scattering problem. Here $H^{(n)}$ is defined by $H^{(n)} := \nabla \times E^{(n)}/i\kappa$.

COROLLARY 2.6.12 *The n-th Fréchet derivative of the solution to the scattering problem (2.6.1) - (2.6.4) with respect to $M = I - N$ at the point M and direction δ solves the equations*

$$\nabla \times E^{(n)} - i\kappa H^{(n)} = 0,$$
$$\nabla \times H^{(n)} + i\kappa N E^{(n)} = i\kappa n\, \delta E^{(n-1)} \qquad (2.6.46)$$

and satisfies the Silver-Müller radiation condition.

Proof. We proceed in the same way as in (2.6.40). $\qquad \square$

Remarks. 1) Note that (2.6.46) can also be obtained directly from the Maxwell equations after the Fréchet differentiability has been established. 2) Under the conditions of Theorem 2.6.7 the equations (2.6.46) have one and only one solution.

3 Uniqueness and stability in inverse scattering

Uniqueness theorems in inverse scattering theory usually investigate the amount of data necessary to determine scattering objects uniquely. Clearly, if we do not have uniqueness, we cannot expect to obtain stable numerical algorithms for the computation of the scattering objects. Thus uniqueness is a question of practical importance.

In this chapter we first investigate the question of uniqueness of the support of obstacle scatterers, given the far field patterns for incident plane waves at a fixed wave number κ. We will show that the shape D of the scatterer $\mathcal{D}$ is uniquely determined by the far field patterns $u^{\infty}(\cdot, d)$ of the scattered fields for all incident plane waves $u^i(\cdot, d)$ with directions of incidence $d \in \Omega$. Uniqueness for penetrable scatterers will be obtained as a consequence of stability, which is the second problem of this section.

The question of stability leads right into the heart of the difficulties of inverse problems. How do errors in the measurements affect the reconstructions? We already indicated that inverse scattering problems are ill-posed, because a radiating solution to the Helmholtz equation in the exterior of a ball does not depend continuously on its far field pattern. We also indicated that with appropriate assumptions on the scatterers under consideration stability can be restored and stability estimates can be derived.

In this section we will derive stability estimates for the reconstruction of the domain D of a scatterer $\mathcal{D}$ from the far field patterns $u^{\infty}(\hat{x}, d)$ for $\hat{x}, d$ in Ω, where for measurements of the far field patterns we will use the practically relevant L^2-norm. A stability estimate consists of a function $F : \mathbb{R}^+ \to \mathbb{R}^+$ with the property

$$F(\delta) \to 0, \quad \delta \to 0, \tag{3.0.1}$$

such that the Hausdorff distance $d(D_1, D_2)$ of two scatterers can be estimated by

$$d(D_1, D_2) \leq F\Big(\|u_1^{\infty}(\cdot, \cdot) - u_2^{\infty}(\cdot, \cdot)\|_{L^2(\Omega \times \Omega)}\Big). \tag{3.0.2}$$

For the convex hulls $\mathcal{H}(D_1)$ and $\mathcal{H}(D_2)$ of impenetrable acoustic or arbitrary electromagnetic scatterers $\mathcal{D}_1$ and $\mathcal{D}_2$ we will derive a logarithmic estimate, i.e.,

$$F(\delta) = \frac{C}{|\ln(\delta)|^c}$$

with constants $C > 0$ and $0 < c < 1$ uniformly for the given classes of scatterers.

Clearly stability implies uniqueness and thus as a consequence of stability we obtain uniqueness statements. In particular, this will yield new uniqueness results for inhomogeneous medium scatterers.

3.1   Acoustic scattering

The following uniqueness theorem for the reconstruction of domains in obstacle scattering was first proven in 1993 by Kirsch and Kress [47], simplifying techniques of Isakav [33]. It will be the starting point for our further considerations, and we will give an even simpler proof than in [47] using mixed reciprocity relations.

THEOREM 3.1.1 *Let $\mathcal{D}_1, \mathcal{D}_2$ be sound-soft or sound-hard scatterers. If the far field patterns $u_1^\infty(\hat{x}, d)$ and $u_2^\infty(\hat{x}, d)$ for scattering of plane waves coincide for all $\hat{x}, d \in \Omega$, then $D_1 = D_2$.*

Proof. Let D_e be the unbounded component of $\mathbb{R}^m \setminus (\overline{D_1 \cup D_2})$. From

$$u_1^\infty(\hat{x}, d) = u_1^\infty(\hat{x}, d), \quad \hat{x}, d \in \Omega,$$

and the Rellich Lemma we obtain

$$u_1^s(z, d) = u_1^s(z, d), \quad z \in D_e, \ d \in \Omega.$$

We use the mixed reciprocity relation (2.1.4) on both sides to derive for the far field patterns of incident point-sources the equality

$$\Phi_1^\infty(d, z) = \Phi_2^\infty(d, z), \quad z \in D_e, \ d \in \Omega.$$

Again we use the Rellich lemma to get

$$\Phi_1^s(x, z) = \Phi_2^s(x, z), \quad x, z \in D_e. \tag{3.1.1}$$

From (3.1.1) and Theorem 2.1.15 we derive $D_1 = D_2$ in the following way. We assume that $\partial D_1 \setminus \overline{D_2} \neq \emptyset$ and $z_0 \in \partial D_1 \setminus \overline{D_2}$. Then

$$\infty > \Phi_2(z_0, z_0) = \lim_{z \to z_0, z \in D_e} \Phi_2(z, z) = \lim_{z \to z_0, z \in D_e} \Phi_1(z, z) = \infty$$

and we obtain a contradiction. In the same way we treat the case $z_0 \in \partial D_2 \setminus \overline{D_1}$. Thus we obtain $D_1 = D_2$ and the proof is complete. $\square$

Since most further results will hold both for scattering from impenetrable and penetrable scatterers, we define the class

$$\mathcal{C} := \mathcal{C}_{obst} \cup \mathcal{C}_m. \qquad (3.1.2)$$

We will now focus on the stability question and develop techniques to prove stability for the reconstruction of the domain D of both impenetrable and penetrable scatterers. Since stability implies uniqueness, we also will obtain uniqueness statements for D. All further results will use the steps introduced in the following two parts of this section.

Approximation of multipoles by plane waves. As the first step, we consider approximations of a multipole by a continuous superposition of plane waves, i.e., by a Herglotz wave function

$$(Hg)(x) := \int_\Omega e^{i\kappa x \cdot d} g(d) \, ds(d), \quad x \in \mathbb{R}^m. \qquad (3.1.3)$$

To obtain these approximations in a uniform way we use the exterior cone condition as follows. With the help of the cone (1.2.10) we define the domain

$$G_{z,p,\rho} := B_{2R_s}\left(z - \frac{p\rho}{\cos(\beta_0)}\right) \setminus \mathrm{co}\left(z - \frac{p\rho}{\cos(\beta_0)}, p, \beta_0\right) \qquad (3.1.4)$$

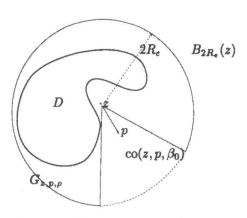

for $z \in \mathbb{R}^m$, $p \in \Omega$ and $\rho > 0$ as shown in Figure 3.1. Here, since the parameter β_0 is kept constant throughout this work, we do not explicitly note the dependence of G on β_0. From the exterior cone condition for $D \subset B_{R_s}(0)$ for each $z \in B \setminus D_\rho$ we obtain a vector $p \in \Omega$, such that $D \subset G_{z,p,\rho}$.

We first consider the operator H from $L^2(\Omega)$ into $L^2(\partial G)$, where G denotes an appropriately chosen domain G with $G \supset \overline{G_{0,p,\rho}}$ and $0 \notin \overline{G}$.

Figure 3.1: Domain of approximation.

LEMMA 3.1.2 *It is possible to choose a domain $G \supset \overline{G_{0,p,\rho}}$ with $0 \notin \overline{G}$ and boundary ∂G of class C^2, such that the homogeneous interior Dirichlet problem*

$$-\Delta u = \kappa^2 u \text{ in } G, \quad u = 0 \text{ on } \partial G \qquad (3.1.5)$$

has only the trivial solution $u = 0$. In this case the operator H has dense range in $L^2(\partial G)$.

Proof. Due to Theorem 4.7 of [58] for the l-th eigenvalue λ_l of the problem (3.1.5), where the eigenvalues are ordered according to their magnitude and multiplicity

$$\lambda_1 \le \lambda_2 \le \dots \le \lambda_l \le \dots,$$

we have the monotonicity property

$$G_1 \overset{\ne}{\subset} G_2 \;\Rightarrow\; \lambda_{1,l} > \lambda_{2,l}, \; l \in I\!\!N. \tag{3.1.6}$$

Thus, if for a domain

$$G_1 \supset G_{0,p,\rho}, \; 0 \notin \overline{G_1}$$

and $l \in I\!\!N$ we have $\lambda_{1,l} = \kappa^2$, we can choose a domain

$$G_2 \supset G_1, \; 0 \notin \overline{G_2},$$

such that $\lambda_{2,l'} \ne \kappa^2$ for all $l' \in I\!\!N$. Then κ^2 is not an eigenvalue for the domain G_2 and every solution u of (3.1.5) with $G := G_2$ must vanish identically.

To show denseness for the range of H in $L^2(\partial G)$ we prove injectivity of the adjoint H^* of the operator H. The adjoint H^* of H is given by

$$(H^*\varphi)(\hat{x}) = \int_{\partial G} e^{-i\kappa\hat{x}\cdot y}\varphi(y)\,ds(y), \; \hat{x} \in \Omega. \tag{3.1.7}$$

The function $H^*\varphi$ is the far field pattern of the single-layer potential

$$(S\varphi)(x) := \int_{\partial G} \Phi(x,y)\varphi(y)\,ds(y), \; x \in I\!\!R^m.$$

We assume $H^*\varphi = 0$ on Ω for $\varphi \in L^2(\partial G)$. Then Rellich's Lemma yields $(S\varphi)(x) = 0$ for $x \in I\!\!R^m \setminus \overline{G}$. Now, from the jump-relations for L^2-densities, Theorem 1.2.4, we get $(I - K^*)\varphi = 0$. We can apply the Fredholm alternative 1.2.6 first in the dual system

$$\big\langle C(\partial D), L^2(\partial D) \big\rangle, \; \langle \varphi, \psi \rangle := \int_D \varphi\bar{\psi}\,dy,$$

and then in the dual system

$$\big\langle L^2(\partial D), L^2(\partial D) \big\rangle, \; \langle \varphi, \psi \rangle := \int_D \varphi\bar{\psi}\,dy,$$

to conclude that the null spaces of $I - K^*$ in $L^2(\partial G)$ and $C(\partial G)$ have the same finite dimension. Since $C(D)$ is a subset of $L^2(D)$, the null spaces coincide and φ is in $C(\partial G)$. Now, continuity of the single-layer potential with continuous density implies that $S\varphi$ solves the homogeneous Dirichlet problem in G. From the first part of the lemma we obtain $S\varphi = 0$ on G. Now, the jump relations of Theorem 1.2.4 yield $\varphi = 0$ and thus the injectivity of H^*. $\qquad\square$

We now complete the first step and show that in the space $C^s(G_{0,p,\rho})$ the function $\Phi_{\mu,q}(\cdot,0)$ can be approximated by a Herglotz wave function.

LEMMA 3.1.3 *Given $\mu, s \in \mathbb{N}_0$, $\rho > 0$ and $\tau > 0$ there is a finite set*

$$\mathcal{G} = \mathcal{G}(\tau, \rho, \mu, s, \beta_0)$$

of densities g in $L^2(\Omega)$, such that for each $p, q \in \Omega$ there is a density $g \in \mathcal{G}$ with

$$\left\| \Phi_{\mu,q}(\cdot,0) - Hg \right\|_{C^s(G_{0,p,\rho})} \leq \tau. \tag{3.1.8}$$

Proof. According to Lemma 3.1.2, given $\epsilon > 0$, we can find $g \in L^2(\Omega)$ such that

$$\left\| \Phi_{\mu,q}(\cdot,0) - Hg \right\|_{L^2(\partial G)} \leq \epsilon. \tag{3.1.9}$$

Let g be the minimum norm solution of (3.1.9), which is unique according to Theorem 1.2.13. We note that both $\Phi_{\mu,q}$ and Hg solve an interior Dirichlet problem for the Helmholtz equation in the domain G and can be represented as the combined acoustic double- and single-layer potential (2.1.13). Differentiating under the integral sign and using the Cauchy-Schwarz inequality we observe that on compact subsets of G the solution of the interior Dirichlet problem depends continuously on the boundary values in $L^2(\partial G)$; i.e., from (3.1.9) with ϵ replaced by ϵ/c with some constant c we obtain (3.1.8).

So far, the density g in (3.1.9) depends on the parameters $p, q \in \Omega$. We observe that for fixed g the norm

$$\left\| \Phi_{\mu,q}(\cdot,0) - Hg \right\|_{C^s(G_{0,p,\rho})}$$

depends continuously on p and q. Thus given $p, q \in \Omega$ and $g \in L^2(\Omega)$ such that

$$\left\| \Phi_{\mu,q}(\cdot,0) - Hg \right\|_{C^s(G_{0,p,\rho})} \leq \frac{\tau}{2} \tag{3.1.10}$$

is satisfied, there is a neighborhood U of p and a neighborhood V of q, such that

$$\left\| \Phi_{\mu,q'}(\cdot,0) - Hg \right\|_{C^s(G_{0,p',\rho})} \leq \tau \tag{3.1.11}$$

is satisfied for all $q' \in U$ and all $p' \in V$. The compact set Ω is covered by a finite number of such domains U and V; i.e., there is a finite set $\mathcal{G}$ of densities g, such that for all $p, q \in \Omega$ the estimate (3.1.8) is satisfied with a density $g \in \mathcal{G}$. $\qquad\qquad\qquad\qquad\qquad\qquad\qquad\qquad\qquad\qquad\quad\square$

Distances between domains estimated by $\Phi_{q,\mu}^s$. In the second step estimates for the difference of the fields $\Phi_{1,\mu,q}^s$ and $\Phi_{2,\mu,q}^s$ for scattering of a multipole from two scatterers $\mathcal{D}_1$ and $\mathcal{D}_2$ will be used to estimate the Hausdorff difference $d(D_1, D_2)$ of the domains D_1 and D_2.

We need to consider six different situations according to the physical properties of the scatterers $\mathcal{D}_1$ and $\mathcal{D}_2$ in two or three dimensions. For each situation we need a corresponding choice of the parameter μ.

Situation	Properties of $\mathcal{D}_1$ and $\mathcal{D}_2$	μ
S1	$\mathcal{D}_1, \mathcal{D}_2$ impenetr. scatterers in $\mathbb{R}^2$	0
S2	$\mathcal{D}_1, \mathcal{D}_2$ impenetr. scatterers in $\mathbb{R}^3$	0
S3	$\mathcal{D}_2$ inhom. medium scatterer and $\mathcal{D}_1$ impenetr. scatterer or vice versa in $\mathbb{R}^2$	$\mu_0 + 2$
S4	$\mathcal{D}_2$ inhom. medium scatterer and $\mathcal{D}_1$ impenetr. scatterer or vice versa in $\mathbb{R}^3$	$\mu_0 + 1$
S5	$\mathcal{D}_1, \mathcal{D}_2$ inhom. medium scatterers in $\mathbb{R}^2$	$\mu_0 + 2$
S6	$\mathcal{D}_1, \mathcal{D}_2$ inhom. medium scatterers in $\mathbb{R}^3$.	$\mu_0 + 1$

Table 3.1

For technical reasons we introduce the set

$$U(D_1, D_2, \rho, \beta_0) := \{ z \in B \setminus (D_{1,\rho} \cup D_{2,\rho}), \qquad\qquad (3.1.12)$$
$$\exists p \in \Omega \text{ such that } \mathrm{co}(z, p, \beta_0) \subset \mathbb{R}^m \setminus (D_{1,\rho} \cup D_{2,\rho}) \}.$$

LEMMA 3.1.4 *We consider scattering of acoustic waves by two scatterers $\mathcal{D}_1, \mathcal{D}_2$ in C and choose $\mu \in \mathbb{N}_0$ according to Table 3.1. Assume that with parameters $\rho, \sigma > 0$ the scattered fields $\Phi_{1,\mu,q}^s$ and $\Phi_{2,\mu,q}^s$ satisfy*

$$\left| \Phi_{1,\mu,q}^s(z, z) - \Phi_{2,\mu,q}^s(z, z) \right| \leq \sigma \qquad\qquad (3.1.13)$$

for all $q \in \Omega$ and $z \in U(D_1, D_2, \rho, \beta_0)$. Then we conclude

$$d(D_1, D_2) \leq F_1(\rho, \sigma) \qquad\qquad (3.1.14)$$

where the function F_1 is defined according to the situations S1 to S6 by

$$F_1(\rho, \sigma) := \begin{cases} \rho + Bb^\sigma \rho^s & S1, \ S5 \ \text{and} \ S6, \\ \rho + \frac{C\rho}{c - \sigma\rho} & S2, \\ \rho + \left(\frac{c}{Cb}|\ln\rho| - \frac{\sigma}{Cb}\right)^{-1/(\mu_0 + m + 2)} & S3 \ \text{and} \ S4 \end{cases}$$

(3.1.15)

with constants B, b, s, c and C uniformly for $\mathcal{D}_1, \mathcal{D}_2 \in \mathcal{C}$.

Proof. We split the proof into two parts.

1. We consider the scattered fields for a point $z \in U(D_1, D_2, \rho, \beta_0)$. From (2.1.48), (2.1.49) and (3.1.13) we obtain for situation S1 of two impenetrable scatterers in the two dimensions the estimate

$$\begin{aligned} C\left|\ln d(z, D_1)\right| + E &\geq \left|\Phi^s_{1,\mu,\nu(z_0)}(z, z)\right| \\ &\geq \left|\Phi^s_{2,\mu,\nu(z_0)}(z, z)\right| - \sigma \qquad (3.1.16) \\ &\geq c\left|\ln d(z, D_2)\right| - \sigma, \end{aligned}$$

which can be transformed into

$$d(z, D_1) \leq B\, b^\sigma\, d(z, D_2)^s \qquad (3.1.17)$$

with constants $B = e^{E/C}, b = e^{1/C}$ and $1 > s = c/C > 0$. In the same way for scattering by impenetrable scatterers in three dimensions (situation S2) we obtain from (2.1.50), (2.1.51) and (3.1.13)

$$\frac{C}{d(z, D_1)} \geq \frac{c}{d(z, D_2)} - \sigma \qquad (3.1.18)$$

and transform it into

$$d(z, D_1) \leq \frac{Cd(z, D_2)}{c - \sigma d(z, D_2)}. \qquad (3.1.19)$$

Consider an inhomogeneous medium scatterer $\mathcal{D}_2$ and an impenetrable scatterer $\mathcal{D}_1$, i.e., the situations S3 and S4. From the definition of the multipoles and the choice of μ we derive

$$\left\|\Phi_{\mu,q}(\cdot, z)\right\|_{C^1(\overline{D_1})} \leq Cd(z, D_1)^{-\mu_0 - 3} \qquad (3.1.20)$$

with some constant C. The scattering map $u^i \mapsto u^s$ from $C^1(\overline{D_1})$ into $C(B \setminus D_{1,\rho})$ is bounded by b/ρ^{m-1} with some constant b uniformly for $\mathcal{D} \in \mathcal{C}_{obst}$. Thus we obtain

$$\left|\Phi^s_{\mu,q}(z, z)\right| \leq C\, b\, |d(z, D_1)|^{-\mu_0 - m - 2}. \qquad (3.1.21)$$

Now, from the lower estimate (2.2.56) we derive

$$C \, b \, |d(z, D_1)|^{-\mu_0 - m - 2} \geq c \, |\ln d(z, D_2)| \, - \, \sigma \tag{3.1.22}$$

with constants $c, C, b > 0$, which yields

$$d(z, D_1) \leq \left(\frac{c}{Cb} \, |\ln d(z, D_2)| \, - \, \frac{\sigma}{Cb}\right)^{-\frac{1}{\mu_0 + m + 2}}. \tag{3.1.23}$$

If D_2 is an impenetrable scatterer and D_1 an inhomogeneous medium, the estimates can be obtained as in (3.1.17) and (3.1.19). Since the functions are dominated by (3.1.23), a bound is again given by (3.1.23). For two inhomogeneous medium scatterers D_1 and D_2 we derive from (2.2.56), (2.2.57) and (3.1.13) the bound (3.1.17)

2. We now come to the second step of the proof. According to the cases S1 to S6 we choose $\epsilon := F_1(\rho, \sigma)$ with F_1 defined in (3.1.15).

First, we consider a point $z_0 \in \partial D_2$, such that for $z' := z_0 + \rho\nu(z_0)$ we have $z' \in U(D_1, D_2, \rho, \beta_0)$. From the triangle inequality, $\rho = d(z', D_2)$ and the first part of the proof we obtain

$$\begin{aligned} d(z_0, D_1) &\leq d(z_0, z') + d(z', D_1) \\ &\leq F_1(\rho, \sigma) = \epsilon. \end{aligned} \tag{3.1.24}$$

Second, we investigate an arbitrary cone

$$\mathrm{co}(z, p, \beta_0) \subset \mathbb{R}^m \setminus \overline{D_{1,\epsilon}}. \tag{3.1.25}$$

To show

$$\mathrm{co}(z, p, \beta_0) \cap D_2 = \emptyset \tag{3.1.26}$$

by contradiction, we assume that

$$\mathrm{co}(z, p, \beta_0) \cap D_2 \neq \emptyset. \tag{3.1.27}$$

Then we have

$$r_1 := \inf \left\{ r > 0 : \left(D_2 \cap \mathrm{co}(z, p, \beta_0)\right) \subset \left(B_r(x) \cap \mathrm{co}(z, p, \beta_e)\right) \right\} > 0,$$

and we can find a point $z_1 \in \partial D_2 \cap \partial B_{r_1}(x) \cap \mathrm{co}(z, p, \beta)$. On the line

$$L = \{z_1 + tp : t \in (0, \infty)\}$$

there is a point z' with $d(z', D_2) = \rho$. Then we have $z' = z_0 + \rho\nu(z_0)$ for some point $z_0 \in \partial D_2$ and $z' \in U(D_1, D_2, \rho, \beta_0)$. Now, the estimate (3.1.24) yields $d(z_0, D_1) \leq \epsilon$. But from (3.1.25) we obtain $\epsilon < d(z_0, D_1)$ and thus a contradiction, i.e., the assumption (3.1.27) is wrong and (3.1.26) is shown.

Finally, we note that for $D_1 \in \mathcal{C}$ and ϵ sufficiently small the open exterior of $D_{1,\epsilon}$ can be covered by cones $\mathrm{co}(z, p, \beta_0)$ with $z \in \mathbb{R}^m \setminus D_{1,\epsilon}$ and $p \in \Omega$, i.e., from (3.1.26) we obtain $D_2 \subset D_{1,\epsilon}$. Since we can go through all arguments with D_1 and D_2 exchanged, we obtain (3.1.14). $\qquad \square$

The backprojection operator Q. We will define and study an operator Q, which is built to approximate the scattered fields of multipoles by a superposition of the far field patterns of the scattered fields of plane waves. To this end we need some further preparations.

a) Consider scattering of an incident field given by a superposition Hg of plane waves (3.1.3) with density $g \in L^2(\Omega)$. Using the scattered field $u^s(\cdot, d)$ and the far field pattern $u^\infty(\cdot, d)$ for the scattering of plane waves, by linearity and boundedness of the scattering maps $u^i \mapsto u^s$ and $u^i \mapsto u^\infty$ the scattered field and far field pattern can be expressed by

$$(H^s g)(x) = \int_\Omega u^s(x, d) g(d) \, ds(d), \quad x \in \mathbb{R}^m \setminus D \qquad (3.1.28)$$

and

$$(H^\infty g)(x) = \int_\Omega u^\infty(\hat{x}, d) g(d) \, ds(d), \quad \hat{x} \in \Omega, \qquad (3.1.29)$$

respectively.

b) Lemma 3.1.3 states the possibility of approximations of multipoles with source-point $x = 0$ on a domain $G_{0,p,\rho}$ by a superposition of plane waves. To approximate a multipole with source-point $x \neq 0$ on $G_{x,p,\rho}$ we consider translations of both the multipole and the Herglotz wave function.

Translations of a Herglotz wave function Hg_0 with a translation vector x can be performed by multiplication of the density $g_0(d)$ with the complex factor $e^{-i\kappa x \cdot d}$. Clearly, the density

$$g(x, d) := e^{-i\kappa x \cdot d} g_0(d) \qquad (3.1.30)$$

of the translated Herglotz wave function has the same norm as the original density g_0. From (3.1.8) by translation we derive

$$\left\| \Phi_{\mu, q}(\cdot, x) - Hg(x, \cdot) \right\|_{C^s(G_{x,p,\rho})} \leq \tau. \qquad (3.1.31)$$

c) The density g defined by (3.1.30) and Lemma 3.1.3 is a function

$$g = g(x, d, p, q, \tau, \rho, \mu, s, \beta_0).$$

We will need the density g with two different sets of values for x, d, μ and τ and for vectors p and q depending on x. We use the abbreviation

$$g_\tau(x, d) := g\Big(x, d, p(x), q(x), \tau, \rho, \mu, s, \beta_0\Big). \qquad (3.1.32)$$

d) To estimate the error of the approximations we need the bound

$$b_{\tau, \rho, \mu, s, \beta_0} := \max\Big\{ \|g\|_{L^2(\Omega)} : \ g \in \mathcal{G}(\tau, \rho, \mu, s, \beta_0) \Big\} \qquad (3.1.33)$$

with the set $\mathcal{G}(\tau,\rho,\mu,s,\beta_0)$ of densities given by Lemma 3.1.3. We use the abbreviation

$$b_\tau = b_{\tau,\rho,\mu,s,\beta_0}$$

or

$$b_{\tau,\rho} = b_{\tau,\rho,\mu,s,\beta_0},$$

if the dependence on ρ,μ,s,β_0 or on μ,s,β_0 is not needed. For the set $\mathcal{G}$ we will write

$$\mathcal{G}_\tau := \mathcal{G}(\tau,\rho,\mu,s,\beta_0).$$

e) For a scatterer $\mathcal{D} \in \mathcal{C}$ and a point $x \in B \setminus D_\rho$, according to the boundedness of D and the exterior cone condition we can find $p \in \Omega$ with

$$D \subset G_{x,p,\rho}. \tag{3.1.34}$$

Thus it is possible to define a function $p : B \to \Omega$, such that with $p = p(x)$ the condition (3.1.34) is satisfied for all $x \in B \setminus D_\rho$. The condition (3.1.34) will be needed to estimate the approximation of $\Phi_{\mu,q}^s(x,z)$ by Qu^∞. Later we will work with two different domains D_1 and D_2. Then for a fixed function p only at points $x \in B$, for which (3.1.34) is satisfied for both D_1 and D_2, the approximation properties of the corresponding operator Q will be valid for both scattered fields $\Phi_{1,\mu,q}^s$ and $\Phi_{2,\mu,q}^s$.

We are now ready to formulate the definition of the operator Q and to investigate its approximation properties.

DEFINITION 3.1.5 (**Backprojection operator Q.**) *Given a set of parameters* $\mu \in I\!N_0$, $s = 1$, $\rho,\tau > 0$ *and functions* $p,q : B \to \Omega$ *with the help of*

$$g_\tau(x,d) = g(x,d,p(x),q(x),\tau,\rho,0,s,\beta_0)$$

and

$$g_\eta(z,\tilde{d}) = g(z,\tilde{d},p(z),q(z),\eta,\rho,\mu,s,\beta_0)$$

specified in (3.1.32) we define the operator $Q : L^2(\Omega \times \Omega) \to L^\infty(B)$ *by*

$$(Qw)(x,z) := \frac{1}{\gamma_m} \int_\Omega \int_\Omega \left\{ g_\tau(x,\tilde{d})g_\eta(z,d) \right\} w(-d,\tilde{d}) \, ds(d)ds(\tilde{d}), \tag{3.1.35}$$

for $x,z \in B$.

THEOREM 3.1.6 (Error estimates for the reconstruction of $\Phi^s_{q,\mu}$ by Q.) *Consider scattering by a sound-soft, sound-hard or inhomogeneous medium scatterer* $\mathcal{D} \in C$. *The error for the approximation of* $\Phi^s_{\mu,q}(x,z)$ *by* Qu^∞ *is estimated by*

$$\left| \Phi^s_{\mu,q}(x,z) - (Qu^\infty)(x,z) \right| \leq c\frac{\eta}{\rho^{(m-1)}} + C\, b_{\eta,\rho,\mu,s,\beta_0}\, \tau \qquad (3.1.36)$$

for all $x, z \in B \setminus D_\rho$, *for which (3.1.34) is satisfied for* $p = p(x)$ *or* $p = p(z)$, *respectively. The constants* c *and* C *depend on* μ, *but not on* $\mathcal{D} \in C$.

Remark. For an appropriate choice of τ and η the error in (3.1.36) can be made arbitrarily small. Given $\epsilon > 0$ we first choose $\eta = \epsilon \rho^{(m-1)}/(2c)$ and then $\tau = \epsilon/(2Cb_\eta)$ to obtain

$$\left| \Phi^s_{\mu,q}(x,z) - (Qu^\infty)(x,z) \right| \leq \epsilon, \quad x,z \in B \setminus D_\rho. \qquad (3.1.37)$$

Proof. First, for $g_\eta(z, \cdot)$ by definition we have

$$\left\| \Phi_{\mu,q}(\cdot,z) - Hg_\eta(z,\cdot) \right\|_{C^1(D)} \leq \eta. \qquad (3.1.38)$$

The scattering map $u^i \mapsto u^s$ is bounded from $C^1(\overline{D})$ into $C(B \setminus D_\rho)$. We use the solutions of the scattering problems as given in Section 2 and estimate the scattered field u^s with the help of the Cauchy-Schwarz inequality to obtain

$$\left\| \Phi^s_{\mu,q}(\cdot,z) - H^s g_\eta(z,\cdot) \right\|_{C(B \setminus D_\rho)} \leq c\frac{\eta}{\rho^{m-1}} \qquad (3.1.39)$$

for all $z \in B \setminus D_\rho$, for which (3.1.34) is satisfied, with some constant c not depending on ρ or $\mathcal{D} \in C$. In the same way for $g_\tau(x, \cdot)$ we derive

$$\left\| \Phi^\infty(\cdot,x) - H^\infty g_\tau(x,\cdot) \right\|_{C(\Omega)} \leq c\,\tau \qquad (3.1.40)$$

for all $x \in B \setminus D_\rho$, for which (3.1.34) is satisfied, with some constant c uniformly for $\mathcal{D} \in C$. We use the mixed reciprocity relations (2.1.4) and (2.2.4) to transform (3.1.40) into

$$\left| u^s(x,d) - \frac{1}{\gamma_m}\left(H^\infty g_\tau(x,\cdot) \right)(-d) \right| \leq \frac{c}{\gamma_m}\,\tau \qquad (3.1.41)$$

for all $d \in \Omega$. Using the Cauchy-Schwarz inequality from (3.1.41) we obtain the estimate

$$\left| \left(H^s g_\eta(z,\cdot) \right)(x) - \left(Qu^\infty \right)(x,z) \right|$$

$$= \left| \int_\Omega \left(u^s(x,d) - \frac{1}{\gamma_m}\int_\Omega u^\infty(-d,\tilde{d})\, g_\tau(x,\tilde{d})\, ds(\tilde{d}) \right) g_\eta(z,d)\, ds(d) \right|$$

$$\leq C \left\| g_\eta(z,\cdot) \right\|_{L^2(\Omega)} \tau \qquad (3.1.42)$$

with some constant C. Now, from (3.1.42) and (3.1.39) we derive

$$
\begin{aligned}
\left|\Phi^s_{\mu,q}(x,z) - \left(Qu^\infty\right)(x,z)\right| &\leq \left|\Phi^s(x,z) - \left(H^s g_\eta(z,\cdot)\right)(x)\right| \\
&\quad + \left|\left(H^s g_\eta(z,\cdot)\right)(x) - \left(Qu^\infty\right)(x,z)\right| \\
&\leq c\,\frac{\eta}{\rho^{m-1}} + C\,\left\|g_\eta(z,\cdot)\right\|_{L^2(\Omega)}\,\tau \quad (3.1.43)
\end{aligned}
$$

with constants $c, C > 0$, which yields (3.1.36). $\qquad\square$

Stability for the reconstruction of D. The operator Q is a bounded operator from $L^2(\Omega \times \Omega)$ into $L^\infty(B)$. We will exploit the behavior of the bounds to derive stability estimates for the reconstruction of the scattered field $\Phi^s_{\mu,q}$ and use these estimates to obtain stability estimates for the reconstruction of the shape D of a scatterer $\mathcal{D}$.

LEMMA 3.1.7 Let $u^\infty_1(\hat{x}, d)$ and $u^\infty_2(\hat{x}, d)$, $\hat{x}, d \in \Omega$, be the far field patterns for scattering of acoustic plane waves from two scatterers $\mathcal{D}_1, \mathcal{D}_2 \in C$. If for some parameter $\delta > 0$ the far field patterns satisfy

$$
\left\|u^\infty_1 - u^\infty_2\right\|_{L^2(\Omega \times \Omega)} \leq \delta, \tag{3.1.44}
$$

then the fields $\Phi^s_{1,\mu,q}$ and $\Phi^s_{2,\mu,q}$, $\mu \in \mathbb{N}_0$, for scattering of multipoles by $\mathcal{D}_1$ and $\mathcal{D}_2$, respectively, satisfy the estimate

$$
\left|\Phi^s_{1,\mu,q}(x,z) - \Phi^s_{2,\mu,q}(x,z)\right| \leq 2c\,\frac{\eta}{\rho^{m-1}} + 2C\,b_\eta\,\tau + \frac{1}{\gamma_m}b_\eta b_\tau\,\delta \tag{3.1.45}
$$

for all $q \in \Omega$, $\rho, \tau, \eta > 0$ and all points $x, z \in U$ defined by (3.1.12), where the constants c, C are given by Theorem 3.1.6.

Proof. We define the function $p : B \to \Omega$ such that the condition (3.1.34) for both scatterers $\mathcal{D}_1$ and $\mathcal{D}_2$ is satisfied in $x \in U$. Then for the operator Q given by (3.1.35) we use (3.1.36) for each scatterer $\mathcal{D}_1$ and $\mathcal{D}_2$ and the Cauchy-Schwarz inequality applied to $Q(u^\infty_{1,\mu,q} - u^\infty_{2,\mu,q})$ to derive from

$$
\begin{aligned}
\left|\Phi^s_{1,\mu,q}(x,z) - \Phi^s_{2,\mu,q}(x,z)\right| &\leq \left|\Phi^s_{1,\mu,q}(x,z) - \left(Qu^\infty_1\right)(x,z)\right| \tag{3.1.46} \\
&\quad + \left|Q\left(u^\infty_1 - u^\infty_2\right)(x,z)\right| + \left|\left(Qu^\infty_1\right)(x,z) - \Phi^s_{2,\mu,q}(x,z)\right|
\end{aligned}
$$

the estimate (3.1.45). $\qquad\square$

It is now possible to use Lemma 3.1.4 to obtain estimates for the Hausdorff distance $d(D_1, D_2)$ between the two scatterers $\mathcal{D}_1$ and $\mathcal{D}_2$ and derive stability estimates for the reconstruction of the shape of scatterers. For positive parameter ρ, τ, η and δ let the function F_2 be given by

$$F_2(\rho, \tau, \eta, \delta) := 2c\frac{\eta}{\rho^{m-1}} + 2C\, b_{\eta,\rho}\, \tau + \frac{1}{\gamma_m} b_{\eta,\rho} b_{\tau,\rho}\, \delta, \qquad (3.1.47)$$

where the bounds $b_{\eta,\rho}$ and $b_{\tau,\rho}$ are defined by (3.1.33) and the constants c and C are chosen according to Theorem 3.1.6. We define

$$F(\delta) := \inf\left\{ F_1(\rho, F_2(\rho, \tau, \eta, \delta)), \ \ \tau, \eta, \rho > 0 \right\} \qquad (3.1.48)$$

with F_1 given by (3.1.15).

The following lemma and corollary will be needed to study the behavior of the function F. For a function $f : (0, \epsilon_0) \times (0, \epsilon_0) \to \mathbb{R}$ we define

$$f^*(s,t) := \sup_{\xi \in [t, \epsilon_0)} f(s, \xi), \qquad (3.1.49)$$

i.e., we build the supremum in the second coordinate.

LEMMA 3.1.8 *For every function $f : (0, \epsilon_0) \to \mathbb{R}$ which satisfies $f(t) \to \infty$ for $t \to 0$ there exists a function $g : (0, \delta_0) \to (0, \epsilon_0)$ with*

$$f^*(g(\delta))\, \delta \ \to\ 0 \quad and \quad g(\delta) \to 0 \quad for\ \delta \to 0. \qquad (3.1.50)$$

Proof. Let f' be a strictly monotone function with $f'(t) \geq f(t)$. On (a, ∞) with $a := f'(\epsilon_0)$ there is an inverse function

$$(f')^{-1} : (a, \infty) \to (0, \epsilon_0)$$

with $(f')^{-1}(f'(s)) = s$ and $f'((f')^{-1}(t)) = t$ for all $s \in (0, \epsilon_0)$ and $t \in (a, \infty)$. Then for the function

$$g : (0, \delta_0) \to (0, \epsilon_0), \quad g(\delta) := (f')^{-1}\left(\frac{1}{\sqrt{\delta}}\right),$$

with δ_0 defined by $(f')^{-1}(1/\sqrt{\delta_0}) = \epsilon_0$ we obtain the behavior (3.1.50). $\square$

COROLLARY 3.1.9 *For every function $f : (0, \epsilon_0) \times (0, \epsilon_0) \to \mathbb{R}$ which satisfies $f(t,s) \to \infty$ for $(t,s) \to 0$ there exists a function $g : (0, \delta_0) \to (0, \epsilon_0) \times (0, \epsilon_0)$, $g = (g_1, g_2)$ with*

$$f^*(g(\delta))\, \delta \ \to\ 0, \ \ g(\delta) \to 0 \quad and \quad \frac{g_1(\delta)}{g_2(\delta)^{m-1}} \to 0 \quad for\ \delta \to 0. \qquad (3.1.51)$$

Proof. Use the preceding Lemma for $\tilde{f}(t) := f^*(t^2, t^{1/(m-1)})$ to obtain a function $\tilde{g}$ and define g by $g_1(\delta) := \tilde{g}(\delta)^2$ and $g_2(\delta) := \tilde{g}(\delta)^{1/(m-1)}$.   □

After these preparations we investigate the behavior of the function F.

LEMMA 3.1.10 *The function F defined by (3.1.48) satisfies*

$$F(\delta) \to 0, \quad \delta \to 0. \tag{3.1.52}$$

Proof. The function $F(\delta)$ is dominated by

$$F_1\Big(h_1(\delta), F_2(h_1(\delta), h_2(\delta), h_3(\delta))\Big)$$

with arbitrary positive functions h_1, h_2 and h_3 on $(0, \delta_0)$. We will show that we can choose h_1, h_2 and h_3 such that

$$F_1\Big(h_1(\delta), F_2(h_1(\delta), h_2(\delta), h_3(\delta))\Big) \to 0, \quad \delta \to 0. \tag{3.1.53}$$

The function F_2 can be decomposed into the sum and product

$$F_2(\rho, \tau, \eta, \delta) = 2c\frac{\eta}{\rho^{m-1}} + b_{\eta,\rho}\left(2C\,\tau + \frac{1}{\gamma_m}b_{\tau,\rho}\,\delta\right).$$

By an application of Corollary 3.1.9 to $f(s,t) := b_{s,t}$ we obtain functions h_4 and h_5 with

$$h_4(\delta) \to 0, \quad \delta \to 0,$$

and

$$h_5(\delta) \to 0, \quad \delta \to 0, \tag{3.1.54}$$

such that the functions

$$h_6(\delta) := 2C\,h_4(\delta) + \frac{1}{\gamma_m}b^*_{h_4,h_5(\delta)}\,\delta$$

and

$$h_7(\delta) := 2c\frac{h_4(\delta)}{h_5(\delta)^{m-1}} + b^*_{h_4(\delta),h_5(\delta)}\,\delta$$

with b^* defined in (3.1.49) satisfy

$$h_6(\delta) \to 0, \quad \delta \to 0,$$

and

$$h_7(\delta) \to 0, \quad \delta \to 0.$$

For

$$
\begin{aligned}
h_1(\delta) &:= \max\{h_5(\delta), h_5(h_6(\delta))\}, \\
h_2(\delta) &:= h_4(\delta), \\
h_3(\delta) &:= h_4(h_6(\delta))
\end{aligned}
$$

we obtain

$$
F_2(h_1(\delta), h_2(\delta), h_3(\delta), \delta)\delta \qquad\qquad (3.1.55)
$$

$$
= \; 2c\frac{h_3(\delta)}{h_1(\delta)^{m-1}} \; + \; b_{h_3(\delta), h_1(\delta)}\left(2Ch_2(\delta) + \frac{1}{\gamma_m}b_{h_2(\delta), h_1(\delta)}\delta\right)
$$

$$
\leq \; 2c\frac{h_4(h_6(\delta))}{h_5(h_6(\delta))^{m-1}}
$$

$$
+ \; b^*_{h_4(h_6(\delta)), h_5(h_6(\delta))}\left(2Ch_4(\delta) + \frac{1}{\gamma_m}b^*_{h_4(\delta), h_5(\delta)}\delta\right)
$$

$$
\to \; 0, \quad \delta \to 0.
$$

Thus we have

$$
F_1\Big(h_1(\delta), F_2(h_1(\delta), h_2(\delta), h_3(\delta), \delta)\Big) \to 0, \quad \delta \to 0 \qquad (3.1.56)
$$

and the proof is complete. $\square$

We now put all previous steps together to obtain the main stability result.

THEOREM 3.1.11 (Stability.) *Let* $u_1^\infty(\hat{x}, d)$ *and* $u_2^\infty(\hat{x}, d)$, $\hat{x}, d \in \Omega$, *be the far field patterns for scattering of acoustic plane waves from two scatterers* $\mathcal{D}_1, \mathcal{D}_2 \in \mathcal{C}$. *We assume that with a nonnegative parameter* δ *the difference between the far field patterns satisfies*

$$
\left\|u_1^\infty - u_2^\infty\right\|_{L^2(\Omega \times \Omega)} \leq \delta. \qquad (3.1.57)
$$

Then with the function F *defined by (3.1.48) we can estimate the Hausdorff distance* $d(\mathcal{D}_1, \mathcal{D}_2)$ *of the two scatterers* $\mathcal{D}_1, \mathcal{D}_2 \in \mathcal{C}$ *by*

$$
d(\mathcal{D}_1, \mathcal{D}_2) \leq F(\delta). \qquad (3.1.58)
$$

Proof. With F_2 defined by (3.1.47) we use Lemma 3.1.7 to derive

$$
\left|\Phi^s_{1,\mu,q}(x, z) - \Phi^s_{2,\mu,q}(x, z)\right| \leq F_2(\rho, \tau, \eta, \delta) \qquad (3.1.59)
$$

for all $q \in \Omega$, $\rho, \tau, \eta > 0$ and all points $x, z \in U$ with U defined by (3.1.12). We can now apply Lemma 3.1.4 to derive

$$d(D_1, D_2) \leq F_1(\rho, F_2(\rho, \tau, \eta, \delta)) \tag{3.1.60}$$

for every choice of parameters $\rho, \tau, \eta > 0$. With the definition (3.1.48) for F from (3.1.60) we obtain (3.1.58). □

Lemma 3.1.10 shows that the estimate (3.1.58) in fact is a stability estimate for the reconstruction of D from the far field patterns for scattering of plane waves. The function F can be calculated or estimated according to the *a priori* knowledge about the unknown scatterers as given by the class $\mathcal{C}$ defined in Definitions 2.1.5 and 2.2.5.

Explicit stability estimates. The natural problem is the investigation of the dependence of $F(\delta)$ on δ. For general angles β_0 we were not yet able to estimate explicitly the behavior of F, but for $\beta_0 = \frac{\pi}{2}$ and for impenetrable domains we will now use the asymptotics of Bessel and Hankel functions to derive a logarithmic bound. More general, we will explicitly estimate the dependence of the function $F(\delta)$ on δ for the estimate

$$d\Big(\mathcal{H}(D_1), \mathcal{H}(D_2)\Big) \leq F(\delta) \tag{3.1.61}$$

for the convex hulls $\mathcal{H}(D_1)$ and $\mathcal{H}(D_2)$ of two scatterers $\mathcal{D}_1, \mathcal{D}_2 \in \mathcal{C}_{obst}$.

THEOREM 3.1.12 *Let* $\mathcal{D}_1, \mathcal{D}_2 \in \mathcal{C}_{obst}$ *be impenetrable scatterers with scattering data* $u_1^\infty(\hat{x}, d)$ *and* $u_2^\infty(\hat{x}, d)$, $\hat{x}, d \in \Omega$. *If*

$$\left\| u_1^\infty - u_2^\infty \right\|_{L^2(\Omega \times \Omega)} \leq \delta, \tag{3.1.62}$$

then we have

$$d\Big(\mathcal{H}(D_1), \mathcal{H}(D_2)\Big) \leq \frac{C}{|\ln \delta|^c} \tag{3.1.63}$$

with constants $C > 0$ *and* $0 < c < 1$ *uniformly for all* $\mathcal{D} \in \mathcal{C}_{obst}$.

Remark. Please note that the statement is independent of the particular scatterer under consideration, i.e., we do not assume that the physical nature of the scatterer or the boundary condition is known in *a priori*, but $\mathcal{D}_1$ and $\mathcal{D}_2$ might be arbitrary scatterers in one of the classes $\mathcal{C}_{obst}$ or $\mathcal{C}_m$.

Proof. We consider the three-dimensional problem. Let $z_0 \in \partial D_2 \cap \partial \mathcal{H}(D_2)$ be a point, such that for $z = z_0 + \nu(z_0)\rho$ we can choose a domain of approximation $G_{(z,p,\rho)}$, $p \in \Omega$, with the angle $\beta_e = \pi/2$ and

$$\overline{D_j} \subset G_{z,p,\rho}, \quad j = 1, 2.$$

Then we can use Lemma 3.1.7 and proceed along the lines of the proof of
Lemma 3.1.4 to obtain

$$d(z_0, D_1) \leq F_1\Big(\rho, F_2(\rho, \tau, \eta, \delta)\Big) \qquad (3.1.64)$$

for every choice of parameters $\rho, \tau, \eta > 0$ with the functions F_1 defined by
(3.1.15) and F_2 defined by (3.1.47). For the special choice

$$\rho = |\ln \delta|^{-\frac{c_1}{2}}, \quad \eta = e^{-|\ln \delta|^{c_2}} \quad \text{and} \quad \tau = e^{-|\ln \delta|^{c_3}} \qquad (3.1.65)$$

with $0 < c_1 < c_2 < 1$, $c_1 + c_2 < 1$ and $0 < c_3 < c_2 - c_1$ we will derive

$$|F_2(\rho, \eta, \epsilon, \delta)| \leq C \qquad (3.1.66)$$

for all sufficiently small $\delta > 0$. Then from (3.1.64), (3.1.65) and (3.1.66)
we obtain the explicit stability estimate (3.1.63).

To prove (3.1.66) we proceed in six steps.

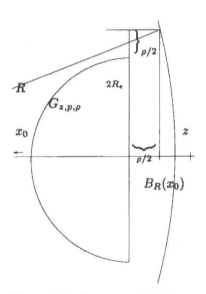

Figure 3.2: Constructed circle.

1. We explicitly construct the domain
G of Lemma 3.1.2. Consider the domain $G_{z,p,\rho}$ and define $G := B_R(x_0)$
by

$$R^2 = (R - \rho/4)^2 + (2R_e + \rho/2)^2,$$
$$x_0 := z - (\tfrac{\rho}{4} + R)p, \qquad (3.1.67)$$

where the first equation yields

$$R := \frac{5\rho}{8} + \frac{2(2R_e)^2}{\rho} + 2R_e. \quad (3.1.68)$$

Then, by straightforward geometric arguments with the help of Figure 3.2 we
derive

$$G_{z,p,\rho} \subset B_R(x_0), \quad d(z, B_R(x_0)) = \frac{\rho}{4}$$

and

$$d(G_{z,p,\rho}, \Omega_R(x_0)) \geq \frac{\rho}{2}$$

for sufficiently small ρ.

2. We need to estimate the bounds b_τ defined in (3.1.33). To this end
we investigate special solutions of the interior Dirichlet problem for the
Helmholtz equation in the domain $B_R(x_0)$. We choose a coordinate system

centered at $x_0 = 0$ and define the special functions

$$u_n(x) := \sum_{k=0}^{n} \sum_{l=-k}^{k} a_k^l \, j_k(\kappa|x|) \, Y_k^{-l}(\hat{x}), \quad x \in \mathbb{R}^3 \qquad (3.1.69)$$

with coefficients $a_k^l := i\kappa h_k^{(1)}(\kappa|z|)Y_k^l(\hat{z})$ for $k = 0, ..., n$, $l = -k, ..., k$. From the expansion

$$\Phi(x, z) = i\kappa \sum_{k=0}^{\infty} \sum_{l=-k}^{k} h_k^{(1)}(\kappa|z|)Y_k^l(\hat{z})j_k(\kappa|x|)Y_k^{-l}(\hat{x}), \quad (3.1.70)$$

$$= \sum_{k=0}^{\infty} \sum_{l=-k}^{k} a_k^l \, j_k(\kappa|x|)Y_k^{-l}(\hat{x}) \qquad (3.1.71)$$

of $\Phi(x, z)$ for $|x| < |z|$ with respect to the spherical harmonics, the addition theorem

$$\sum_{m=-k}^{k} Y_k^l(\hat{z}) \, Y_k^{-l}(\hat{x}) = \frac{2k+1}{4\pi} P_k(\cos(\theta)) \qquad (3.1.72)$$

with the Legendre Polynomial P_k, where θ denotes the angle between $\hat{z}$ and $\hat{x}$, the asymptotic behavior (1.2.36) and (1.2.37) of the Hankel and Bessel functions and $|P_k(t)| \leq 1$ for $t \in [-1, 1]$, we obtain

$$\left\| \Phi(\cdot, z) - u_n \right\|_{L^2(\Omega_R(x_0))} \qquad (3.1.73)$$

$$= \left\| \kappa \sum_{k=n+1}^{\infty} \sum_{m=-k}^{k} h_k^{(1)}(\kappa|z|) \, Y_k^l(\hat{z}) \, j_k(\kappa R) \, Y_k^{-l}(\hat{x}) \right\|_{L^2(\Omega)}$$

$$= \left\| \kappa \sum_{k=n+1}^{\infty} h_k^{(1)}(\kappa|z|) \, j_k(\kappa R) \, \frac{2k+1}{4\pi} P_k(\cos(\theta)) \right\|_{L^2(\Omega)}$$

$$\leq \frac{c_0}{|z|} \sum_{k=n+1}^{\infty} q^k = \frac{c_0}{|z|} \frac{q^{n+1}}{1-q} \qquad (3.1.74)$$

with $q := R/|z| < 1$ and a constant c_0 not depending on R, $|z|$ or n. Since for compact subsets $\overline{G_{z,p,\rho}}$ of $B_R(x_0)$ the solution of the interior Dirichlet problem in $B_R(x_0)$ with L^2-boundary data on $\Omega_R(x_0)$ defines a bounded mapping from $L^2(\Omega_R(x_0))$ into $C(\overline{G_{z,p,\rho}})$, we obtain a factor λ such that

$$\left\| \Phi(\cdot, z) - u_n \right\|_{C(\overline{G_{z,p,\rho}})} \leq \lambda \left\| \Phi(\cdot, z) - u_n \right\|_{L^2(\Omega_R(x_0))} \qquad (3.1.75)$$

$$\leq \frac{\lambda c_0}{|z|} \frac{q^{n+1}}{1-q}. \qquad (3.1.76)$$

The factor λ is a function of z, R and ρ.

3. To evaluate the dependence of λ on ρ let u be a solution to the interior Dirichlet problem for the Helmholtz equation in $B_R(x_0)$. We will show

$$|u(x)| \leq \frac{C}{\rho^2} \|u\|_{L^2(\Omega_R(x_0))} \qquad (3.1.77)$$

for $x \in B_{R-\rho/2}(x_0)$, i.e., $\lambda \leq C\rho^{-2}$ with some constant C. We present a proof by means of spherical harmonics and Bessel functions. Again using a coordinate system centered at $x_0 = 0$, from Green's formula and the expansion (3.1.70) we observe that

$$u(x) = \sum_{k=0}^{\infty} \sum_{l=-k}^{k} a_k^l j_k(\kappa|x|) Y_k^l(\hat{x}), \quad x \in \overline{B_R(x_0)}, \qquad (3.1.78)$$

where the sum (3.1.78) converges uniformly on compact subsets of $B_R(x_0)$. We use the Cauchy-Schwarz inequality to calculate for $|u(x)|$ with $x = r\hat{x} \in B_{R-\rho/2}(x_0)$ the estimate

$$|u(x)|^2 = \left| \sum_{k=0}^{\infty} \sum_{l=-k}^{k} a_k^m j_k(\kappa|x|) Y_k^l(\hat{x}) \right|^2 \qquad (3.1.79)$$

$$\leq \left(\sum_{k=0}^{\infty} \sum_{l=-k}^{k} \left| a_k^l j_k(\kappa R) \right|^2 \right) \left(\sum_{k=0}^{\infty} \sum_{l=-k}^{k} \left| Y_k^l(\hat{x}) \frac{j_k(\kappa r)}{j_k(\kappa R)} \right|^2 \right).$$

For the first term of (3.1.79) we observe that

$$\sum_{k=0}^{\infty} \sum_{l=-k}^{k} \left| a_k^l j_k(\kappa R) \right|^2 = \| u \|_{L^2(\Omega_R)}^2 < \infty. \qquad (3.1.80)$$

For the second term of (3.1.79) we use the estimate

$$\left| Y_k^l(\hat{x}) \right| \leq C k^{1/2} \|Y_k^l\|_{L^2(\Omega)}, \quad \hat{x} \in \Omega, \qquad (3.1.81)$$

for spherical harmonics of order $k \in I\!N$ (see Chapter X, Lemma 6.1 of [60]) and the asymptotic behavior of the spherical Bessel functions to derive

$$\sum_{k=1}^{\infty} \sum_{l=-k}^{k} \left| Y_k^l(\hat{x}) \frac{j_k(\kappa r)}{j_k(\kappa R)} \right|^2 \qquad (3.1.82)$$

$$\leq \sum_{k=0}^{\infty} \sum_{l=-k}^{k} |k| \left| \frac{j_k(\kappa r)}{j_k(\kappa R)} \right|^2 \leq c \sum_{k=0}^{\infty} k \, q^{2k} \leq c \sum_{k=0}^{\infty} (2k+1) \, q^{2k}$$

$$= c \frac{d}{dq} \sum_{k=0}^{\infty} q^{2k+1} = c \frac{d}{dq} \left(\frac{q}{1-q^2} \right) = c \left(\frac{1}{1-q^2} + \frac{2q^2}{(1-q^2)^2} \right)$$

with $q := \frac{r}{R}$ and a constant c. From (3.1.68) and $0 \le r \le R - \rho$ we obtain

$$c\rho^2 \le 1 - q \le 1, \quad |\ln q| \ge B\rho^2 \tag{3.1.83}$$

for sufficiently small $\rho > 0$ with constants c, C and B. We now use $1 - q^2 = (1-q)(1+q)$ and $1 + q \ge 1$ to estimate (3.1.82) by $c\rho^{-4}$ with a constant c. We take the square root of (3.1.79) to obtain (3.1.77), i.e., we have proven $\lambda \le C\rho^{-2}$ with a constant C not depending on ρ.

4. For the function

$$g_n(\hat{x}) = \sum_{k=0}^{n} \sum_{l=-k}^{k} \frac{1}{4\pi i^k} \, a_k^l \, Y_k^{-l}(\hat{x}), \quad \hat{x} \in \Omega, \tag{3.1.84}$$

we calculate $u_n = Hg_n$. With n chosen as

$$n := \left[\frac{\ln(\tau \frac{(1-q)|z|}{\lambda c_0})}{\ln(q)} \right] - 1, \tag{3.1.85}$$

where $[a]$ denotes the smallest integer larger than $a \in I\!\!R$, by straightforward calculation from the estimate (3.1.76) we derive

$$\left\| \Phi(\cdot, z) - Hg_n \right\|_{C^1(\overline{G_{z,\nu,\rho}})} \le \tau. \tag{3.1.86}$$

With the help of (3.1.83) and $\lambda \le c\rho^{-2}$ we derive for n the estimate

$$n \le C \frac{|\ln(c\tau\rho^3)|}{\rho^2} \tag{3.1.87}$$

with constants C, c.

5. We now estimate the norm of the function g_n with n given by (3.1.85). We calculate

$$\begin{aligned}
\left\| g_n \right\|_{L^2(\Omega)}^2 &= \left(\frac{1}{4\pi} \right)^2 \sum_{k=0}^{n} \sum_{l=-k}^{k} |a_k^l|^2 \\
&\le c_1 \sum_{k=0}^{n} (2k+1) \left(\frac{2k}{e\kappa|z|} \right)^{2k} \\
&\le c_1 \, (2n+1)(n+1) \left(\frac{2n}{e\kappa|z|} \right)^{2n} \tag{3.1.88}
\end{aligned}$$

with a constant c_1. We insert (3.1.85) into (3.1.88) to obtain

$$\left\| g_n \right\|_{L^2(\Omega)} \le C \left(\frac{|\ln(c\tau\rho^3)|}{\rho^2} \right)^2 \left(\frac{C|\ln(c\tau\rho^3)|}{\rho} \right)^{C\frac{|\ln(c\tau\rho^3)|}{\rho^2}} \tag{3.1.89}$$

with constants c and C not depending on ρ.

6. We need to investigate the set of densities $\mathcal{G}$ which we constructed in Lemma 3.1.3. Since we chose $g \in \mathcal{G}$ to be minimum norm solutions of (3.1.9), the norm $\|g_n\|_{L^2(\Omega)}$ is an upper bound for b_τ. Finally, the bound (3.1.66) for $|F_2(\rho, \eta, \tau, \delta)|$ can be obtained by straightforward (but lengthy) calculation from definition (3.1.47) of F_2, the choice of parameters (3.1.65) and the estimate (3.1.89) for the bounds b_τ and b_η.

The proof in two dimensions can be worked out in the same way with obvious modifications. $\square$

From stability, as a corollary we obtain uniqueness of the support of penetrable or impenetrable scatterers.

COROLLARY 3.1.13 (**Uniqueness.**) *For acoustic scatterers $D \in C$ the domain D is uniquely determined by the far field pattern for scattering of all plane waves.*

3.2   Electromagnetic scattering

In this section we will prove stability estimates for inverse electromagnetic scattering. Consider the electromagnetic Herglotz pair

$$(Vg)(x,p) := \int_\Omega \frac{i}{\kappa}(p \cdot \nabla_x)\nabla_x e^{i\kappa x \cdot d}g(d)\, ds(d), \quad \frac{1}{i\kappa}\mathrm{curl}\, Vg \quad (3.2.1)$$

with a density $g \in L^2(\Omega)$. With the aid of

$$\nabla \times \nabla \times (pw) = (p \cdot \nabla)\nabla\, w, \quad p \in \Omega, \quad (3.2.2)$$

for solutions w of the Helmholtz equation and

$$E_{pl}^i(x,d,p) = i\kappa(d \times p) \times d e^{i\kappa x \cdot d} = \frac{i}{\kappa}\nabla_x \times \nabla_x \times (pe^{i\kappa x \cdot d}) \quad (3.2.3)$$

we derive

$$(Vg)(x,p) = \int_\Omega E_{pl}^i(x,d,p)g(d)ds(d), \quad x \in \mathbb{R}^3. \quad (3.2.4)$$

By linearity and continuity of the direct scattering problem the incident electromagnetic wave

$$E^i := Vg, \quad H^i := \frac{1}{i\kappa}\mathrm{curl}\, Vg$$

has the scattered field

$$(V^s g)(x,p) := \int_\Omega E_{pl}^s(x,d,p)g(d)ds(d), \quad \frac{1}{i\kappa}\text{curl } (V^s g)(\cdot,p) \quad (3.2.5)$$

and far field pattern

$$(V^\infty g)(x,p) := \int_\Omega E_{pl}^\infty(x,d,p)g(d)ds(d), \quad \nu \times (V^\infty g)(\cdot,p). \quad (3.2.6)$$

For fixed $\rho > 0$ we will construct an approximation

$$E_{edp}^i(x,z,p) \approx (Vg(z,\cdot))(x,p), \quad x \in \overline{D}, \quad (3.2.7)$$

for an electric dipole uniformly for $z \in B \setminus D_\rho$. Since both plane waves and multipoles in acoustic and electromagnetic scattering are strongly related to each other, we will be able to use the results of the acoustic parts. From Lemma 3.1.3 we first derive the following lemma.

LEMMA 3.2.1 *Given* $\rho,\tau > 0$ *there exists a density* $g \in L^2(\Omega)$, *such that for each* $p,q \in \Omega$ *there is an orthogonal matrix* M *with*

$$\left\| E_{edp}^i(\cdot,0,q) - \left(Vg(M^{-1}\cdot)\right)(\cdot,q) \right\|_{C^2(G_{0,p,\rho})} \leq \tau. \quad (3.2.8)$$

Proof. We use Lemma 3.1.3 for $s = 4$ and $\mu = 0$. For the special case $\mu = 0$ the multipole $\Phi_{0,q} = \Phi$ has rotational symmetry. Thus for $p \in \Omega$ we obtain the approximation of $\Phi(\cdot,0)$ by Hg on $G_{0,p,\rho}$ from the approximation on $G_{0,e_1,\rho}$, $e_1 = (1,0,0)$, by a simple rotation of the Herglotz wave function. The domain of approximation $G_{0,p,\rho}$ is obtained from the domain of approximation $G_{0,e_1,\rho}$ by rotation with an orthogonal matrix M satisfying $p = Me_1$. Clearly, the rotated Herglotz wave function $v(x) := (Hg)(M^{-1}x)$ approximates the point-source $\Phi(\cdot,0)$ on $G_{0,p,\rho}$. We calculate

$$\begin{aligned} v(x) &= \int_\Omega e^{i\kappa M^{-1}x \cdot d} g(d)\, ds(d) \\ &= \int_\Omega e^{i\kappa x \cdot Md} g(d)\, ds(d) \\ &= \int_\Omega e^{i\kappa x \cdot \tilde{d}} g(M^{-1}\tilde{d})\, ds(d). \quad (3.2.9) \end{aligned}$$

To obtain the statement of the lemma we now apply the differential operator $\frac{i}{\kappa}(q \cdot \nabla)\nabla$ to both $\Phi(\cdot,0)$ and $Hg(M^{-1}\cdot)$. $\square$

Given the density g of the preceding lemma and an orthogonal matrix function M on B we define the translated and rotated density by

$$g(x, d, M, \tau, \rho, \beta_0) := e^{-i\kappa x \cdot d} g(M^{-1}(x)d), \quad d \in \Omega. \qquad (3.2.10)$$

We use the abbreviations

$$g_\tau(x, d) = g(x, d, M, \tau, \rho, \beta_0),$$

$$b_{\tau,\rho} := \|g_\tau(x, \cdot)\|_{L^2(\Omega)},$$

and

$$b_\tau = b_{\tau,\rho},$$

where $b_{\tau,\rho}$ is well defined, since the norm of $g_\tau(x, \cdot)$ is independent of $x \in \mathbb{R}^3$.

DEFINITION 3.2.2 *Given a set of parameters $\mu \in \mathbb{N}_0$, $s = 4$, $\rho, \tau, \eta > 0$, $\gamma = \frac{1}{4\pi}$ and an orthogonal matrix function M on B with the help of*

$$g_\tau(x, d) = g(x, d, M, \tau, \rho, 0, s, \beta_0)$$

and

$$g_\eta(z, \tilde{d}) = g(z, \tilde{d}, M, \eta, \rho, \mu, s, \beta_0)$$

we define the operator $Q : L^2(\Omega \times \Omega, \mathbb{R}^3) \to L^\infty(B, \mathbb{R}^3)$ by

$$(Qw)(x, z) := \frac{1}{\gamma} \int_\Omega \int_\Omega \left\{ g_\tau(x, d) g_\eta(z, \tilde{d}) \right\} w(-d, \tilde{d}) \, ds(d) ds(\tilde{d}) \qquad (3.2.11)$$

for $x, z \in B$.

The Operator (3.2.11) is basically the operator given by (3.1.35), but now applied to vector-valued functions. It can be used to construct E^s_{edp} from the knowledge of E^∞_{pl}. Again, we observe the strong relationship between acoustic and electromagnetic scattering. We use the class $\mathcal{C}$ of electromagnetic scatterers defined by

$$\mathcal{C} := \mathcal{C}_{pc} \cup \mathcal{C}_{elm}. \qquad (3.2.12)$$

THEOREM 3.2.3 *Consider electromagnetic scattering by a perfect conductor or an inhomogeneous medium scatterer $\mathcal{D} \in \mathcal{C}$. The error for the approximation of E^s_{edp} by QE^∞_{pl} satisfies*

$$\left| \tilde{q} \cdot E^s_{edp}(x, z, q) - (q \cdot QE^\infty_{pl}(\cdot, \cdot, \tilde{q}))(x, z) \right| \le c\frac{\eta}{\rho^3} + Cb_{\eta,\rho}\tau \qquad (3.2.13)$$

for all $q, \tilde{q} \in \Omega$ and all points $x, z \in B \setminus D_\rho$, for which (3.1.34) is satisfied with $p = M(x)e_1$ or $p = M(z)e_1$, respectively. The constants c and C hold uniformly for $\mathcal{D} \in \mathcal{C}$.

Proof. For $\eta > 0$ and $q \in \Omega$ by definition of the kernel g_η we have

$$\left\|E^i_{edp}(\cdot, z, q) - \left(Vg_\eta(z, \cdot)\right)(\cdot, q)\right\|_{C^a(\overline{D})} \leq \eta \qquad (3.2.14)$$

for all $z \in B \setminus D_\rho$, for which (3.1.34) is satisfied with $p = M(z)e_1$.

For the perfect conductor we use the boundedness of the integral operator $(I + M + iNPS_0^2)^{-1}$, the singularity of the kernels of the potential P_E and the Cauchy-Schwartz inequality to estimate the corresponding scattered fields by

$$\left\|E^s_{edp}(\cdot, z, q) - \left(V^s g_\eta(z, \cdot)\right)(\cdot, q)\right\|_{C(B \setminus \overline{D_\rho})} \leq c \frac{\eta}{\rho^3} \qquad (3.2.15)$$

with some constant c not depending on ρ or $\mathcal{D} \in \mathcal{C}$. If the scatterer is an inhomogeneous medium we obtain an analogous estimate by consideration of the volume integral equation and the potential T_e.

In the same way for a second independent parameter τ, $\tilde{q} \in \Omega$ and all points $x \in B \setminus D_\rho$, for which (3.1.34) is satisfied with $p = M(x)e_1$, we derive

$$\left\|E^\infty_{edp}(\cdot, x, \tilde{q}) - \left(V^\infty g_\tau(x, \cdot)\right)(\cdot, \tilde{q})\right\|_{C(\Omega)} \leq c\tau \qquad (3.2.16)$$

with some constant c. With the help of the mixed reciprocity relations Theorems 2.3.4 and 2.4.4 we transform (3.2.16) into

$$\left|\tilde{q} \cdot E^s_{pl}(x, -d, q) - \left(q \cdot \frac{1}{\gamma}V^\infty g_\tau(x, \cdot)\right)(d, \tilde{q})\right| \leq \frac{c}{\gamma}\tau \qquad (3.2.17)$$

with some constant c. We use (3.2.17) to estimate

$$\left|\left(\tilde{q} \cdot V^s g_\eta(z, \cdot)\right)(x, q) - \left(q \cdot QE^\infty(\cdot, \cdot, \tilde{q})\right)(x, z)\right|$$

$$= \left|\int_\Omega \left(\tilde{q} \cdot E^s_{pl}(x, d, q)\right.\right.$$

$$\left.\left. - \frac{1}{\gamma}\int_\Omega q \cdot E^\infty(-d, \tilde{d}, \tilde{q})g_\tau(x, \tilde{d}) \, ds(\tilde{d})\right)g_\eta(z, d) \, ds(d)\right|$$

$$\leq C\left\|g_\eta(z, \cdot)\right\|_{L^2(\Omega)}\tau \qquad (3.2.18)$$

with some constant C. Now from (3.2.18) and (3.2.15) we derive

$$\left|\tilde{q} \cdot E^s_{edp}(x, z, q) - \left(q \cdot QE^\infty(\cdot, \cdot, \tilde{q})\right)(x, z)\right|$$

$$\leq c\frac{\eta}{\rho^3} + C\left\|g_\eta(z, \cdot)\right\|_{L^2(\Omega)}\tau$$

with constants $c, C > 0$, and the proof is complete.   □

We will use the preceding theorem to estimate the difference between the scattered fields of electric dipoles for two scatterers $\mathcal{D}_1$ and $\mathcal{D}_2$ in C.

LEMMA 3.2.4 *Let $E_{pl,1}^{\infty}(\hat{x}, d, q)$ and $E_{pl,2}^{\infty}(\hat{x}, d, q)$ for $\hat{x}, d, q \in \Omega$ be the far field patterns for scattering of electromagnetic plane waves from two scatterers $\mathcal{D}_1, \mathcal{D}_2$ in C. If for some parameter $\delta > 0$ the far field patterns satisfy*

$$\left\| E_{pl,1}^{\infty}(\cdot, \cdot, p) - E_{pl,2}^{\infty}(\cdot, \cdot, p) \right\|_{L^2(\Omega \times \Omega)} \leq \delta, \ p \in \Omega, \tag{3.2.19}$$

then with the constants c, C given by Theorem 3.2.3 we have

$$\left| E_{edp,1}^{s}(x, z, q) - E_{edp,2}^{s}(x, z, q) \right| \tag{3.2.20}$$

$$\leq 2c\frac{\eta}{\rho^3} + 2Cb_{\eta,\rho}\,\tau + \frac{1}{\gamma}b_{\eta,\rho}\,b_{\tau,\rho}\,\delta$$

for all $q \in \Omega$, $\rho, \tau, \eta > 0$ and all points $x, z \in U$, where U is defined by (3.1.12).

Proof. Let M be an orthogonal matrix function on B, such that for all $x \in U$ with $p = M(x)e_1$ the condition (3.1.34) is satisfied for both scatterers $\mathcal{D}_1$ and $\mathcal{D}_2$. We use (3.2.13) for each scatterer $\mathcal{D}_1$ and $\mathcal{D}_2$. Then as in (3.1.46) the estimate (3.2.20) is obtained with the help of the Cauchy-Schwarz inequality. $\qquad \square$

We now prove the electromagnetic counterpart of Lemma 3.1.4. From estimates for the difference of $E_{edp,1}^{s}$ and $E_{edp,2}^{s}$ for scattering by two scatterers $\mathcal{D}_1$ and $\mathcal{D}_2$ we derive estimates for the Hausdorff distance $d(\mathcal{D}_1, \mathcal{D}_2)$ of the scatterers. Again we have to consider different situations according to the physical properties of the scatterers $\mathcal{D}_1$ and $\mathcal{D}_2$.

Situation	Properties of $\mathcal{D}_1$ and $\mathcal{D}_2$
S1	$\mathcal{D}_1, \mathcal{D}_2$ are impenetrable scatterers
S2	$\mathcal{D}_2$ is an inhomogeneous medium scatterer and $\mathcal{D}_1$ is an impenetrable scatterer or vice versa
S3	$\mathcal{D}_1, \mathcal{D}_2$ are inhomogeneous medium scatterers

Table 3.2

LEMMA 3.2.5 *We consider scattering of electromagnetic waves by two scatterers $\mathcal{D}_1, \mathcal{D}_2 \in C$. Assume that with parameters $\sigma, \rho > 0$ the scattered fields*

$E^{\bullet}_{edp,1}$ and $E^{\bullet}_{edp,2}$ *for scattering of electric dipoles by* $\mathcal{D}_1$ *or* $\mathcal{D}_2$, *respectively,
satisfy*

$$\left| E^{\bullet}_{edp,1}(z,z,q) - E^{\bullet}_{edp,2}(z,z,q) \right| \leq \sigma \qquad (3.2.21)$$

for all $q \in \Omega$ *and for points* $z \in B \setminus (D_{1,\rho} \cup D_{2,\rho})$, *for which a cone*
$co(z,p_z,\beta_0)$, $p_z \in \Omega$, *in the exterior of* $D_{1,\rho} \cup D_{2,\rho}$ *exists. Then we conclude*

$$d(D_1, D_2) \leq F_1(\rho,\sigma), \qquad (3.2.22)$$

where the function F_1 *is defined according to the situations S1 to S3 by*

$$F_1(\rho,\sigma) := \begin{cases} \rho + \frac{C\rho}{(c-\sigma\rho^3)^{1/3}}, & S1, \\ \rho + \left(\frac{C\rho}{c-\sigma\rho}\right)^{1/3}, & S2, \\ \rho + \frac{C\rho}{c-\sigma\rho}, & S3 \end{cases} \qquad (3.2.23)$$

with constants c *and* C *not depending on the scatterers* $\mathcal{D}_1, \mathcal{D}_2 \in C$.

Proof. We split the proof in two parts. First, consider the scattered
fields at a special point z in $B \setminus (\overline{D_1 \cup D_2})$, such that $z = z_0 + \rho\nu(z_0)$ with
$z_0 \in \partial D_2$ and $\rho = d(z, D_2)$ sufficiently small and there exists a cone

$$co(z,p,\beta_0) \subset \mathbb{R}^m \setminus (\overline{D_{1,\rho} \cup D_{2,\rho}}), \quad p \in \Omega.$$

Then from (2.3.33), (2.3.34) and (3.2.21) for situation S1 of two impenetrable scatterers we obtain the estimate

$$\frac{C}{|d(z,D_1)|^3} \geq \left| E^{\bullet}_{edp,1}(z,z,p) \right|$$

$$\geq \left| E^{\bullet}_{edp,1}(z,z,p) \right| - \sigma \qquad (3.2.24)$$

$$\geq \frac{c}{|d(z,D_2)|^3} - \sigma, \qquad (3.2.25)$$

which can be transformed into

$$d(z,D_1) \leq \frac{Cd(z,D_2)}{\left(c - \sigma d(z,D_2)^3\right)^{1/3}} \qquad (3.2.26)$$

with constants C and c. In the same way for situation S2 we use (2.3.33),
(2.3.34), (2.4.20) and (2.4.21) to derive

$$d(z,D_1) \leq \left(\frac{Cd(z,D_2)}{c - \sigma d(z,D_2)}\right)^{\frac{1}{3}}. \qquad (3.2.27)$$

For situation S3 we estimate

$$d(z, D_1) \leq \frac{Cd(z, D_2)}{c - \sigma d(z, D_2)}. \tag{3.2.28}$$

The second part of the proof is literally the same as in Lemma 3.1.4 and thus from the estimates (3.2.26), (3.2.27) and (3.2.28) we obtain Lemma 3.2.5. □

Let the function F_2 be given by

$$F_2(\rho, \tau, \eta, \delta) := 2c\frac{\eta}{\rho^3} + 2Cb_{\eta,\rho}\tau + \frac{1}{\gamma}b_{\eta,\rho}\, b_{\tau,\rho}\, \delta \tag{3.2.29}$$

for $\rho, \tau, \eta, \delta > 0$, where the constants c, C are chosen according to Theorem 3.2.3. Then we define the function

$$F(\delta) := \inf\left\{F_1(\rho, F_2(\rho, \tau, \eta, \delta)), \ \tau, \eta, \rho > 0\right\} \tag{3.2.30}$$

with F_1 given by (3.2.23). The functions F for acoustic and electromagnetic scattering differ by constants and by the definitions of the function $F_1(\rho, \sigma)$. The proof and statement of Lemma 3.1.10 is literally the same for the electromagnetic case (3.2.30), i.e., we have

$$F(\delta) \to 0, \ \delta \to 0. \tag{3.2.31}$$

We now collect all previous results to obtain the main stability result for electromagnetic scattering by a perfect conductor or an inhomogeneous electromagnetic medium.

THEOREM 3.2.6 (Stability estimate.) *Let $E_{pl,1}^\infty(\hat{x}, d, p)$, $E_{pl,2}^\infty(\hat{x}, d, p)$ for $\hat{x}, d, p \in \Omega$ be the far field patterns for scattering of electromagnetic plane waves from two scatterers $D_1, D_2 \in C$. We assume that with a non-negative parameter δ the difference between the far field patterns satisfies*

$$\left\|E_{pl,1}^\infty(\cdot, \cdot, p) - E_{pl,2}^\infty(\cdot, \cdot, p)\right\|_{L^2(\Omega \times \Omega)} \leq \delta \tag{3.2.32}$$

for all $p \in \Omega$. Then with the function F defined by (3.2.30) we can estimate the Hausdorff distance $d(D_1, D_2)$ of the domains D_1 and D_2 by

$$d(D_1, D_2) \leq F(\delta). \tag{3.2.33}$$

Proof. With F_2 defined by (3.2.29) we use Lemma 3.2.4 to derive

$$\left|E_{edp,1}^s(x, z, p) - E_{edp,2}^s(x, z, p)\right| \leq F_2(\rho, \tau, \eta, \delta) \tag{3.2.34}$$

for all $p \in \Omega$, $\rho, \tau, \eta > 0$ and all points $x, z \in U$ with U defined by (3.1.12). We now apply Lemma 3.2.5 to derive (3.2.33). □

We would like to explicitly estimate the behavior of the function $F(\delta)$ for $\delta \to 0$. For the convex hulls we will now derive a logarithmic estimate both for a perfect conductor and an inhomogeneous electromagnetic medium.

THEOREM 3.2.7 (**Explicit stability estimate.**) *Let* $\mathcal{D}_1$, $\mathcal{D}_2 \in \mathcal{C}$ *be electromagnetic scatterers with scattering data* $E^{\infty}_{pl,1}(\hat{x}, d, p)$ *and* $E^{\infty}_{pl,2}(\hat{x}, d, p)$, $\hat{x}, d, p \in \Omega$. *Then*

$$\left\| E^{\infty}_{pl,1}(\cdot, \cdot, p) - E^{\infty}_{pl,2}(\cdot, \cdot, p) \right\|_{L^2(\Omega \times \Omega)} \leq \delta \qquad (3.2.35)$$

for all $p \in \Omega$ *yields*

$$d\Big(\mathcal{H}(\mathcal{D}_1), \mathcal{H}(\mathcal{D}_2) \Big) \leq \frac{C}{|\ln \delta|^c} \qquad (3.2.36)$$

with constants $C > 0$ *and* $0 < c < 1$.

Proof. The theorem can be proven analogously to Theorem 3.1.12. Here we paraphrase the proof and point out the places where changes have to be made.

The explicit estimate (3.1.63) is obtained by explicitly estimating the behavior of the functions $F_2(\rho, \tau, \eta, \delta)$, and $F_1(\rho, \sigma)$. To this end the domain of approximation $G_{z,p,\rho}$ is placed in a ball $B_R(x_0)$. In the ball $B_R(x_0)$ an expansion of both the point-source $\Phi(\cdot, z)$ and the incident Herglotz wave function with respect to spherical harmonics and (spherical) Bessel functions is used. With the help of the asymptotic behavior of the spherical Bessel functions, the norm of the minimum norm solution g with discrepancy τ of

$$(Hg)(x) = \Phi(x, z), \quad x \in \Omega_R(x_0),$$

with $H : L^2(\Omega) \to L^2(\Omega_R(x_0))$ is estimated in (3.1.89).

For the electromagnetic cases we first note that the incident electric dipole is obtained by

$$E^i_{edp}(x, z, p) = \frac{i}{\kappa}(p \cdot \nabla_z)\nabla_z \Phi(x, z) \qquad (3.2.37)$$

as a second derivative of the acoustic point-source and that the electric field

$$(Vg)(x, p) = \frac{i}{\kappa}(p \cdot \nabla_z)\nabla_z \int_\Omega e^{i\kappa x \cdot d} g(d) \, ds(d) \qquad (3.2.38)$$

of the electromagnetic Herglotz pair is obtained by an application of the same differential operator to the acoustic Herglotz wave function Hg. Thus to obtain an approximation of E^i_{cdp} by Vg instead of equation (3.1.75) we have to derive the estimate

$$\left\| \Phi(\cdot, z) - u_n \right\|_{C^4(\overline{G_{s,y,\rho}})} \leq \lambda \left\| \Phi(\cdot, z) - u_n \right\|_{L^2(\Omega_R(x_0))} \qquad (3.2.39)$$

and investigate the behavior of the corresponding constant λ with respect to variations of ρ. Instead of a singularity of second order here we obtain $\lambda \leq C\rho^{-6}$. Then we use the special choice (3.1.65) to derive (3.2.36) in the same way as in Theorem 3.1.12. $\qquad\qquad\square$

From stability as a corollary we obtain uniqueness of the support of penetrable or impenetrable scatterers.

COROLLARY 3.2.8 *For electromagnetic scatterers $\mathcal{D} \in C$ the domain D is uniquely determined by the far field pattern for scattering of all plane waves.*

4   The case of finite data

In this chapter we will investigate uniqueness and stability of inverse problems in the case where only a finite number of measurements of the far field patterns $u^\infty(\cdot, d)$ are given for a finite number of incident plane waves.

Since we search for domains in a space of infinite dimension, in this case in general we will not obtain a full uniqueness or stability result. We introduce a concept to treat the situation appropriately, which we call ϵ-uniqueness or ϵ-stability, respectively.

First, we show that for given $\epsilon > 0$ there are integers n_i, n_o in $I\!\!N$, such that, if for n_i incident plane waves the far field patterns for two scatterers $\mathcal{D}_1$ and $\mathcal{D}_2$ coincide for n_o observation directions, the Hausdorff distance $d(D_1, D_2)$ of the domains D_1 and D_2 satisfies

$$d(D_1, D_2) \leq \epsilon.$$

The distance $d(D_1, D_2)$ tends to zero, if n_i and n_o tend to infinity. This is a kind of continuity statement: with more measurements we obtain better reconstructions, and in the limit $n_i, n_o \to \infty$ we obtain precise reconstructions. We call the concept ϵ-uniqueness in analogy to the ϵ-δ-formulation of continuity.

We would like to point out the difference of ϵ-uniqueness and stability. Stability investigates the continuity of the mapping from the data space into the space of domains. Stability implies full uniqueness. In contrast to stability, ϵ-uniqueness investigates a sequence of finite data-spaces for which uniqueness is not necessarily satisfied.

Second, we will investigate stability for the case of finitely many measurements and for a finite number of incident plane waves. In this case, where we do not have uniqueness, it cannot be possible to obtain full stability results. We develop a concept of ϵ-stability as follows. Given $\epsilon > 0$ it is possible to find $n_o, n_i \in I\!\!N$ and a function $F_{(n_0, n_i)} : I\!\!R^+ \to I\!\!R^+$ with the behavior

$$\limsup_{\delta \to 0} F_{(n_0, n_i)}(\delta) \leq \epsilon \qquad (4.0.1)$$

such that the Hausdorff distance $d(D_1, D_2)$ of the domains D_1 and D_2 of the scatterers $\mathcal{D}_1$ and $\mathcal{D}_2$ can be estimated by

$$d(D_1, D_2) \leq F_{(n_0, n_i)}\left(\|u_1^\infty(\cdot, \cdot) - u_2^\infty(\cdot, \cdot)\|_{L^2(\Omega_{n_i} \times \Omega_{n_o})} \right). \qquad (4.0.2)$$

Clearly, each function $F_{(n_0, n_i)}$ which satisfies (4.0.1) and (4.0.2) provides an ϵ-stability estimate for the reconstruction of the shapes of scatterers.

4.1   Finite data in inverse acoustic scattering

The proof of the uniqueness theorem 3.1.1 is based on the application of
Rellich's lemma and it cannot be applied to the case where the far field
pattern is known only for a finite number of observation points and incident
waves. Since Rellich's lemma includes an analyticity argument, it is not
possible to use approximations to treat a finite data set. We now develop
modified techniques to derive the results of ϵ-uniqueness, which in the limit-
case $\epsilon \to 0$ also yield the above uniqueness results.

A discrete backprojection operator. We will replace the role of
Rellich's lemma by an operator $Q_{(n_o,n_i)}$ for the approximate reconstruction
of point-sources $\Phi^s(x,z)$ and, more general, the scattered field $\Phi^s_{\mu,q}(x,z)$
of multipoles. We will obtain

$$\Phi^s_{q,\mu}(x,z) \approx \Big(Q_{(n_o,n_i)}u^\infty_{(n_o,n_i)}\Big)(x,z), \quad x,z \in B \setminus D_\rho,$$

on a set $B \setminus D_\rho$ with $\rho > 0$, where for $d_j, d_k \in \Omega$

$$u^\infty_{(n_o,n_i)} := \Big(u^\infty(d_j,d_k)\Big)_{j=1,\dots,n_o,k=1,\dots,n_i} \quad \in \mathbb{C}^{n_o \cdot n_i} \tag{4.1.1}$$

denotes a finite set of measured far field patterns. We consider subsets of
the unit sphere

$$\Omega_n := \{d_j : j = 1,\dots n\} \quad \subset \Omega \tag{4.1.2}$$

with $d_i \neq d_j$ for $i \neq j$, $d_i, d_j \in \Omega_n$. For simplicity, for the sequence
$(\Omega_n)_{n \in \mathbb{N}}$ we demand the denseness property

$$d(\hat{x},\Omega_n) \to 0, \quad n \to \infty \tag{4.1.3}$$

for all $\hat{x} \in \Omega$, the symmetry property

$$d \in \Omega_n \;\Rightarrow\; -d \in \Omega_n \tag{4.1.4}$$

if n is even and the monotonicity property

$$\Omega_{n'} \subset \Omega_n \text{ for } n > n'. \tag{4.1.5}$$

We also assume that the d_j are distributed uniformly in the sense that the
sets $\tilde{V}_j$ defined by

$$\tilde{V}_j := \{\hat{x} \in \Omega, d(\hat{x},d_j) \leq d(\hat{x},d_l) \; \forall \, l \neq j\}. \tag{4.1.6}$$

satisfy

$$\frac{c_m}{2n} \leq \int_{\tilde{V}_j} ds \leq \frac{2c_m}{n}, \quad j = 1,\dots,n, \tag{4.1.7}$$

where c_m is the surface of the ball or unit sphere, respectively, given by

$$c_m = \begin{cases} 2\pi, & m = 2 \\ 4\pi, & m = 3. \end{cases} \qquad (4.1.8)$$

The function space $L^2(\Omega_n)$ is defined as the space of functions

$$w : \Omega_n \to \mathbb{C}$$

equipped with the norm

$$\|w\|_{L^2(\Omega_n)} := \left(\frac{c_m}{n} \sum_{k=1}^{n} |w(d_k)|^2 \right)^{\frac{1}{2}}.$$

The mapping

$$\iota : w \mapsto (w(d_k))_{k=1,\ldots,n} \in \mathbb{C}^n$$

is a norm isomorphism from the space $L^2(\Omega_n)$ onto the space $\mathbb{C}^n$ equipped with the norm

$$\|a\|_{L^2(\mathbb{C}^n)} := \left(\frac{c_m}{n} \sum_{k=1}^{n} |a_k|^2 \right)^{\frac{1}{2}}. \qquad (4.1.9)$$

In the same way $L^2(\Omega_{n_o} \times \Omega_{n_i})$ with the norm

$$\|w\|_{L^2(\Omega_{n_o} \times \Omega_{n_i})} := \left(\frac{c_m}{n_o} \frac{c_m}{n_i} \sum_{k=1}^{n_o} \sum_{j=1}^{n_i} |w(d_k, d_j)|^2 \right)^{\frac{1}{2}}$$

is defined and shown to be isomorphic to $\mathbb{C}^{n_o \cdot n_i}$ equipped with the norm (4.1.9), where c_m has to be replaced by c_m^2. Usually, we will identify the two spaces and treat $u_{(n_o,n_i)}^\infty$ (defined in (4.1.1)) as an element of $L^2(\Omega_{n_o} \times \Omega_{n_i})$. By $\tilde{A} := \iota^{-1} \circ A \circ \iota$ operators $A : \mathbb{C}^n \to Y$ can be considered as operators

$$\tilde{A} : L^2(\Omega_{n_o} \times \Omega_{n_i}) \to Y.$$

We will usually identify A and $\tilde{A}$.

To construct the approximation operator $Q_{(n_o,n_i)}$ in principle we have two possibilities. First, we may discretize the continuous operator Q. Second, we may approximate point-sources by a finite superposition of plane waves and proceed analogously to the derivation of the operator Q. Here, we will choose the second approach.

Finite superpositions of plane waves. A finite superposition of plane waves is given by the finite Herglotz wave function

$$(H_n a)(x) := \frac{c_m}{n} \sum_{j=1}^{n} e^{i\kappa x \cdot d_j} a_j, \quad x \in \mathbb{R}^m \tag{4.1.10}$$

with density vector $a \in \mathbb{C}^n$ and c_m given by (4.1.8). For scattering of the functions $H_n a$ from a scatterer $\mathcal{D}$ we can exploit the linearity of the scattering problem. If $H_n a$ is the incident field, the corresponding scattered field and its far field pattern are given by

$$(H_n^s a)(x) = \frac{c_m}{n} \sum_{j=1}^{n} u^s(x, d_j) a_j, \quad x \in \mathbb{R}^m \setminus D, \tag{4.1.11}$$

and

$$(H_n^\infty a)(\hat{x}) = \frac{c_m}{n} \sum_{j=1}^{n} u^\infty(x, d_j) a_j, \quad \hat{x} \in \Omega. \tag{4.1.12}$$

In the next two lemmas we investigate the approximation of a multipole $\Phi_{\mu,q}$ of order μ by a finite superposition of plane waves $H_n g$ on the domains $G_{z,p,\rho}$ introduced in (3.1.4).

LEMMA 4.1.1 *For fixed $\mu, s \in \mathbb{N}_0$ and $\rho > 0$ the error*

$$E(n) := \sup_{p,q \in \Omega} \inf_{a \in \mathbb{C}^n} \left\| \Phi_{\mu,q}(\cdot, z) - H_n a \right\|_{C^s(G_{z,p,\rho})} \tag{4.1.13}$$

for an approximation of the multipole of order μ by a finite superposition of plane waves is independent of $z \in \mathbb{R}^m$ and satisfies

$$\lim_{n \to \infty} E(n) = 0. \tag{4.1.14}$$

Define

$$b(\epsilon, n) := \sup_{p,q \in \Omega} \inf_{a \in \mathbb{C}^n} \left\{ \|a\|_{L^2(\Omega_n)} \middle| \ \|\Phi_{\mu,q}(\cdot, z) - H_n a\|_{C^s(G_{z,p,\rho})} \leq \epsilon \right\} \tag{4.1.15}$$

if

$$\left\{ a \in \mathbb{C}^n \middle| \ \|\Phi_{\mu,q}(\cdot, z) - H_n a\|_{C^s(G_{z,p,\rho})} \leq \epsilon \right\} \neq \emptyset \ \forall p, q \in \Omega$$

and $b(\epsilon, n) := 0$ otherwise. For fixed $\epsilon > 0$ the function $b(\epsilon, \cdot)$ is bounded.

Proof. As for the continuous case we use the fact that $G_{z,p,\rho}$ is obtained from $G_{0,p,\rho}$ by translation. If we translate a finite Herglotz wave function (4.1.10) by $z \in \mathbb{R}^m$, the result is again a finite Herglotz wave function with density vector

$$\tilde{a}_j := e^{-i\kappa z \cdot d_j} a_j, \quad j = 1, .., n. \tag{4.1.16}$$

Hence, an approximation of $\Phi(\cdot, z)$ on $\overline{D} \subset G_{z,q,\rho}$ can be derived from an approximation of $\Phi(\cdot, 0)$ on $G_{0,q,\rho}$.

Because of (4.1.16) the error $E(n)$ is independent of z and we can restrict our investigation to the case $z = 0$. We will use an approximation argument to derive the statements from the continuous case.

For $\epsilon > 0$ consider the finite set $\mathcal{G}(\epsilon/3, \rho, \mu, s, \beta_0)$ given by Lemma 3.1.3. By the definition of L^2 a function $g \in \mathcal{G} \subset L^2(\Omega)$ can be approximated by a function $\tilde{g} \in C^1(\Omega)$ such that

$$\left\| Hg - H\tilde{g} \right\|_{C^*(G_{0,p,\rho})} \leq \frac{\epsilon}{3}.$$

For the density $\tilde{g} \in C^1(\Omega)$ we can apply the standard convergence theorems for quadrature rules to approximate the integral $H\tilde{g}$ by a finite sum $H_n a$ with error

$$\left\| H\tilde{g} - H_n a \right\|_{C^*(G_{0,p,\rho})} \leq \frac{\epsilon}{3},$$

where $a \in \mathbb{C}^n$ is given by $a_j := \alpha_j^{(n)} \tilde{g}(d_j)$, $j = 1, ..., n$, with appropriate weights $\alpha_j^{(n)}$. For the weights we assume $|\alpha_j^{(n)}| \leq c$, $j = 1, ..., n$, $n \in \mathbb{N}$ with some constant c, which is valid for example for the approximate computation of integrals by Riemann sums. Hence, given $\epsilon > 0$ we can find $n \in \mathbb{N}$, such that for every pair $p, q \in \Omega$ there is $a \in \mathbb{C}^n$ with

$$\left\| \Phi_{\mu,q}(\cdot, 0) - H_n a \right\|_{C^*(G_{0,p,\rho})} \tag{4.1.17}$$

$$\leq \left\| \Phi_{\mu,q}(\cdot, 0) - Hg \right\|_{C^*(G_{0,p,\rho})} \tag{4.1.18}$$

$$+ \left\| Hg - H\tilde{g} \right\|_{C^*(G_{0,p,\rho})} + \left\| H\tilde{g} - H_n a \right\|_{C^*(G_{0,p,\rho})}$$

$$\leq \epsilon,$$

where $g \in \mathcal{G}(\epsilon/3, \rho, \mu, s, \beta_0)$ has to be chosen appropriately. Clearly, the function on the left-hand side of (4.1.17) is an upper bound for the function $E(n)$ defined in (4.1.13). By (4.1.5) the function E is monotonous. Thus we obtain $E(n) \to 0$ for $n \to \infty$. For the norm of a we estimate for sufficiently large n

$$\|a\|_{L^2(\Omega_n)}^2 = \frac{c_m}{n} \sum_{j=1}^{n} |\alpha_j^{(n)} \tilde{g}(d_j)|^2 \tag{4.1.19}$$

$$\leq \quad c \frac{c_m}{n} \sum_{j=1}^{n} |\alpha_j^{(n)}| \, |\bar{g}(d_j)|^2 \; \to \; c \, \|g\|_{L^2(\Omega)}^2, \quad n \to \infty;$$

i.e., the norm of a is bounded uniformly for $n \in I\!N$. Thus we obtain the boundedness of $b(\epsilon, \cdot)$ and the proof of Lemma 4.1.1 is complete.   □

For $\mu, s \in I\!N_0$ and $\rho, \tau > 0$ according to the behavior (4.1.14) of $E(n)$ there is an even integer $n \in I\!N$ such that

$$\sup_{p,q \in \Omega} \; \inf_{a \in \mathbb{C}^n} \left\{ \left\| \Phi_{\mu,q}(\cdot, 0) - H_n a \right\|_{C^s(G_{0,p,\rho})} \right\} \leq \frac{\tau}{2} \qquad (4.1.20)$$

is satisfied.

LEMMA 4.1.2 *Given $\mu, s \in I\!N_0$, $\rho, \tau > 0$ and an even integer $n \in I\!N$ such that (4.1.20) is satisfied, there is a finite set*

$$\mathcal{G} = \mathcal{G}(n, \tau, \rho, \mu, s, \beta_0) \subset \mathbb{C}^n$$

such that for each $p, q \in \Omega$ there is a vector $a \in \mathcal{G}$ with

$$\left\| \Phi_{\mu,q}(\cdot, 0) - H_n a \right\|_{C^s(G_{0,p,\rho})} \leq \tau. \qquad (4.1.21)$$

If for μ, s, ρ, τ and n condition (4.1.20) is not satisfied, we define

$$\mathcal{G}(n, \tau, \rho, \mu, s, \beta_0) := \{0\}.$$

Proof. Consider $p, q \in \Omega$ and $a \in \mathbb{C}^n$ with

$$\left\| \Phi_{\mu,q}(\cdot, 0) - H_n a \right\|_{C^s(G_{0,p,\rho})} \leq \frac{\tau}{2}. \qquad (4.1.22)$$

Since for fixed $a \in \mathbb{C}^n$ the function on the left-hand side of (4.1.22) depends continuously on p and q, by compactness of Ω as in the proof of Lemma 3.1.3 we obtain a finite set $\mathcal{G}$ of vectors $a \in \mathbb{C}^n$, such that for $p, q \in \Omega$ there is a vector $a \in \mathcal{G}$ which satisfies the estimate (4.1.24).   □

As shown in (4.1.16), an approximation of $\Phi_{\mu,q}(\cdot, 0)$ on $G_{0,p,\rho}$ yields an approximation of $\Phi_{\mu,q}(\cdot, x)$ on $G_{x,p,\rho}$. For a vector a in $\mathbb{C}^n$ we define the vector function $a(\cdot) : B \to \mathbb{C}^n$ by

$$a_j(x) = e^{-i\kappa x \cdot d_j} a_j, \quad j = 1, ..., n, \quad x \in B, \qquad (4.1.23)$$

where d_j is given by (4.1.2). From (4.1.21) by translation we derive

$$\left\| \Phi_{\mu,q}(\cdot, x) - H_n a(x) \right\|_{C^s(x,p,\rho)} \le \tau. \qquad (4.1.24)$$

Following Lemma 4.1.2 and (4.1.23) the vector a is a function

$$a = a(x, p, q, \tau, \rho, \mu, s, \beta_0). \qquad (4.1.25)$$

We will need the vector with two different sets of values for x, τ and μ and with vectors p, q depending on x. To indicate the dependence we will use the notation

$$a_\tau(x) = a(x, p(x), q(x), \tau, \rho, \mu, s, \beta_0). \qquad (4.1.26)$$

The j-th component of $a_\tau(x) \in \mathbb{C}^n$ is denoted by

$$a_{\tau,j}(x).$$

For the set $\mathcal{G}$ we write as in the continuous case

$$\mathcal{G}_\tau = \mathcal{G}(n, \tau, \rho, \mu, s, \beta_0). \qquad (4.1.27)$$

We are now prepared to define the operator $Q_{(n_o, n_i)}$.

DEFINITION 4.1.3 (**Discrete backprojection operator.**) *Given a set of parameters $\mu \in \mathbb{N}_0$, $s = 1$, $\rho, \tau, \eta > 0$, even integers $n_o, n_i \in \mathbb{N}$, for which (4.1.20) is satisfied for $(0, \tau, n_o)$, (μ, η, n_i), and functions*

$$p, q : B \to \Omega$$

with the help of

$$a_\tau(x) = a(x, p(x), q(x), \tau, \rho, 0, s, \beta_0)$$

and

$$a_\eta(z) = a(z, p(z), q(z), \tau, \rho, \mu, s, \beta_0)$$

we define the operator

$$Q_{(n_o, n_i)} : L^2(\Omega_{n_o} \times \Omega_{n_i}) \to L^\infty(B)$$

by

$$(Q_{(n_o,n_i)} w)(x, z) := \frac{1}{\gamma_m} \frac{c_m}{n_o} \frac{c_m}{n_i} \sum_{j=1}^{n_o} \sum_{k=1}^{n_i} a_{\tau,j}(x) a_{\eta,k}(z) w(-d_k, d_j) \qquad (4.1.28)$$

for $x, z \in B$.

THEOREM 4.1.4 *Consider scattering by a sound-soft, sound-hard or inhomogeneous medium scatterer $\mathcal{D} \in \mathcal{C}$. For all $x, z \in B \setminus D_\rho$, for which (3.1.34) is satisfied for $p = p(x)$ or $p = p(z)$, respectively, the error for the approximation of $\Phi^s_{\mu,q}(x, z)$ by $Q_{(n_o, n_i)} u^\infty_{(n_o, n_i)}$ is estimated by*

$$\left| \Phi^s_{\mu,q}(x, z) - \left(Q_{(n_o,n_i)} u^\infty_{(n_o,n_i)} \right)(x, z) \right| \leq c \frac{\eta}{\rho^{m-1}} + C \left\| a_\eta(z) \right\|_{L^2(\Omega_{n_i})} \tau$$

$$\tag{4.1.29}$$

uniformly for $\mathcal{D} \in \mathcal{C}$ with constants c and C depending on μ.

Proof. We start with the estimate

$$\left\| \Phi_{\mu,q}(\cdot, z) - H_{n_i} a_\eta(z) \right\|_{C^1(D)} \leq \eta \tag{4.1.30}$$

derived from (4.1.24). The scattering map $u^i \mapsto u^s$ is bounded from $C^1(\overline{D})$ into $C(B \setminus D_\rho)$. Estimating the combined single- and double-layer potential with the help of the Cauchy-Schwarz inequality we derive a constant c, such that (4.1.30) yields

$$\left\| \Phi^s_{\mu,q}(\cdot, z) - H^s_{n_i} a_\eta(z) \right\|_{C(B \setminus D_\rho)} \leq \frac{c}{\rho^{m-1}} \eta \tag{4.1.31}$$

uniformly for all scatterers $\mathcal{D} \in \mathcal{C}$. We exploit the estimate (4.1.24) a second time, now for τ, n_o and $\mu = 0$, to obtain a constant c with

$$\left\| \Phi^\infty(\cdot, x) - H^\infty_{n_o} a_\tau(x) \right\|_{C(\Omega)} \leq c \tau \tag{4.1.32}$$

uniformly for all scatterers $\mathcal{D} \in \mathcal{C}$. We use the mixed reciprocity relation (2.1.4) to transform (4.1.32) into

$$\left| u^s(x, d) - \frac{1}{\gamma_m} \left(H^\infty_{n_o} a_\tau(x) \right)(-d) \right| \leq \frac{c}{\gamma_m} \tau \tag{4.1.33}$$

for all $d \in \Omega$. We now insert the approximation (4.1.33) for $u^s(x, -d)$ into

$$\left(H^s_{n_i} a_\eta(z) \right)(x) = \frac{c_m}{n_i} \sum_{k=1}^{n_i} u^s(x, d_k) a_{\eta,k}(z).$$

From the Cauchy-Schwarz inequality we obtain the estimate

$$\left| \left(H^s_{n_i} a_\eta(z) \right)(x) - \left(Q_{(n_o,n_i)} u^\infty_{(n_o,n_i)} \right)(x, z) \right|$$

$$= \left| \frac{c_m}{n_i} \sum_{j=k}^{n_i} \left(u^s(x, d_k) - \frac{c_m}{\gamma_m n_o} \sum_{j=1}^{n_o} u^\infty(-d_k, d_j) a_{\tau,j}(x) \right) a_{\eta,k}(z) \right|$$

$$\leq C \left\| a_\eta(z) \right\|_{L^2(\Omega_{n_i})} \tau \tag{4.1.34}$$

with some constant C uniformly for scatterers $\mathcal{D} \in \mathcal{C}$. We can now use (4.1.34) and (4.1.31) to estimate the distance between $\Phi_{n,q}^s$ and the reconstruction $Q_{(n_o,n_i)} u_{(n_o,n_i)}^\infty$. We calculate

$$\left| \Phi_{\mu,q}^s(x,z) - \left(Q_{(n_o,n_i)} u_{(n_o,n_i)}^\infty \right)(x,z) \right| \tag{4.1.35}$$

$$\leq \left| \Phi_{\mu,q}^s(x,z) - \left(H_n^s a_\eta(z) \right)(x) \right|$$

$$+ \left| \left(H_n^s a_\eta(z) \right)(x) - \left(Q_{(n_o,n_i)} u_{(n_o,n_i)}^\infty \right)(x,z) \right|$$

$$\leq \frac{c}{\rho^{m-1}} \eta + C \left\| a_\eta(z) \right\|_{L^2(\Omega_{n_i})} \tau \tag{4.1.36}$$

with constants c and C uniformly for scatterers $\mathcal{D} \in \mathcal{C}$. $\square$

In the finite data case it is of interest to explicitly formulate the approximation properties of $Q_{(n_o,n_i)}$ in a corollary.

COROLLARY 4.1.5 *For scattering by a sound-soft, sound-hard or inhomogeneous medium scatterer $\mathcal{D} \in \mathcal{C}$ we consider the approximation of $\Phi_{\mu,q}^s$ by $Q_{(n_o,n_i)} u_{(n_o,n_i)}^\infty$. Given $\mu \in \mathbb{N}_0$, $\rho, \sigma > 0$ and functions $p, q : B \to \Omega$ there are parameter $\tau, \eta > 0$ and even integers $n_i, n_o \in \mathbb{N}$, such that*

$$\left| \Phi_{\mu,q}^s(x,z) - \left(Q_{(n_o,n_i)} u_{(n_o,n_i)}^\infty \right)(x,z) \right| \leq \sigma, \tag{4.1.37}$$

for all $x, z \in B \backslash D_\rho$, for which (3.1.34) is satisfied for $p = p(x)$ or $p = p(z)$, respectively.

Proof. Given $\rho, \sigma > 0$ for $\eta = \sigma/(2c)$ we use Lemma 4.1.2 and first obtain $n_i \in \mathbb{N}$ such that (4.1.20) is satisfied with (τ, n) replaced by (η, n_i). We then construct the function $a_\eta(\cdot)$ as in (4.1.26). In the same way for $\tau = \sigma/(2C\|a_\eta(x)\|)$ and $\mu = 0$ we get $n_o \in \mathbb{N}$ to obtain the estimate (4.1.20) with n replaced by n_o. Now an application of Theorem 4.1.4 yields (4.1.37). $\square$

We can use the operator $Q_{(n_o,n_i)}$ to obtain estimates for the difference of the scattered fields $\Phi_{1,\mu,q}^s$ or $\Phi_{2,\mu,q}^s$ by scatterers $\mathcal{D}_1$ or $\mathcal{D}_2$, respectively, from the knowledge of a finite data set of far field patterns for scattering of plane waves.

LEMMA 4.1.6 *Let $u_1^\infty(\hat{x}, d)$ and $u_2^\infty(\hat{x}, d)$, $\hat{x}, d \in \Omega$, be the far field patterns for scattering of acoustic plane waves from scatterers $\mathcal{D}_1, \mathcal{D}_2 \in \mathcal{C}$. Given $\mu \in \mathbb{N}_0$, $\rho > 0$ and $\sigma > 0$ there are even integers $n_i, n_o \in \mathbb{N}$, such that*

$$u_1^\infty(\hat{x}, d) = u_2^\infty(\hat{x}, d), \quad \hat{x} \in \Omega_{n_o}, \, d \in \Omega_{n_i}, \tag{4.1.38}$$

yields

$$\left| \Phi^s_{1,\mu,q}(x,z) - \Phi^s_{2,\mu,q}(x,z) \right| \leq \sigma \qquad (4.1.39)$$

for all $q \in \Omega$ and all points $x, z \in U$, where U is defined by (3.1.12).

Proof. First we remark that we can choose a function $p : B \to \Omega$ such that (3.1.34) is satisfied for both scatterers $\mathcal{D}_1$ and $\mathcal{D}_2$ and for all $x \in U$. We choose an arbitrary constant function $q(\cdot) : B \to \Omega$. With $Q_{(n_o,n_i)}$ given by (4.1.28) we estimate

$$\left| \Phi^s_{1,\mu,q}(x,z) - \Phi^s_{2,\mu,q}(x,z) \right|$$
$$\leq \left| \Phi^s_{1,\mu,q}(x,z) - \left(Q_{(n_o,n_i)} u^\infty_{1,(n_o,n_i)} \right)(x,z) \right|$$
$$+ \left| \left(Q_{(n_o,n_i)} u^\infty_{1,(n_o,n_i)} \right)(x,z) - \left(Q_{(n_o,n_i)} u^\infty_{2,(n_o,n_i)} \right)(x,z) \right|$$
$$+ \left| \left(Q_{(n_o,n_i)} u^\infty_{2,(n_o,n_i)} \right)(x,z) - \Phi^s_{2,\mu,q}(x,z) \right|. \qquad (4.1.40)$$

According to Corollary 4.1.5 given $\sigma/2$ we obtain $\tau, \eta > 0$, $n_o, n_i \in \mathbb{N}$, such that the first and the last term of (4.1.40) are bounded by $\sigma/2$ for all points $x, z \in U$, where according to Lemma 4.1.2 we may choose n_i, n_o and τ, η uniformly for $q \in \Omega$. From (4.1.38) we derive that the central term of (4.1.40) is zero; thus we obtain the estimate (4.1.39).   □

Uniqueness for finite data. We are now prepared to formulate the main theorem on ϵ-uniqueness for the reconstruction of the shape of a sound-soft, a sound-hard or an inhomogeneous medium scatterer $\mathcal{D} \in \mathcal{C}$ from a finite data set $u^\infty_{(n_o,n_i)}$ of scattered plane waves.

THEOREM 4.1.7 (ϵ-Uniqueness.) *Let $u^\infty_1(\hat{x},d)$ and $u^\infty_2(\hat{x},d)$, $\hat{x}, d \in \Omega$, be the far field patterns for scattering of plane waves from two acoustic scatterers $\mathcal{D}_1, \mathcal{D}_2$ in $\mathcal{C}$. Given $\epsilon > 0$ there is $n_o, n_i \in \mathbb{N}$, such that*

$$u^\infty_1(\hat{x},d) = u^\infty_2(\hat{x},d), \quad \hat{x} \in \Omega_{n_o}, \ d \in \Omega_{n_i}, \qquad (4.1.41)$$

yields the estimate

$$d(D_1, D_2) \leq \epsilon \qquad (4.1.42)$$

for the Hausdorff distance $d(D_1, D_2)$ of D_1 and D_2.

Proof. Consider one of the situations S1 to S6 from Table 3.1 on page 126. Given $\epsilon > 0$ we choose ρ and σ such that $\epsilon = F_1(\rho, \sigma)$ with F_1 given by (3.1.15). We use Lemma 4.1.6 to obtain integers n_i and n_o such that

$$\left| \Phi^s_{1,\mu,q}(z,z) - \Phi^s_{2,\mu,q}(z,z) \right| \leq \sigma \qquad (4.1.43)$$

is satisfied for all $q \in \Omega$ and for all points $z \in U$ with U defined by (3.1.12). Then the estimate (4.1.42) for the chosen situation is given by (3.1.14) of Lemma 3.1.4. To obtain the statement (4.1.42) for arbitrary situations we take the maximum of all n_i, n_o for S1 to S6. □

We have shown that a sufficiently large finite number of measurements of the far field patterns determine the boundary of an acoustic scatterer up to a given error ϵ in the Hausdorff distance. This is independent of the physical properties of the scatterer, i.e., the scatterer may be an impenetrable sound-soft or sound-hard obstacle or an inhomogeneous medium scatterer.

If more information about the physical properties of the scatterer is known, we may obtain better estimates for the dependence of n_o and n_i on ϵ; i.e., less measurements are necessary to determine the boundary up to an error ϵ. This is due to the fact that we do not have to build a maximum of all bounds for the different possible situations S1 to S6. The observation indicates that the ill-posedness of an inverse problem is influenced by the amount of information given for reconstructions.

As a simple consequence of ϵ-uniqueness we obtain the uniqueness statements of Corollary 3.1.13 for the support of either impenetrable or inhomogeneous medium scatterers $\mathcal{D} \in \mathcal{C}$.

Stability in the case of finite data. The second theme of this section is stability in the case of finite data. In this case we do not have uniqueness and thus it cannot be possible to obtain full stability statements. We will show that it is still possible to derive statements close to stability which we refer to as ϵ-stability.

The concept of ϵ-stability is closely related both to full stability and to ϵ-uniqueness as introduced in Chapter 3. To emphasize and highlight the connections we will give two proofs for the main theorem. The first approach does not build on stability estimates, but derives the statements of ϵ-stability in a way close to ϵ-uniqueness. The second approach shows how the statements of ϵ-uniqueness and ϵ-stability can be derived from full stability.

The main tool of the first approach is the operator $Q_{(n_0,n_i)}$ defined in (4.1.28). We need some further preparations. To estimate the norm of $Q_{(n_0,n_i)}$ we use the bound

$$b_{n,\tau,\rho} := \max\left\{\|a\|_{L^2(\mathbb{C}^n)} : a \in \mathcal{G}(n,\tau,\rho,\mu,s,\beta_0)\right\} \qquad (4.1.44)$$

where the norm $\|\cdot\|_{L^2(\mathbb{C}^n)}$ is given by (4.1.9) and the set $\mathcal{G}(n,\tau,\rho,\mu,s,\beta_0)$ by (4.1.27). In Lemma 4.1.1 it has been shown that for fixed $\rho,\tau > 0$ the constant $b_{n,\tau,\rho}$ is bounded uniformly for $n_o, n_i \in I\!N$. We define

$$F_2(n_o, n_i, \rho, \tau, \eta, \delta) \quad := \quad 2c\frac{\eta}{\rho^{m-1}} + 2C\, b_{n_i,\eta,\rho}\,\tau \qquad (4.1.45)$$
$$+ \frac{1}{\gamma_m}b_{n_i,\eta,\rho}b_{n_o,\tau,\rho}\,\delta$$

and

$$F_{(n_o,n_i)}(\delta) := \inf \left\{ F_1\Big(\rho, F_2(n_o, n_i, \rho, \tau, \eta, \delta)\Big)\Big|\, \rho, \tau, \eta > 0 \quad (4.1.46)\right.$$
$$\left. \text{for which } (4.1.20) \text{ is satisfied for } (\mu, \eta, n_i) \text{ and } (0, \tau, n_o) \right\}$$

with the function F_1 given by (3.1.15) according to the situations S1 to S6. We first study the behavior of $F_{(n_o,n_i)}(\delta)$ for $\delta \to 0$.

LEMMA 4.1.8 *Given* $\epsilon > 0$ *there is* $n_o, n_i \in I\!N$, *such that the function* $F_{(n_o,n_i)}$ *defined by (4.1.46) satisfies*

$$\limsup_{\delta \to 0} F_{(n_o,n_i)}(\delta) \leq \epsilon. \qquad (4.1.47)$$

Proof. For $n_o, n_i \in I\!N$ the function $F_{(n_o,n_i)}(\delta)$ is dominated by

$$F_1\Big(\rho_0, F_2(n_o, n_i, \rho_0, \tau_0, \eta_0, \delta)\Big)$$

for all positive parameters ρ_0, τ_0 and η_0, for which the condition (4.1.20) is satisfied. We show that for $\epsilon > 0$ we can find $n_o, n_i \in I\!N$ and parameters ρ_0, τ_0 and η_0, such that

$$\limsup_{\delta \to 0} F_1\Big(\rho_0, F_2(n_o, n_i, \rho_0, \tau_0, \eta_0, \delta)\Big) \leq \epsilon \qquad (4.1.48)$$

and (4.1.20) is satisfied for (μ, η_0, n_i), $(0, \tau_0, n_o)$ and ρ_0. We proceed in two steps.

1. The function F_2 can be decomposed into the sum and product

$$F_2(n_o, n_i, \rho, \tau, \eta, \delta) = 2c\frac{\eta}{\rho^{m-1}} + b_{n_i,\eta,\rho}\left(2C\,\tau + \frac{1}{\gamma_m}b_{n_o,\tau,\rho}\,\delta\right).$$

Since for fixed $\rho, \tau > 0$ the constant $b_{n_o,\tau,\rho}$ is bounded independently of $n_o \in I\!N$, by an application of Corollary 3.1.9 to $f(t,s) := \sup_{n \in I\!N} b_{n,t,s}$ we obtain functions h_4 and h_5 with

$$h_4(\delta) \to 0, \quad \delta \to 0,$$

and

$$h_5(\delta) \to 0, \quad \delta \to 0, \tag{4.1.49}$$

such that the functions

$$h_6(\delta) := 2C \, h_4(\delta) + \frac{1}{\gamma_m} b^*_{n_o, h_4, h_5(\delta)} \, \delta$$

and

$$h_7(\delta) := 2c \frac{h_4(\delta)}{h_5(\delta)^{m-1}} + b^*_{n_i, h_4(\delta), h_5(\delta)} \, \delta$$

with b^* defined in (3.1.49) satisfy

$$h_6(\delta) \to 0, \quad \delta \to 0,$$

and

$$h_7(\delta) \to 0, \quad \delta \to 0,$$

uniformly for $n_i, n_o \in I\!N$. For

$$\begin{aligned}
h_1(\delta) &:= \max\{h_5(\delta), h_5(h_6(\delta))\}, \\
h_2(\delta) &:= h_4(\delta) \\
h_3(\delta) &:= h_4(h_6(\delta))
\end{aligned}$$

we obtain

$$F_2(n_o, n_i, h_1(\delta), h_2(\delta), h_3(\delta), \delta)\delta \tag{4.1.50}$$

$$= 2c \frac{h_3(\delta)}{h_1(\delta)^{m-1}}$$

$$\quad + b_{n_i, h_5(\delta), h_1(\delta)} \left(2C h_2(\delta) + \frac{1}{\gamma_m} b_{n_o, h_2(\delta), h_1(\delta)} \delta \right)$$

$$\leq 2c \frac{h_4(h_6(\delta))}{h_5(h_6(\delta))^{m-1}}$$

$$\quad + b^*_{n_i, h_4(h_6(\delta)), h_5(h_6(\delta))} \left(2C h_4(\delta) + \frac{1}{\gamma_m} b^*_{n_o, h_4(\delta), h_5(\delta)} \delta \right)$$

$$\to 0, \quad \delta \to 0.$$

Thus we have

$$F_1 \Big(h_1(\delta), F_2(n_o, n_i, h_1(\delta), h_2(\delta), h_3(\delta), \delta) \Big) \to 0, \quad \delta \to 0, \tag{4.1.51}$$

uniformly for all $n_o, n_i \in I\!N$.

2. For $\epsilon > 0$ from (4.1.51) with the explicit form of F_2 and F_1 we get parameters $\rho_0, \tau_0, \eta_0 > 0$ and some constant C, such that we have

$$\sup_{n_o, n_i \in I\!N} F_2(n_o, n_i, \rho_0, \tau_0, \eta_0, \delta) \leq C$$

and

$$F_1(\rho_0, \sigma) \leq \epsilon$$

for all δ sufficiently small and $0 < \sigma \leq C$. This yields

$$\limsup_{\delta \to 0} \sup_{n_o, n_i \in I\!N} F_1\left(\rho_0, F_2(n_o, n_i, \rho_0, \tau_0, \eta_0, \delta)\right) \leq \epsilon.$$

We now explicitly take into account condition (4.1.20). For ρ_0, τ_0 and η_0 following Lemma 4.1.1 we choose $n_o, n_i \in I\!N$ such that (4.1.20) is satisfied for (μ, η_0, n_i) and $(0, \tau_0, n_0)$. Then we obtain (4.1.48) and thus (4.1.47). $\square$

To prove the following Theorem on ϵ-stability we will use the approximation of the fields Φ^s by the operator $Q_{(n_o, n_i)}$ applied to

$$u^\infty_{(n_o, n_i)} = \left(u^\infty(d_j, d_k)\right)_{j=1,..,n_0, k=1,...,n_i} \in L^2(\Omega_{n_o} \times \Omega_{n_i})$$

to derive an estimate for the Hausdorff distance $d(D_1, D_2)$ between the domains of two scatterers $\mathcal{D}_1$ and $\mathcal{D}_2$ in $\mathcal{C}$.

THEOREM 4.1.9 (ϵ-**Stability**.) *Let $u^\infty_1(\hat{x}, d)$ and $u^\infty_2(\hat{x}, d)$, $\hat{x}, d \in \Omega$, be the far field patterns for scattering of acoustic plane waves from two scatterers $\mathcal{D}_1, \mathcal{D}_2$ in $\mathcal{C}$. Given $\epsilon > 0$ there are even integers $n_o, n_i \in I\!N$ such that, if for a nonnegative parameter δ the far field patterns satisfy*

$$\left\|u^\infty_{1(n_o, n_i)} - u^\infty_{2(n_o, n_i)}\right\|_{L^2(\Omega_{n_o} \times \Omega_{n_i})} \leq \delta, \qquad (4.1.52)$$

the Hausdorff distance $d(D_1, D_2)$ of the two scatterers can be estimated by

$$d(D_1, D_2) \leq F_{(n_o, n_i)}(\delta). \qquad (4.1.53)$$

Proof. Let $\mu \in I\!N_0$ be given according to the situations S1 to S6 defined in Table 3.1. For $\epsilon > 0$ following Lemma 4.1.8 there are even integers $n_o, n_i \in I\!N$, such that the function $F_{(n_o, n_i)}$ satisfies (4.1.47). We now keep n_i, n_o fixed and apply Theorem 4.1.2 to each choice of $\rho, \tau, \eta > 0$, for which the condition (4.1.20) is satisfied. Then for a function $p : B \to \Omega$ and $q \in \Omega$ an application of Theorem 4.1.4 to the scatterer $\mathcal{D}_j$, $j = 1, 2$, yields

$$\left|\Phi^s_{j,\mu,q}(x, z) - \left(Q_{(n_o, n_i)} u^\infty_{j,(n_o, n_i)}\right)(x, z)\right| \leq c\,\frac{\eta}{\rho^{m-1}} + C\, b_{n_i, \eta, \rho}\, \tau \quad (4.1.54)$$

for all $x, z \in B$, for which (3.1.34) is satisfied. We use (4.1.54) for $\mathcal{D}_1$ and $\mathcal{D}_2$ and the Cauchy-Schwarz inequality applied to

$$Q(u^\infty_{1,\mu,q} - u^\infty_{2,\mu,q})$$

to derive from

$$\left|\Phi_{1,\mu,q}^s(x,z) - \Phi_{2,\mu,q}^s(x,z)\right| \leq \left|\Phi_{1,\mu,q}^s(x,z) - \left(Qu_1^\infty\right)(x,z)\right| \quad (4.1.55)$$
$$+ \left|Q\left(u_1^\infty - u_2^\infty\right)(x,z)\right| + \left|\left(Qu_1^\infty\right)(x,z) - \Phi_{2,\mu,q}^s(x,z)\right|$$

the estimate

$$\left|\Phi_{1,\mu,q}^s(x,z) - \Phi_{2,\mu,q}^s(x,z)\right| \leq F_2(n_o, n_i, \rho, \tau, \eta, \delta) \quad (4.1.56)$$

for all $q \in \Omega$ and all points $x, z \in B \setminus \left(D_{1,\rho} \cup D_{2,\rho}\right)$, for which cones $co(x, p_x, \beta_0)$ and $co(z, p_z, \beta_0)$, $p_x, p_z \in \Omega$, in the exterior of $D_{1,\rho} \cup D_{2,\rho}$ exist. Now we apply Lemma 3.1.4 to obtain

$$d(D_1, D_2) \leq F_1(\rho, F_2(n_o, n_i, \rho, \tau, \eta, \delta)). \quad (4.1.57)$$

Since this is valid for each choice of the parameters ρ, τ, η satisfying (4.1.20), we obtain the estimate (4.1.53). □

A second approach to ϵ-stability and explicit estimates. We now come to the second approach to prove the estimates of ϵ-stability. For Ω_n given by (4.1.2) we know that

$$\tau(n) := \max_{x \in \Omega} \min_{j=1,..,n} d(x, d_j) \to 0, \quad n \to \infty. \quad (4.1.58)$$

We will prove Theorem 4.1.9 with the function $F_{(n_o, n_i)}$ replaced by

$$F_{(n_o, n_i)}(\delta) := F\left(16\pi^2\left(\pi C_\infty(\tau(n_o) + \tau(n_i)) + \delta\right)\right), \quad \delta \geq 0, \quad (4.1.59)$$

where F is given by (3.1.48).

From Theorem 2.1.14 for impenetrable scatterers and Theorem 2.2.7 for the inhomogeneous medium scatterer we obtain that the set of far field patterns $u^\infty(\cdot, d) \in C^1(\Omega)$ is bounded uniformly for $d \in \Omega$ and scatterers $\mathcal{D} \in \mathcal{C}$. By the same arguments this can be seen to be true for the surface gradient $\text{Grad}_d\, u^\infty(\cdot, d)$. Thus the functions $u^\infty(\cdot, \cdot)$ are bounded in $C^1(\Omega \times \Omega)$ by a constant C_∞ uniformly for all scatterers $\mathcal{D} \in \mathcal{C}$. Using the mean value theorem we have for $n_o, n_i \in I\!N$ and $j = 1, 2$

$$\left|u_j^\infty(\hat{x}, \hat{y}) - u_j^\infty(d_l, d_k)\right| \leq \pi \|u_j^\infty\|_{C^1(\Omega \times \Omega)}\left(d(\hat{x}, d_l) + d(\hat{y}, d_k)\right). \quad (4.1.60)$$

We use a disjoint partition $\Omega = \bigcup_{j=1,\dots,n} V_j$ of the unit sphere with $d_j \in V_j$ and

$$\overline{V_j} = \{\hat{x} \in \Omega, d(\hat{x}, d_j) \le d(\hat{x}, d_l)\ \forall\, l \ne j\}. \tag{4.1.61}$$

Then by $2ab \le a^2 + b^2$ and

$$\begin{aligned}(a+b+c)^2 &= a^2 + b^2 + c^2 + 2ab + 2ac + 2bc \\ &\le 3(a^2 + b^2 + c^2)\end{aligned}$$

we estimate for each $k = 1, \dots, n_i, l = 1, \dots, n_o$

$$\begin{aligned}&\left\|u_1^\infty(\hat{x}, \hat{y}) - u_2^\infty(\hat{x}, \hat{y})\right\|_{L^2(V_l \times V_k)}^2 \\ &\le 3\Big(\left\|u_1^\infty(\hat{x}, \hat{y}) - u_1^\infty(d_l, d_k)\right\|_{L^2(V_l \times V_k)}^2 \\ &\quad + \left\|u_1^\infty(d_l, d_k) - u_2^\infty(d_l, d_k)\right\|_{L^2(V_l \times V_k)}^2 \\ &\quad + \left\|u_2^\infty(d_l, d_k) - u_2^\infty(\hat{x}, \hat{y})\right\|_{L^2(V_l \times V_k)}^2\Big).\end{aligned} \tag{4.1.62}$$

We also note that we have

$$\begin{aligned}&\sum_{k,l=1}^{n_i,n_o} \left\|u_1^\infty(d_l, d_k) - u_2^\infty(d_l, d_k)\right\|_{L^2(V_l \times V_k)}^2 \\ &\le \sum_{k,l=1}^{n_i,n_o} \left|u_1^\infty(d_l, d_k) - u_2^\infty(d_l, d_k)\right|^2 \int_{V_l \times V_k} ds\, ds \\ &\le 4\delta^2.\end{aligned} \tag{4.1.63}$$

From (4.1.60) we derive

$$\begin{aligned}&\sum_{k,l=1}^{n_i,n_o} \left\|u_1^\infty(\hat{x}, \hat{y}) - u_1^\infty(d_l, d_k)\right\|_{L^2(V_l \times V_k)}^2 \\ &\le c_m^2 \left(\pi C_\infty(\tau(n_i) + \tau(n_o))\right)^2.\end{aligned} \tag{4.1.64}$$

With the help of (4.1.63) and (4.1.64) using $c_m \le 4\pi$ and $\sqrt{a^2 + b^2} \le a + b$ from (4.1.62) we now derive

$$\begin{aligned}\left\|u_1^\infty - u_2^\infty\right\|_{L^2(\Omega \times \Omega)} &= \Big(\sum_{k=1}^{n_i}\sum_{l=1}^{n_o} \left\|u_1^\infty - u_2^\infty\right\|_{L^2(V_l \times V_k)}^2\Big)^{1/2} \\ &\le \Big(6c_m^2(\pi C_\infty(\tau(n_i) + \tau(n_o))^2 + 12\delta^2\Big)^{1/2} \\ &\le 16\pi^2(\pi C_\infty(\tau(n_i) + \tau(n_o)) + \delta),\end{aligned} \tag{4.1.65}$$

where we did not use the best possible estimate for the square roots and sums under consideration, but some rough upper bound. Given $\epsilon > 0$ because of (4.1.58) and the behavior (3.1.52) of F it is possible to choose $n \in \mathbb{N}$ such that

$$F\Big(16\pi^3 C_\infty(\tau(n) + \tau(n))\Big) \leq \epsilon.$$

Then for $F_{(n_0, n_i)}$ defined by (4.1.59) we calculate (4.1.47). Finally, from (4.1.65) and Theorem 3.1.11 we derive (4.1.53) with $F_{(n_o, n_i)}$ defined by (4.1.59).

It is an important question to explicitly determine the behavior of $F_{(n_o, n_i)}(\delta)$ for $\delta \to 0$ and for $n_i, n_0 \to \infty$. For the reconstruction of the convex hull $\mathcal{H}(\mathcal{D})$ of a scatterer $\mathcal{D} \in \mathcal{D}$ we will now answer this question.

THEOREM 4.1.10 *Let* $u_1^\infty(\hat{x}, d)$ *and* $u_2^\infty(\hat{x}, d)$, $\hat{x}, d \in \Omega$, *be the far field patterns for scattering of acoustic plane waves from two scatterers* $\mathcal{D}_1, \mathcal{D}_2 \in \mathcal{C}$. *Given* $\epsilon > 0$ *there is* $n_o, n_i \in \mathbb{N}$ *such that, if for a nonnegative parameter* δ *the far field patterns satisfy*

$$\left\| u_{1(n_o, n_i)}^\infty - u_{2(n_o, n_i)}^\infty \right\|_{L^2(\Omega_{n_o} \times \Omega_{n_i})} \leq \delta, \qquad (4.1.66)$$

the Hausdorff distance $d(\mathcal{H}(\mathcal{D}_1), \mathcal{H}(\mathcal{D}_2))$ *of the convex hulls of two scatterers* $\mathcal{D}_1, \mathcal{D}_2 \in \mathcal{C}$ *can be estimated by*

$$d(\mathcal{H}(\mathcal{D}_1), \mathcal{H}(\mathcal{D}_2)) \leq \frac{C}{\left| \ln \left[16\pi^2 \Big(\pi C_\infty(\tau(n_o) + \tau(n_i)) + \delta \Big) \right] \right|^c} \qquad (4.1.67)$$

with constants $C > 0$, $0 < c < 1$ *and* C_∞.

Proof. We proceed analogously to the second proof of Theorem 4.1.9, where the role of Theorem 3.1.11 has to be replaced by Theorem 3.1.12.

□

4.2   Inverse electromagnetic scattering

For electromagnetic scattering to obtain uniqueness and ϵ-uniqueness we can in principle proceed in the same way as for acoustic scattering. For diversity here we use a second approach to ϵ-uniqueness via the results of stability. The following theorem includes scattering from a perfect conductor or from an inhomogeneous electromagnetic medium.

THEOREM 4.2.1 (ϵ-uniqueness.) *Let* $E_{pl,1}^\infty(\hat{x}, d, q)$ *and* $E_{pl,2}^\infty(\hat{x}, d, q)$ *be the electric far field patterns for scattering of plane waves from two electromagnetic scatterers* $\mathcal{D}_1, \mathcal{D}_2 \in \mathcal{C}$. *Given* $\epsilon > 0$ *there is* $n_o, n_i, n_{pol} \in \mathbb{N}$, *such that*

$$E_{pl,1}^\infty(\hat{x}, d, q) = E_{pl,2}^\infty(\hat{x}, d, q), \quad \hat{x} \in \Omega_{n_o}, \, d \in \Omega_{n_i}, \, q \in \Omega_{n_{pol}} \qquad (4.2.1)$$

yields the estimate

$$d(\mathcal{D}_1, \mathcal{D}_2) \leq \epsilon \qquad (4.2.2)$$

for the Hausdorff distance $d(\mathcal{D}_1, \mathcal{D}_2)$ *of* $\mathcal{D}_1$ *and* $\mathcal{D}_2$.

Proof. We base our proof on the stability estimates. A stability estimate consists of a function $F : \mathbb{R}^+ \to \mathbb{R}^+$

$$F(\delta) \to 0, \quad \delta \to 0, \qquad (4.2.3)$$

such that for two scatterers $\mathcal{D}_1$ and $\mathcal{D}_2$ in $\mathcal{C}$ with electric far field patterns $E_{pl,1}^\infty(\hat{x}, d, q)$ and $E_{pl,2}^\infty(\hat{x}, d, q)$, respectively, the estimate

$$\left\| E_{pl,1}^\infty(\cdot, \cdot, q) - E_{pl,2}^\infty(\cdot, \cdot, q) \right\|_{L^2(\Omega \times \Omega)} \leq \delta, \quad q \in \Omega, \qquad (4.2.4)$$

yields

$$d(\mathcal{D}_1, \mathcal{D}_2) \leq F(\delta). \qquad (4.2.5)$$

From Theorem 2.3.11 for the perfect conductor and Theorem 2.4.7 for the inhomogeneous electromagnetic medium we obtain bounds for the far field patterns $E_{pl,j}^\infty(\cdot, \cdot, \cdot)$, $j = 1, 2$, in $C^1(\Omega \times \Omega \times \Omega)$ uniformly for scatterers $\mathcal{D}_j \in \mathcal{C}$. Thus given $\epsilon > 0$ and δ with $F(\delta) = \epsilon$, we can find n_o, n_i and n_{pol} in $\mathbb{N}$ such that

$$E_{pl,1}^\infty(\hat{x}, d, q) = E_{pl,2}^\infty(\hat{x}, d, q), \quad \hat{x} \in \Omega_{n_o}, \, d \in \Omega_{n_i}, \, q \in \Omega_{n_{pol}} \qquad (4.2.6)$$

yields

$$\left\| E_{pl,1}^\infty(\cdot, \cdot, q) - E_{pl,2}^\infty(\cdot, \cdot, q) \right\|_{L^2(\Omega \times \Omega)} \leq \delta \qquad (4.2.7)$$

for all $q \in \Omega$. Now from (4.2.7) and (4.2.5) we obtain (4.2.2) and the proof is complete. $\square$

The brevity of the preceding proof compared to the lengthy estimates for the acoustic case may be surprising at a first glance. Of course, the main work to obtain estimates has to be done at some place. Here it is hidden in the stability estimate (4.2.5), proven in the preceding chapter.

In a similar way it is possible to derive the results of ϵ-stability for the above inverse electromagnetic scattering problems. To avoid repetitions we omit the presentation.

5 The point-source method and applications

In the preceding sections we used point-sources to obtain uniqueness and stability results for inverse scattering problems. We will now investigate the reconstruction of the domain D of scatterers $\mathcal{D}$ from the algorithmical and numerical viewpoint.

As a starting point we develop a point-source method to reconstruct the scattered field $u^s(z, d)$ in the exterior of a scatterer $\mathcal{D} \in \mathcal{C}$. More explicitly, given a scatterer $\mathcal{D} \in \mathcal{C}$ and $\rho, \tau > 0$ we will construct a kernel

$$g_\tau(z, \hat{x}), \quad z \in B, \quad \hat{x} \in \Omega,$$

such that the reconstruction of the scattered field $u^s(z, d)$ by

$$(Au^\infty(\cdot, d))(z) := \int_\Omega g_\tau(z, \hat{x}) u^\infty(\hat{x}, d) ds(\hat{x}), \quad z \in B \setminus D_\rho$$

satisfies the error estimate

$$\left\| u^s(\cdot, d) - Au^\infty(\cdot, d) \right\|_{L^\infty(B \setminus D_\rho)} \leq \tau. \tag{5.0.1}$$

Analogous estimates will be derived for electromagnetic scattering from a perfect conductor or an inhomogeneous electromagnetic medium. We will prove the convergence of the point-source method to reconstruct u^s, i.e., given a family of measured far field patterns u_δ^∞, $\delta \in (0, 1)$, with

$$\|u^\infty(\cdot, d) - u_\delta^\infty\|_{L^2(\Omega)} \leq \delta$$

we will show that we can choose $\tau = \tau(\delta)$ such that

$$\left\| u^s(\cdot, d) - Au_\delta^\infty \right\|_{L^\infty(B \setminus D_\rho)} \to 0, \quad \delta \to 0.$$

The operator A is a kind of backprojection operator as used for the Backus-Gilbert or mollifier methods; see [16], [43], and [59].

In a second step we will use the scattered field u^s, the known incident field u^i and the boundary conditions of an impenetrable scatterer $\mathcal{D}$ for reconstructions of the domain D of $\mathcal{D}$. Numerical examples are provided for acoustic scattering in three dimensions.

A different approach to the point-source method, which does not use the reciprocity relations, has been developed in [74], [75] and [76].

5.1   Reconstruction of acoustic scatterers

The central idea of the point-source method is the reconstruction of the scattered field u^s using a backprojection operator A. The main ingredients for the definition of the operator and the derivation of the error estimates of Theorem 5.1.2 are mixed reciprocity relations and the superposition of point-sources by plane waves. Please note that the reconstruction of the scattered field is obtained for scatterers with different physical properties (impenetrable or penetrable scatterers) in a uniform way.

DEFINITION 5.1.1 *For* $\mu = 0, s = 1$, $\rho, \tau > 0$ *and a function* $p : B \to \Omega$ *let the density* $g_\tau(z, \cdot)$ *be given by (3.1.32). We define the operator*

$$A : L^2(\Omega) \to L^\infty(B)$$

by

$$(Aw)(z) := \frac{1}{\gamma_m} \int_\Omega g_\tau(z, d) w(-d) \, ds(d), \quad z \in B. \qquad (5.1.1)$$

Error estimates for reconstructions of u^s. The error for the reconstruction of a scattered field $u^s(\cdot, d)$ from its far field pattern $u^\infty(\cdot, d)$ is estimated in the following theorem.

THEOREM 5.1.2 *Consider scattering by a sound-soft, sound-hard or inhomogeneous medium scatterer* $\mathcal{D} \in \mathcal{C}$. *For* $\mu = 0, s = 1$, $\rho, \tau > 0$ *and a vector field* $p : B \to \Omega$ *with*

$$D \subset G_{x,p(x),\rho}, \quad x \in B \setminus D_\rho, \qquad (5.1.2)$$

let the operator A be given by (5.1.1). Then for fixed $d \in \Omega$ and a measured far field pattern $u_\delta^\infty(\cdot, d)$ with

$$\left\| u^\infty(\cdot, d) - u_\delta^\infty(\cdot, d) \right\|_{L^2(\Omega)} \leq \delta \qquad (5.1.3)$$

the error for the approximation of $u^s(\cdot, d)$ by $Au_\delta^\infty(\cdot, d)$ satisfies

$$\left| u^s(z, d) - (Au_\delta^\infty(\cdot, d))(z) \right| \leq c\tau + Cb_{\tau,\rho}\, \delta, \quad z \in B \setminus D_\rho, \qquad (5.1.4)$$

with the constant $b_{\tau,\rho}$ defined by (3.1.33) and constants $c, C > 0$ uniformly for scatterers $\mathcal{D} \in \mathcal{C}$.

Remark. The elements of the following proof have already been used to investigate the reconstruction properties of the operator Q.

Proof. First, for $\tau > 0$ and $z \in B \setminus D_\rho$ by definition of g we have

$$\left\| \Phi_{\mu,q}(\cdot, z) - H g_\tau(z, \cdot) \right\|_{C^1(D)} \leq \tau. \tag{5.1.5}$$

The scattering map maps the incident field in $C^1(\overline{D})$ boundedly onto the far field pattern of the scattered field in $L^2(\Omega)$. Thus we obtain a constant c such that

$$\left\| \Phi^\infty(\cdot, z) - H^\infty g_\tau(z, \cdot) \right\|_{C(\Omega)} \leq c\tau. \tag{5.1.6}$$

We use the mixed reciprocity relations (2.1.4) and (2.2.4) to transform (5.1.6) into

$$\left| u^s(z, d) - \frac{1}{\gamma_m} \Big(H^\infty g_\tau(z, \cdot) \Big)(-d) \right| \leq \frac{c}{\gamma_m} \tau \tag{5.1.7}$$

for all $d \in \Omega$. From the decomposition

$$\left| u^s(z, d) - \Big(A u_\delta^\infty(\cdot, d) \Big)(z) \right|$$
$$\leq \left| u^s(z, d) - \frac{1}{\gamma_m} \Big(H^\infty g_\tau(z, \cdot) \Big)(-d) \right| \tag{5.1.8}$$
$$+ \left| \frac{1}{\gamma_m} \int_\Omega g_\tau(z, \tilde{d}) \Big\{ u^\infty(-\tilde{d}, d) - u_\delta^\infty(-\tilde{d}, d) \Big\} \, ds(\tilde{d}) \right|$$

with the help of the Cauchy-Schwarz inequality we derive (5.1.4). $\qquad \square$

We use the operator A to reconstruct the field $u^s(\cdot, d)$ from a measured far field pattern u_δ^∞. Given the error δ in the measurements, the error of the reconstructions is given by (5.1.4). For a family of measurements u_δ^∞, $\delta \in (0, 1)$, with

$$\| u_\delta^\infty - u^\infty(\cdot, d) \|_{L^2(\Omega)} \to 0, \ \delta \to 0,$$

by an application of Lemma 3.1.8 there is a function h with

$$h(\delta) \to 0, \ \delta \to 0,$$

such that A defined with $\tau = h(\delta)$ satisfies

$$\left\| u^s(\cdot, d) - A u_\delta^\infty \right\|_{C(B \setminus D_\rho)} \to 0, \ \delta \to 0.$$

We have proven the convergence of the point-source method for the reconstruction of the scattered field $u^s(\cdot, d)$.

Efficient calculation of A. The kernel $g_\tau(z, \cdot)$ of the operator A can be calculated according to the *a priori* knowledge about the unknown scatterers $\mathcal{D}$ given by the class $\mathcal{C}$ of scatterers (see Definitions 2.1.5 and 2.2.5). In a first step the set of densities $\mathcal{G}$ has to be computed according to Lemma 3.1.3 . Then $g_\tau(z, \cdot)$ is given b y (3.1.30). Clearly the computation of the function $g_0 \in L^2(\Omega)$ with

$$\left\| \Phi(\cdot, 0) - H g_0 \right\|_{C^s(G_{0,p,\rho})} \leq \tau \qquad (5.1.9)$$

for each $p \in \Omega$ and $s = 1$ is not effective from a computational point of view. We now describe a possibility to obtain $g_\tau(z, \cdot)$ by the computation of only one solution $g_0 \in L^2(\Omega)$ of (5.1.9).

LEMMA 5.1.3 *For a vector $p_0 \in \Omega$ let g_0 be a solution of (5.1.9) and let $p_1 \in \Omega$ be given. Then the minimum norm solution g_1 for (5.1.9) with p_0 replac d by p_1 can be obtained from g_0 by r otation*

$$g_1(d) := g_0(M^{-1}d) \qquad (5.1.10)$$

with an orthogonal matrix M which satisfies $p_1 = M p_0$.

Proof. The proof is obtained by a simple rotation of the domain of approximation and the Herglotz wav e function from the fact that the rotated Herglotz wave function

$$\int_\Omega e^{i\kappa M^{-1}x\cdot d} g(d) ds(d) \;=\; \int_\Omega e^{i\kappa x\cdot M d} g(d) ds(d)$$
$$=\; \int_\Omega e^{i\kappa x\cdot \tilde{d}} g(M^{-1}\tilde{d}) ds(\tilde{d}) \qquad (5.1.11)$$

is again a Herglotz wav e function with the rotated density $g(M^{-1}d)$. $\qquad \square$

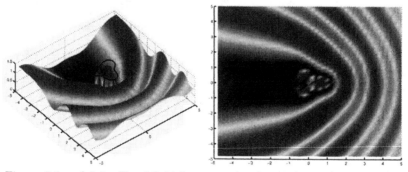

Figure 5.1 and 5.2: Total field for scattering by an obstacle.

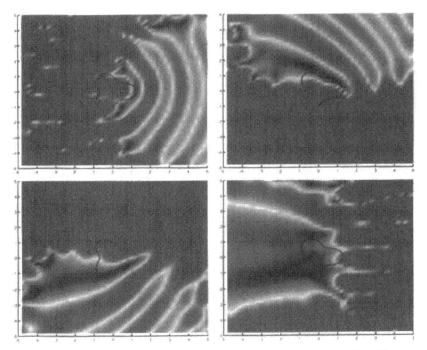

Figures 5.3-5.6: Reconstructions using the point-source method for wave number $\kappa = 2$ with different fixed orientations of the domain of approximation. We see different enlightened areas by the reconstruction method.

Reconstruction of sound-soft scatterers $\mathcal{D}$. We now investigate the reconstruction of an unknown impenetrable scatterer $\mathcal{D}$ from the knowledge of the far field pattern for scattering of a plane wave. As described in Theorem 5.1.2 it is possible to use the operator A to reconstruct $u^s(\cdot, d)$ in the exterior of $\mathcal{D}$.

We first consider the reconstruction of a sound-soft scatterer. In this case we can use the boundary condition (2.1.4) to find the unknown scatterer as a minimum curve of the total field

$$u^s(\cdot, d) + u^i(\cdot, d). \qquad (5.1.12)$$

But so far the construction of the operator A to compute an approximation for u^s is built on the knowledge of D. The construction of the kernel $g_\tau(z, \cdot)$ of A to obtain the approximation (5.1.4) assumes the knowledge of adequate directions $p = p(z)$ such that

$$\overline{D} \subset G_{z, p(z), \rho}, \quad z \in B \setminus D_\rho, \qquad (5.1.13)$$

is satisfied, i.e., it assumes the knowledge of the unknown scatterer. Clearly we do not know the scatterer and a knowledge for the choice of $p(z)$ is not available, when we start the algorithm. This gives rise to different strategies to choose $p(z)$ in a multistep procedure.

We now describe one simple strategy to reconstruct u^s and p in several steps. We start with a number of fixed directions $p(z) = p_\xi$ for $\xi = 1, ..., N$, and compute the corresponding operators A_ξ and the fields

$$\left(A_\xi u_\delta^\infty\right)(z) + u^i(z, d). \tag{5.1.14}$$

For each $\xi = 1, ..., N$ we search for parts of the unknown boundary ∂D of the domain D as the minimum curve of (5.1.14) and obtain a first approximation ∂D_1 to the boundary ∂D. In each further step we adapt the choice of $p(z)$ according to the reconstruction ∂D_n of the n-th step to achieve the condition (5.1.13) for further points $z \in B$ and obtain an approximation ∂D_{n+1}. A stopping criterion is provided by the condition (5.1.13), which has to be satisfied for the current choice of p and the current reconstruction D_n. Efficient adaptive algorithms for the choice of p will have to be a part of further research (see also [21]).

Reconstruction of sound-hard scatterers $\mathcal{D}$. For the sound-hard scatterer we have to modify the approach according to the different boundary condition

$$\frac{\partial}{\partial \nu}\left(u^s(\cdot, d) + u^i(\cdot, d)\right) = 0. \tag{5.1.15}$$

So far, we described the reconstruction of u^s. But to use the normal derivative in (5.1.15) the reconstruction of ∇u^s is needed.

As a consequence of the linearity and boundedness of the scattering operator $S : u^i \mapsto u^\infty$ for scattering of point-sources $\Phi(\cdot, z)$ the operator S and the differentiation ∇_z with respect to the source point can be exchanged. This is a simple consequence of the linearity and boundedness of S. Thus we have

$$\begin{aligned}
\nabla_z \Phi^\infty(-d, z) &= \nabla_z S\left(\Phi(\cdot, z)\right)(-d) \\
&= S\left(\nabla_z \Phi(\cdot, z)\right)(-d) \\
&= -S\left(\nabla \Phi(\cdot, z)\right)(-d).
\end{aligned} \tag{5.1.16}$$

Due to Lemma 3.1.3 there is an approximation of $\nabla \Phi(\cdot, z)$ by a Herglotz wave function

$$\nabla_y \int_\Omega e^{i\kappa y \cdot d} g_\tau(z, d) \, ds(d) = \int_\Omega i\kappa d \, e^{i\kappa y \cdot d} g_\tau(z, d) \, ds(d), \quad y \in B \setminus D_\rho. \tag{5.1.17}$$

We define

$$(A'w)(z) := \frac{1}{\gamma_m} \int_\Omega (-i\kappa d) \, g_\tau(z,d) \, w(-d) \, ds(d), \quad z \in B \setminus D_\rho \quad (5.1.18)$$

and obtain an approximation of $\nabla u^s(\cdot, d)$ by $A'u^\infty(\cdot, d)$ with the error estimate

$$\left| \nabla_z u^s(z,d) - (A'u_\delta^\infty(\cdot, d))(z) \right| \leq c\tau + Cb_{\tau,\rho} \, \delta, \quad (5.1.19)$$

analogous to (5.1.4). For the reconstruction of a scatterer $\mathcal{D}$ with sound-hard boundary condition we can now proceed analogously to the sound-soft case, where the sound-soft boundary condition (5.1.12) has to be replaced by the sound-hard condition (5.1.15).

Numerical examples. In the following we show numerical results by Giebermann and Potthast [21], who used the point-source method for the reconstruction of scatterers in three dimensions. The measured data consist of the far field patterns for six different incident plane waves at 256 observation points.

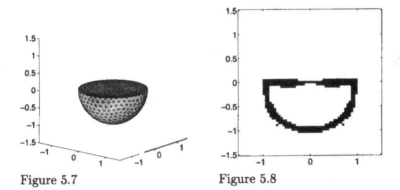

Figure 5.7   Figure 5.8

Figure 5.8 shows a cut of the reconstruction in the x-z-plane in the second second step of the reconstruction algorithm. The first step is given by Figure 5.9 to Figure 5.14, i.e., the minimum curves for different waves and orientations of the domain of approximation. With the direction of incidence d and $p = -d$ we show the minimum curves of $|u_{approx}|$ for $d = (0,0,-1)$ in Figure 5.9, $d = (0,0,1)$ in Figure 5.10, $d = (0,1,0)$ in Figure 5.11, $d = (-1,0,0)$ in Figure 5.12, $d = (0,-1,0)$ in Figure 5.13 and $d = (0,0,1)$ in Figure 5.14.

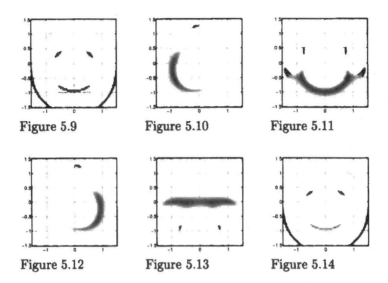

Figure 5.9   Figure 5.10   Figure 5.11

Figure 5.12   Figure 5.13   Figure 5.14

Instead of a further minimization here we built the union of the minimum points and then removed those points from the union, which according to the orientation of the domain of approximation in comparison with the union of minimum points for the six steps could not be part of the boundary of the scatterer. The union of the minimum points and the result of the removal-step can be seen in Figure 5.15, 5.16 and 5.8.

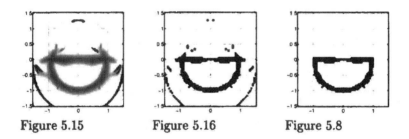

Figure 5.15   Figure 5.16   Figure 5.8

In a second example we want to show that we can reconstruct obstacles which consist out of several separate components. Figure 5.17 shows two balls which are reconstructed with the point-source method. Here we again use six different waves and show the second step, i.e., the union of the minimum points, which is this time shown in a full three-dimensional plot in Figure 5.18 without an application of a removal-step.

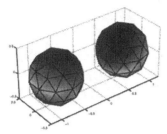

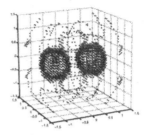

Figure 5.17   Figure 5.18

5.2 Inverse electromagnetic scattering

The operator A defined by (5.1.1) for the reconstruction of the scattered acoustic fields can be used also for the reconstruction of the scattered electromagnetic field $E_{pl}^s(z, d, p)$, $z \in B \setminus D_\rho$, from its far field pattern $E_{pl}^\infty(\cdot, d, p)$. For the electromagnetic case we have to choose the kernel $g_\tau(z, \cdot)$ such that the estimates (3.1.31) or (5.1.9), respectively, are satisfied with $s = 4$. The following theorem investigates not only the perfect conductor, but also an inhomogeneous electromagnetic medium.

THEOREM 5.2.1 *Consider the scattering of electromagnetic waves by a perfect conductor or an inhomogeneous medium scatterer $D \in C$. For $\mu = 0, s = 4$, $\rho, \tau > 0$ and a vector field $p : B \to \Omega$ satisfying (5.1.2) the operator A is defined by (5.1.1). Let the measured far field pattern E_δ^∞ satisfy the data error estimate*

$$\left\| E_{pl}^\infty(\cdot, d, p) - E_\delta^\infty \right\|_{L^2(\Omega, \mathbb{R}^3)} \leq \delta. \qquad (5.2.1)$$

Then the error for the approximation of $E_{pl}^s(\cdot, d, p)$ by AE_δ^∞ is estimated by

$$\left| E^s(z, d, p) - \left(AE_\delta^\infty \right)(z) \right| \leq c\tau + C \|g_0\|_{L^2(\Omega)} \delta, \quad z \in B \setminus D_\rho, \qquad (5.2.2)$$

with constants $c, C > 0$ and g_0 defined by Lemma 5.1.3.

Proof. Given $\rho, \tau > 0$ and $z \in B \setminus D_\rho$ by the definition of g_τ we have

$$\left\| E_{edp}^i(\cdot, z, q) - \int_\Omega E_{pl}^i(\cdot, \tilde{d}, p) g_\tau(z, \tilde{d}) \, ds(\tilde{d}) \right\|_{C^2(\overline{D})}$$

$$= \left\| (q \cdot \nabla) \nabla \left(\Phi(\cdot, z) - \int_\Omega e^{i\kappa \cdot \tilde{d}} g_\tau(z, \tilde{d}) \, ds(\tilde{d}) \right) \right\|_{C^2(\overline{D})}$$

$$\leq \left\| \Phi(\cdot,z) - \int_\Omega e^{i\kappa\cdot\tilde{d}} g_\tau(z,\tilde{d})\, ds(\tilde{d}) \right\|_{C^4(\overline{D})}$$

$$\leq \tau \tag{5.2.3}$$

for all $q \in \Omega$. The mapping of the incident electric field in $C^2(\overline{D})$ onto the far field pattern of the scattered electric field in $C^1(\Omega)$ is bounded uniformly for scatterers $\mathcal{D} \in \mathcal{C}$. Thus there is a constant c such that

$$\left\| E_{edp}^\infty(\cdot,z,q) - \int_\Omega E_{pl}^\infty(\cdot,\tilde{d},p) g_\tau(z,\tilde{d})\, ds(\tilde{d}) \right\|_{C^1(\Omega)} \leq c\,\tau. \tag{5.2.4}$$

With the help of the reciprocity relations (2.3.20), (2.3.22), (2.4.7) and (2.4.8) from (5.2.4) we obtain the estimate

$$\left| E_{pl}^s(z,d,q) - \frac{1}{\gamma}\int_\Omega E_{pl}^\infty(-\tilde{d},d,p) g_\tau(z,\tilde{d})\, ds(\tilde{d}) \right|_{C^1(\Omega)} \leq \frac{c}{\gamma}\,\tau \tag{5.2.5}$$

for $z \in B \setminus D_\rho$. Using the Cauchy-Schwarz inequality we calculate the norm of the operator A and estimate

$$\begin{aligned}
\left| E^s(z,d,p) - \left(AE_\delta^\infty\right)(z) \right| &\leq \left| E^s(z,d,p) - \left(AE_{pl}^\infty(\cdot,d,p)\right)(z) \right| \\
&\quad + \left| A\left(E_{pl}^\infty(\cdot,d,p) - E_\delta^\infty\right)(z) \right| \\
&\leq \frac{c}{\gamma}\,\tau + \frac{4\pi}{\gamma}\,\|g_0\|_{L^2(\Omega)}\delta. \tag{5.2.6}
\end{aligned}$$

This completes the proof.   □

We can use the preceding theorem to formulate a point-source method for the reconstruction of a perfect conductor $\mathcal{D}$ from the knowledge of a measured far field pattern E_δ^∞ for an incident plane wave $E_{pl}^i(\cdot,d,p)$. According to (5.2.2) for an appropriate choice of the parameters ρ,τ and the function $p: B \to \Omega$ the function AE_δ^∞ approximates the scattered field $E_{pl}^s(\cdot,d,p)$ on $B \setminus D_\rho$. We can use the boundary condition (2.3.5) to search for parts Λ of the unknown boundary ∂D of the scatterer $\mathcal{D}$ as a surface where

$$\left| \nu_\Lambda(\cdot) \times \left(E_{pl}^i(\cdot,d,p) + AE_\delta^\infty \right) \right|$$

is small. Here ν_Λ is a function $B \to \Omega$, which coincides with the normal vector ν to Λ on Λ.

We face similar problems as for the reconstruction of sound-hard acoustic scatterers. First, the choice of $g_\tau(z,\cdot)$ involves the orientation $p \in \Omega$ of the domain of approximation $G_{z,p,\rho}$, which has to be chosen such that the unknown scatterer $\mathcal{D}$ is in the interior of $G_{z,p(z),\rho}$. Second, the boundary condition involves the normal vector to the unknown surface.

For the reconstruction of a perfect conductor the strategy, which we suggested for the acoustic inverse scattering problems, can in principle be used to overcome the problems and to reconstruct the scatterer in a multistep procedure.

5.3 Reconstruction of the boundary values of scattered fields

We want to study the reconstruction of the boundary values of a solution to an exterior Dirichlet problem from the far field pattern of the solution. To this end we consider the boundary values given by a point-source $\Phi(\cdot, z)$ with source point z on the boundary ∂D of the domain D. Since in $\mathbb{R}^3$ the singularity of the point-source is $1/r$ the function $\Phi(\cdot, z)$ for $z \in \partial D$ is not in $L^2(\partial D)$, but it is an element of $L^p(\partial D)$ for $p \in (1, 2)$. Thus we need to work in L^p-spaces and collect some basic results. Here we follow [75] and [69].

We consider the single-layer potential with density $\varphi \in L^q(\partial D)$ for $q \in (2, \infty)$ and its far field pattern

$$(S^\infty \varphi)(x) := \int_{\partial D} \Phi_{true}^\infty(x, y)\varphi(y)\, ds(y), \quad x \in \mathbb{R}^m, \tag{5.3.1}$$

and the double-layer potential with density φ with far field pattern

$$(K^\infty \varphi)(\hat{x}) := \int_{\partial D} \frac{\Phi_{true}^\infty}{\partial \nu(y)}\varphi(y)\, ds(y), \quad x \in \mathbb{R}^m, \tag{5.3.2}$$

where $\Phi_{true}^\infty(\cdot, z)$ denotes the far field pattern of the point-source $\Phi(\cdot, z)$ itself, in contrast to the far field pattern $\Phi^\infty(\cdot, z)$ of the scattered field for an incident point-source.

The Herglotz operator $H : L^2(\Omega) \to L^p(\partial D)$, $p \in (1, 2)$ or $H : L^2(\Omega) \to C(\partial D)$ is defined by (3.1.3). We briefly summarize some basic properties of the single- and double-layer operators in L^p-spaces.

LEMMA 5.3.1 *The single- and double-layer potential operators are bounded compact mappings in $L^p(\partial D)$ for $p \in (1, \infty)$.*

Proof. For $p = 2$ the results may be derived using the theorem of Lax and are well known; see for example [8]. For $p \in (2, \infty)$ and with

$$\frac{1}{p} + \frac{1}{q} = 1, \tag{5.3.3}$$

from the Hölder inequality we immediately obtain

$$\left\| \int_{\partial D} k(\cdot, y)\varphi(y) ds(y)\right\|_p \leq \sup_{x \in \partial D} \|k(x, \cdot)\|_q \|\varphi\|_p, \qquad (5.3.4)$$

where the weakly singular kernel $k(x, \cdot)$ is in $L^q(\partial D)$ for each $x \in \partial D$ since $q \in (1,2)$. For $p \in (1,2)$ with q given by (5.3.3) we use the identity

$$\|w(\cdot)\|_p = \sup_{\psi \in L^q, \|\psi\|_q = 1} \left| \int_{\partial D} w(y)\psi(y) ds(y)\right|, \qquad (5.3.5)$$

applied to $w(x) = \int_{\partial D} k(x, y)\varphi(y) ds(y)$ with a weakly singular kernel k to estimate

$$\left\| \int_{\partial D} k(\cdot, y)\varphi(y) ds(y)\right\|_p$$

$$\leq \sup_{\psi \in L^q, \|\psi\|_q = 1} \left| \int_{\partial D} \int_{\partial D} k(x, y)\varphi(y) ds(y)\psi(x) ds(x)\right|$$

$$= \sup_{\psi \in L^q, \|\psi\|_q = 1} \left| \int_{\partial D} \left(\int_{\partial D} k(x, y)\psi(x) ds(x)\right) \varphi(y) ds(y)\right|$$

$$\leq |\partial D|^{1/q} \sup_{\psi \in L^q, \|\psi\|_q = 1} \sup_{y \in \partial D} \|k(\cdot, y)\|_p \cdot \|\psi\|_q \|\varphi\|_p$$

$$= |\partial D|^{1/q} \sup_{y \in \partial D} \|k(\cdot, y)\|_p \cdot \|\varphi\|_p, \qquad (5.3.6)$$

where we used the Hölder inequality twice. This proves the boundedness of the operators in the L^p-spaces. The operators are compact since they are limits of finite dimensional operators which can be constructed from finite dimensional approximations of the kernel $\Phi(x, y)$ in $L^p(\partial D)$. □

We next note that the jump-relations are satisfied in an L^p-sense, i.e., we have

$$\lim_{h \to 0} \int_{\partial D} \left|(S\varphi)(x \pm h\nu(x)) - (S\varphi)(x)\right|^p ds(x) = 0, \qquad (5.3.7)$$

$$\lim_{h \to 0} \int_{\partial D} \left|\frac{\partial(S\varphi)}{\partial\nu(x)}(x \pm h\nu(x)) - (K^*\varphi)(x) \pm \varphi(x)\right|^p ds(x) = 0 \qquad (5.3.8)$$

for the single-layer potential and

$$\lim_{h \to 0} \int_{\partial D} \left|(K\varphi)(x \pm h\nu(x)) - (K\varphi)(x) \mp \varphi(x)\right|^p ds(x) = 0 \qquad (5.3.9)$$

for the double-layer potential. A proof is obtained in the same way as in (5.3.6).

For the solution of the exterior Dirichlet problem we use a combined single- and double-layer approach (2.1.13). The combined potential (2.1.13) solves the exterior Dirichlet problem with boundary values f, if the density φ is a solution to the integral equation

$$\varphi + K\varphi - iS\varphi = 2f. \qquad (5.3.10)$$

We have stated in Theorem 2.1.2 that the integral equation has a unique solution $\varphi \in C(\partial D)$. We now extend the analysis to densities $f \in L^p(\partial D)$ for $p \in (1,2)$.

THEOREM 5.3.2 *The exterior Dirichlet problem for the Helmholtz equation with boundary values $f \in L^p(\partial D)$, $p \in (1,2)$, and boundary condition in the L^p-sense has a unique solution and the solution depends continuously on the boundary data with respect to uniform convergence of the solution on closed subsets of $\mathbb{R}^m \setminus \overline{D}$.*

Proof. We apply the Fredholm Alternative Theorem in the dual systems $\langle C(\partial D), L^q(\partial D) \rangle$ and then in the dual system $\langle L^p(\partial D), L^q(\partial D) \rangle$, both equipped with the canonical duality pairing

$$(\varphi, \psi) \;=\; \int_{\partial D} \varphi(x)\overline{\psi(x)}\, ds(x) \qquad (5.3.11)$$

and q is chosen according to (5.3.3). □

We need to study the denseness of the Herglotz wave operator defined by (3.1.3) from $L^2(\partial D)$ into $L^p(\partial D)$.

LEMMA 5.3.3 *Assume that the homogeneous interior Dirichlet problem for the domain D has only the trivial solution. Then the Herglotz wave operator*

$$H : L^2(\Omega) \to L^p(\partial D)$$

has dense range in $L^p(\partial D)$ for each $p \in (1,2)$.

Proof. To prove denseness of H we need to show that

$$\int_{\partial D} (H\varphi)(y)\psi(y)\, ds(y) = 0 \text{ for all } \varphi \in L^2(\Omega) \qquad (5.3.12)$$

implies $\psi = 0$ for $\psi \in L^q(\partial D)$ where q satisfies (5.3.3). The adjoint of the operator H is given by (3.1.7). We consider $\psi \in L^q(\partial D)$ and assume that (5.3.12) is satisfied. Then we obtain $H^*\psi = 0$ from

$$0 \;=\; \int_{\partial D} (H\varphi)(y)\psi(y)\, ds(y) \;=\; \int_{\Omega} \varphi(d)(H^*\psi)(d)\, ds(d)$$

for all $\varphi \in L^2(\Omega)$. Because of $H^*\psi = S^\infty \psi$, the single-layer potential $S\psi$ with density ψ has vanishing farfield pattern $S^\infty \psi = 0$. We conclude $S\psi = 0$ in $I\!\!R^m \setminus \overline{D}$ using the Rellich lemma. Since we have the imbedding

$$L^q(\partial D) \subset L^2(\partial D) \tag{5.3.13}$$

for $2 < q$ and the norm estimate

$$\frac{1}{|\partial D|^{1/2}} \|\varphi\|_{L^2(\partial D)} \leq \frac{1}{|\partial D|^{1/q}} \|\varphi\|_{L^q(\partial D)}, \tag{5.3.14}$$

which is valid for bounded domains D, we might use jump relations for densities $\psi \in L^2(\partial D)$ to derive $(I - K^*)\psi = 0$ on ∂D. Using the Fredholm Alternative Theorem it is straightforward to show that the nullspaces of $I - K^*$ in $C(\partial D)$ and $L^2(\partial D)$ coincide. Thus we obtain $\psi \in C(\partial D)$. Now, by the continuity of the single-layer potential we argue that the function $S\psi$ solves the interior Dirichlet problem for the domain D with homogeneous boundary conditions and, since this problem has only the trivial solution, vanishes identically. As a last step we again use the jump relations for the derivative of the single-layer potential and obtain $\psi = 0$ on ∂D. This concludes our proof.   □

The reconstruction of boundary values from the far field pattern. We now investigate the reconstruction of the boundary values of a solution to the exterior Dirichlet problem from the knowledge of the far field pattern of the solution, i.e., the inversion of the operator $\mathcal{F}$ which maps f onto u^∞. We will study four methods: the point-source method, minimum norm solutions, the Tikhonov regularization, and the spectral cut-off for the inversion of $\mathcal{F}$. The results will be needed in Section 5.4 as regularization procedures for Newton's method. But especially the bounds obtained by the point-source method are of interest for themselves and build a more general tool for the investigation of other than Newton's method; see for example the results in [70] on stability.

The point-source method. For the reconstruction of the boundary values $f \in C^{1,\alpha}(\partial D)$ of a solution u^s to the exterior Dirichlet problem for a domain D from its far field pattern u^∞ we will use the following approximation procedure consisting of seven steps. For simplicity we first assume that κ^2 is not an interior Dirichlet eigenvalue for the domain D. Later remarks will show how this condition may be removed.

Step 1. First we approximate the fundamental solution $\Phi(\cdot, z) \in L^p(\partial D)$, $p \in (1, 2)$, for a source point $z \in \partial D$ by a superposition of plane waves.

This is possible because of the denseness as stated in Lemma 5.3.3. We use the notation

$$\Phi(x, z) \approx \int_\Omega e^{i\kappa z \cdot d} g(z, d) \, ds(d) \tag{5.3.15}$$

as abbreviation for the statement: for all $\epsilon > 0$ and $z \in \partial D$ there is a density $g(z, \cdot) \in L^2(\Omega)$ such that

$$\left\| \Phi(x, z) - \int_\Omega e^{i\kappa z \cdot d} g(z, d) \, ds(d) \right\|_p \leq \epsilon. \tag{5.3.16}$$

Step 2. We consider the exterior Dirichlet problem with boundary values in $L^p(\partial D)$ to obtain a corresponding estimate for the solution to the boundary value problem, i.e., we derive

$$\Phi^s(x, z) \approx \int_\Omega u^s(x, d) g(z, d) \, ds(d). \tag{5.3.17}$$

Step 3. We use the mixed reciprocity relation (2.1.4) to obtain

$$\Phi^s(x, z) \approx \frac{1}{\gamma_m} \int_\Omega \Phi^\infty(-d, x) g(z, d) \, ds(d), \tag{5.3.18}$$

in the sense of (5.3.16) for $z \in \partial D$, where the constant γ_m depends on the dimension m and is given by (2.1.18).

Step 4. We now represent the boundary values $f \in C^{1,\alpha}(\partial D)$ by a single-layer representation

$$
\begin{aligned}
f(z) &= \int_{\partial D} \Phi(z, y) \varphi(y) ds(y) \\
&= \int_{\partial D} \Phi^s(z, y) \varphi(y) ds(y), \quad z \in \partial D,
\end{aligned}
\tag{5.3.19}
$$

with density $\varphi \in C^{0,\alpha}(\partial D)$. By Theorems 3.30 and 2.23 and equation (3.45) of [7] for two times continuously differentiable f the integral equation

$$S\varphi = f \tag{5.3.20}$$

has a unique solution

$$\varphi = -T(I - K)^{-1}(I + K)^{-1} f \in C^{0,\alpha}(\partial D) \tag{5.3.21}$$

which depends continuously on the $C^{1,\alpha}$-norm of f.

Step 5. Using the reciprocity relation $\Phi^s(x, y) = \Phi^s(y, x)$ we transform the representation (5.3.19) into

$$f(z) = \int_{\partial D} \Phi^s(y, z)\varphi(y)ds(y), \quad z \in \partial D. \qquad (5.3.22)$$

Step 6. From (5.3.19) we conclude that the far field pattern u^∞ of the solution u^s to the exterior Dirichlet problem with boundary function f is given by

$$u^\infty(\hat{x}) = \int_{\partial D} \Phi^\infty(\hat{x}, y)\varphi(y)ds(y), \quad \hat{x} \in \Omega. \qquad (5.3.23)$$

Step 7. We insert the approximation (5.3.18) into (5.3.22) to derive

$$\begin{aligned}
f(z) &\approx \frac{1}{\gamma_m} \int_{\partial D} \int_{\Omega} \Phi^\infty(-d, y)g(z, d)\varphi(y) \, ds(d) \, ds(y) \\
&= \frac{1}{\gamma_m} \int_{\Omega} \left(\int_{\partial D} \Phi^\infty(-d, y)\varphi(y) \, ds(y) \right) g(z, d) \, ds(d) \\
&= \frac{1}{\gamma_m} \int_{\Omega} u^\infty(\hat{x})g(z, d) \, ds(d), \qquad (5.3.24)
\end{aligned}$$

in the sense that

$$\left| f(z) - \frac{1}{\gamma_m} \int_{\Omega} u^\infty(\hat{x})g(z, d) \, ds(d) \right| \leq C\|\varphi\|_q \epsilon \qquad (5.3.25)$$

with some constant C, where q is given by (5.3.3). The estimate (5.3.25) is obtained as indicated by (5.3.24) from (5.3.18) by

$$\begin{aligned}
&\left| f(z) - \frac{1}{\gamma_m} \int_{\Omega} u^\infty(\hat{x})g(z, d) \, ds(d) \right| \\
&\leq \left| \int_{\partial D} \left\{ \Phi^s(y, z) - \int_{\Omega} u^s(x, d)g(z, d) \, ds(d) \right\} \varphi(y)ds(y) \right| \\
&\leq C\epsilon\|\varphi\|_q. \qquad (5.3.26)
\end{aligned}$$

We define the backprojection operator A_ϵ by (5.1.1). This operator maps the far field pattern of a solution to the exterior Dirichlet problem onto the boundary values of the scattered field. We use the notation S_q^{-1} for the operator which maps the boundary values f in $C^{1,\alpha}(\partial D)$ onto the solution φ of $S\varphi = f$ considered as an element of $L^q(\partial D)$. Clearly, the operator is bounded from $C^{1,\alpha}(\partial D)$ into $L^q(\partial D)$ with a bound b. Bounds for the reconstruction of the boundary values are now obtained from (5.3.25) by

$$
\begin{aligned}
\|f\|_{C(\partial D)} &\leq \|f - A_\epsilon u^\infty\|_{C(\partial D)} + \|A_\epsilon u^\infty\|_{C(\partial D)} \\
&\leq Cb\|f\|_{C^{1,\alpha}(\partial D)}\epsilon + \|A_\epsilon u^\infty\|_{C(\partial D)} \quad\quad (5.3.27) \\
&\leq Cb\|f\|_{C^{1,\alpha}(\partial D)}\epsilon + \|A_\epsilon\|_{L^2(\Omega),C(\partial D)} \cdot \|u^\infty\|_{L^2(\Omega)}.
\end{aligned}
$$

If we know an *a priori* bound

$$
\|f\|_{C^{1,\alpha}(\partial D)} \leq L \quad\quad (5.3.28)
$$

on the $C^{1,\alpha}$-norm of the boundary values f, then we can estimate the norm of the boundary values by the norm of the far field pattern u^∞

$$
\|f\|_{C(\partial D)} \leq CbL\epsilon + \|A_\epsilon\|_{L^2(\Omega),C(\partial D)} \cdot \|u^\infty\|_{L^2(\Omega)}. \quad\quad (5.3.29)
$$

If $\|u^\infty\| \leq \delta$ with $\delta \to 0$, then we may choose ϵ depending on δ such that

$$
\epsilon \to 0 \quad \text{and} \quad \|A_\epsilon\|\delta \to 0 \quad \text{for} \quad \delta \to 0 \quad\quad (5.3.30)
$$

and thus

$$
\|f\| \leq CbL\epsilon + \|A_\epsilon\|\delta \to 0 \text{ for } \delta \to 0, \quad\quad (5.3.31)
$$

i.e., we obtain continuity for the reconstruction of the boundary values f from the far field pattern u^∞.

Remark 1: We can remove the condition on the domain D that the interior Dirichlet problem needs to be uniquely solvable, when the solution u^s can be extended into a neighbourhood of the boundary ∂D. In this case due to the monotonicity property of the Dirichlet eigenvalues for the negative Laplacian (see for example [58]) we can find a domain $D_i \subset D$ for which the interior homogeneous Dirichlet problem is uniquely solvable and for this domain all the above arguments with D replaced by D_i are true and we are able to reconstruct the values of u^s on ∂D_i and by continuity of the exterior Dirichlet problem also on ∂D.

Remark 2: If the function f is the trace of an analytic function $-u^i$ on ∂D and ∂D is analytic, we can extend the solution u^s into a neighbourhood $B \setminus \overline{D_i}$ of ∂D with some ball B with $D \subset B$ by the Cauchy-Kowalewski Theorem (see for example [17]).

Remark 3: Clearly, we obtain the continuity of the reconstruction under the above conditions by quite general arguments which yield the boundedness of the inverse of continuous operators defined on compact subsets of Banach spaces (see for example Theorem 7.1 of [7]). But here we have fully constructed the inverse of the mapping and we may obtain explicit estimates for the bounds by the norm of the operator A_ϵ as worked out in [70].

Reconstruction of boundary values by minimum norm solutions, Tikhonov regularization and spectral cut-off. We have described a solution of the exterior Dirichlet problem by means of integral equations in Chapter 2. According to (5.3.10), (5.3.1) and (5.3.2) the operator $\mathcal{F}$, which maps the boundary values f onto the far field pattern of the solution, is given by

$$\mathcal{F} = 2(K^\infty - iS^\infty)\left(I + K - iS\right)^{-1}. \tag{5.3.32}$$

To study convergence of regularization methods for the inversion of $\mathcal{F}$ we need some of its basic properties. Since we would like to obtain estimates in $C(\partial D)$ we work with the Hilbert space $H^2(\partial D)$. The norm $\|\cdot\|_{H^2(\partial D)}$ is stronger than $\|\cdot\|_{C(\partial D)}$.

LEMMA 5.3.4 *The operator* $\mathcal{F} : H^2(\partial D) \to L^2(\Omega)$ *is an injective compact linear operator with dense range.*

Proof. Clearly the operator is linear and since both S^∞ and K^∞ have smooth kernel it is compact. To show injectivity we assume that $\mathcal{F}(f) = 0$ for some function $f \in C(\partial D) \supset H^2(\partial D)$. By Rellich's lemma we obtain $u = 0$ in $\mathbb{R}^m \setminus \overline{D}$ for the field u given by (2.1.13) with density $\varphi = (I + K - iS)^{-1}2f$. Jump relations now yield

$$-u_- = \varphi, \quad -\frac{\partial u_-}{\partial \nu} = i\varphi \text{ on } \partial D. \tag{5.3.33}$$

By Green's theorem we obtain

$$i\int_{\partial D} |\varphi|^2 ds = \int_{\partial D} \bar{u}_- \frac{\partial u_-}{\partial \nu} ds = \int_D \left\{ |\operatorname{grad} u|^2 - \kappa^2 |u|^2 \right\} dx \tag{5.3.34}$$

and taking the imaginary part of this equation we obtain $\varphi = 0$, i.e., the injectivity of $\mathcal{F}$.

To show that $\mathcal{F}$ has dense range we first prove injectivity of the adjoint operator as an operator from $L^2(\Omega)$ into $L^2(\partial D)$. Since the adjoint $\mathcal{F}^*$ of $\mathcal{F}$ is given by

$$(I + K^* + iS)^{-1}(K^\infty - iS^\infty)^* \tag{5.3.35}$$

and $(I + K^* + iS)^{-1}$ is boundedly invertible in $L^2(\partial D)$ by the Fredholm Alternative Theorem we need to prove injectivity of $(K^\infty - iS^\infty)^*$. Let $0 = (K^\infty - iS^\infty)^*\psi$ for some function $\psi \in L^2(\Omega)$. Then the Herglotz wave function

$$v(x) = \int_\Omega e^{i\kappa x \cdot d}\psi(d)ds(d) \tag{5.3.36}$$

is a solution to the Helmholtz equation in D with boundary values

$$\left(\frac{\partial}{\partial\nu(x)} + i\right)v(x) = 0. \tag{5.3.37}$$

In the same way as in (5.3.34) we obtain $v(x) = 0$ for $x \in D$ and then $v \equiv 0$ in $\mathbb{R}^m$ by analyticity. Now, Theorem 3.15 of [8] yields $\psi = 0$ and the denseness of $\mathcal{F}$ from $L^2(\partial D)$ into $L^2(\Omega)$. Since $H^2(\partial D)$ is dense in $L^2(\partial D)$, we conclude the denseness of $\mathcal{F}$ from $H^2(\partial D)$ into $L^2(\partial D)$ and the proof is complete. $\qquad\square$

From the above Lemma 5.3.4 we obtain convergence properties for the reconstruction of boundary values using the methods under consideration.

THEOREM 5.3.5 *Let $u_\delta^\infty \in L^2(\Omega)$ with $\|u_\delta^\infty - u^\infty\| \leq \delta \leq u_\delta^\infty$ and u_δ be the minimum norm solution with discrepancy δ. Then there holds the convergence*

$$u_\delta \to \mathcal{F}^{-1}u^\infty, \quad \delta \to 0 \tag{5.3.38}$$

in $C(\partial D)$. For the Tikhonov regularization R_α with a choice of the regularization parameter $\alpha(\delta)$ such that

$$\alpha(\delta) \to 0, \quad \frac{\delta^2}{\alpha(\delta)} \to 0, \quad \text{for } \delta \to 0 \tag{5.3.39}$$

for $u_\delta := R_{\alpha(\delta)}u_\delta^\infty$ we obtain (5.3.38). Under the conditions (5.3.39) the spectral cut-off $u_\delta := R_\alpha u_\delta^\infty$ defined by (1.2.81) satisfies (5.3.38). For a compact set U of boundaries of class C^2 the convergence (5.3.38) is uniform on U.

Proof. The statements follow from Theorems 1.2.11, 1.2.12 and 1.2.13. For the uniformity of the reconstructions for $\partial D \in U$ we remark that the operators $\mathcal{F}$ and R_α depend continuously on the boundary with respect to the C^2-norm for the set of domains U. We obtain uniform convergence in (5.3.38) for ∂D in the compact set U. $\qquad\square$

5.4   Convergence of a regularized Newton method

Let F be a Fréchet differentiable operator defined on a subset of a normed space X into a normed space Y. Let us briefly review the simple proof for the convergence of Newton's method for the iterative solution of an equation of the form

$$F(x) = 0 \qquad (5.4.1)$$

by the iteration scheme

$$r_{n+1} = r_n - \left(F'(r_n)\right)^{-1} F(r_n). \qquad (5.4.2)$$

We assume that $F'(r_n)$ is injective and has dense range, i.e., the inverse operator $(F'(r_n))^{-1}$ is well defined on a dense subset of Y. Clearly, in general we will not have $F(r_n) \in \mathcal{R}(F'(r_n))$, but for some heuristic considerations we assume that this condition is satisfied. Let r_* be a solution of (5.4.1), i.e., a value with $F(r_*) = 0$. Then we have

$$\left(F'(r_n)\right)^{-1} F(r_*) = 0. \qquad (5.4.3)$$

We now calculate an estimate for the error in the $(n+1)$st iteration from the error in the nth iteration. We have

$$
\begin{aligned}
r_{n+1} - r_* &= r_n - \left(F'(r_n)\right)^{-1} F(r_n) - r_* \\
&= r_n - r_* - \left(F'(r_n)\right)^{-1} \left(F(r_n) - F(r_*)\right) \qquad (5.4.4) \\
&= \left(F'(r_n)\right)^{-1} \left\{ F'(r_n)(r_n - r_*) - \left(F(r_n) - F(r_*)\right) \right\}.
\end{aligned}
$$

For two-times continuously Fréchet differentiable functions F the estimate

$$\left\| F'(r_n)(r_n - r_*) - \left(F(r_n) - F(r_*)\right) \right\| = O\left(\|r_n - r_*\|^2\right) \qquad (5.4.5)$$

is well known. If $\left(F'(r_n)\right)^{-1}$ would be bounded, we immediately could derive the typical superlinear convergence of Newton's scheme. But for the solution of the exterior Dirichlet problem we have a compact operator F and in this case also the derivative F' will be compact and cannot have a bounded inverse. We will now show how we can prove convergence of a regularized Newton method for the case of the inverse scattering problems.

To solve the inverse domain problem by Newton's method we have to solve the equation

$$\mathcal{F}(D) = u_*^\infty \qquad (5.4.6)$$

where $\mathcal{F}(D)$ is defined by (5.3.32) with boundary values $f = -u^i|_{\partial D}$. We consider a domain ∂D_0 with boundary of class C^3 which is sufficiently close to the solution of the inverse problem. Then the normal vector ν_0 as a function on ∂D_0 is C^2-smooth and with the help of a function $r \in C^2(\partial D_0)$ defined on the boundary ∂D_0 all surfaces ∂D in a neighbourhood V of ∂D_0 can be represented by

$$\partial D(r) = \{x + r(x)\nu_0(x) : \quad x \in \partial D_0\}. \tag{5.4.7}$$

In this framework we are looking for the zero r_* of the function

$$\begin{aligned} F(r) \quad &:= \quad \mathcal{F}(D(r)) - u_*^\infty \\ &= \quad \mathcal{F}(D(r)) - \mathcal{F}(D(r_*)). \end{aligned} \tag{5.4.8}$$

For the following arguments we assume that the solution u_*^s to the exterior Dirichlet problem for the domain D_* can be extended as a solution of the Helmholtz equation into the exterior $\mathbb{R}^n \backslash D_i$ of some domain D_i with $\overline{D_i} \subset D_*$. As mentioned above for analytic domains D_* this assumption is satisfied. Then for each boundary r in some neighborhood of r_* the total field $u_*^s|_{\partial D(r)}$ and its normal derivative $\frac{\partial u_*^s}{\partial \nu}$ are well defined. We use the notation f for the boundary values, u_*^s for the exterior solution with boundary values $f = -u^i$ on $D(r_*)$, u^s for the exterior solution with boundary values $f = -u^i$ on $D(r)$, $u_* = u^i + u_*^s$ and $u = u^i + u^s$. The solution for the domain $D(r_n)$ is denoted by u_n^s and the corresponding far field pattern by u_n^∞.

It is worked out in [44] and [78] that the calculation of the Fréchet derivative $F'(r)h$ is equivalent to the evaluation of the far field pattern of an exterior Dirichlet problem with boundary values

$$-h\frac{\partial u}{\partial \nu_0} \quad \text{on} \quad \partial D(r). \tag{5.4.9}$$

The boundary values for the exterior solution with far field pattern $F(r_n) = \mathcal{F}(D(r_n)) - \mathcal{F}(D(r_*))$ defined in (5.4.8) are given by

$$u_n^s|_{\partial D(r_n)} - u_*^s|_{\partial D(r_n)} \quad = \quad -u^i|_{\partial D(r_n)} - u_*^s|_{\partial D(r_n)} \quad = \quad -u_*|_{\partial D(r_n)}.$$

Then Newton's method for the solution of

$$F(r) = 0 \tag{5.4.10}$$

with F defined by (5.4.8) consists in the iteration scheme

$$\begin{aligned} r_{n+1} \quad &= \quad r_n - \left(F'(r_n)\right)^{-1} F(r_n) \\ &= \quad r_n - \left(u_*|_{\partial D(r_n)} \Big/ \frac{\partial u_n}{\partial \nu_0}\right). \end{aligned} \tag{5.4.11}$$

Here the operator $/$ denotes the pointwise quotient of the two functions under consideration. We have proven the following lemma.

LEMMA 5.4.1 *Let $D(r_*)$ be a solution to the inverse scattering problem described by $F(r) = 0$ with analytic surface $\partial D_* = \partial D(r_*)$. Then there is a neighbourhood of ∂D_* such that the Newton scheme to find the zeros of F is given by (5.4.11).*

Remark. The lemma provides an interpretation of Newton's method in terms of the total fields u and u_*, respectively, at the boundary of the domain under consideration.   □

For a better understanding of our convergence analysis we will now first formulate a corresponding local problem and solve it by a Newton scheme, which is strongly related to the above solution of the far field equation. This local Newton scheme is obtained for example when solving the inverse obstacle problem using the point-source method as introduced in Section 5.1 when Newton updates are used to find the zeros of the total field u_*. We define the nonlinear function

$$\begin{aligned} \tilde{F}(\tilde{r}) \; &:= \; -u^i|_{\partial D(\tilde{r})} - u_*^s|_{\partial D(\tilde{r})} \\ &= \; -u_*(x + \tilde{r}(x)\nu_0(x)), \quad x \in \partial D_0. \end{aligned} \tag{5.4.12}$$

Its Fréchet derivative is given by

$$\tilde{F}'(\tilde{r})h \; = \; -h\frac{\partial u_*}{\partial \nu_0}. \tag{5.4.13}$$

Thus the Newton updates to find the zeros of $\tilde{F}(\tilde{r}) = 0$ are

$$\begin{aligned} \tilde{r}_{n+1} \; &= \; \tilde{r}_n - \left(\tilde{F}'(\tilde{r}_n)\right)^{-1}\tilde{F}(\tilde{r}_n) \\ &= \; \tilde{r}_n - \left(u_*|_{\partial D(\tilde{r}_n)} \, / \, \frac{\partial u_*}{\partial \nu_0}\right). \end{aligned} \tag{5.4.14}$$

Under the condition

$$|\frac{\partial u_*}{\partial \nu}(x)| \; > 0, x \in \partial D_*. \tag{5.4.15}$$

superlinear convergence of the local Newton scheme (5.4.14) is obtained as in (5.4.5) by straightforward calculation. Since for perfect data in principle we are able to perfectly reconstruct the field u_* under the conditon (5.4.15) we obtain superlinear convergence for the determination of the unknown scatterer for the perfect scheme (5.4.14) and thus for the perfect point-source method with Newton updates.

For the further argument we will assume that (5.4.15) is satisfied. Note that for the potential theoretic case and a domain D which is a circle this condition is satisfied. Thus in the three-dimensional case (where the solution of the exterior Dirichlet problem depends continuously on the wave number κ) by continuity of the solutions with respect to κ for the problems under consideration the condition (5.4.15) is satisfied for small κ and domains close to a circle. Here, we will focus on the development of our main line of reasoning and leave further investigations of this condition to future research.

From (5.4.15) by the continuity of the solution to the exterior Dirichlet problem with respect to variations of the domain we conclude that

$$|\frac{\partial u}{\partial \tilde{\nu}}(x)| > 0, x \in \partial D(r) \tag{5.4.16}$$

for all functions r in a neighbourhood of r_* and all vector fields $\tilde{\nu}$ in a neighbourhood of the normal vectors ν of r.

We first investigate the perfect Newton scheme (5.4.11). The function u_n is the solution to an exterior Dirichlet problem; $\frac{\partial u_n}{\partial \nu_0}$ is the derivative of u_n at the boundary. Studying the regularity properties of the solution to boundary value problems with boundaries of class $C^{n,\alpha}$ we observe that the regularity of the derivative $\frac{\partial u_n}{\partial \nu_0}$ is one order less than the regularity of the boundary under consideration. Thus according to (5.4.11) we lose one order of regularity in each Newton step.

Note that this cannot be remedied by considering smooth surfaces. Even if we start with sufficiently smooth boundaries, the C^2-norm of r_{n+1} in (5.4.11) may become very large and we will not be able to derive convergence results in this case. We need to apply a regularization to control the C^2 norm of the boundaries.

We have seen that the perfect Newton scheme is ill-posed and, in general, it will not converge without regularization. Thus, as a second step we consider a regularized version of the scheme. We use quasi-solutions to regularize the local ill-posedness of the perfect Newton scheme as follows. First we calculate an update $\tilde{r}_{n+1}$ by the perfect Newton scheme

$$\begin{aligned}
\tilde{r}_{n+1} &= r_n - \left(F'(r_n)\right)^{-1} F(r_n) \\
&= r_n - \left(u_* |_{\partial D(r_n)} / \frac{\partial u_n}{\partial \nu_0}\right). \tag{5.4.17}
\end{aligned}$$

Then we regularize $\tilde{r}_{n+1}$ to obtain a C^3-boundary r_{n+1} with norm bounded by some global constant C_0. The function $r_{n+1} \in C^3(\partial D_0)$ with

$$\|r_{n+1}\|_{C^3(\partial D_0)} \leq C_0$$

is chosen such that

$$\|r_{n+1} - \tilde{r}_{n+1}\|_{C(\partial D_0)} \le \|r - \tilde{r}_{n+1}\|_{C(\partial D_0)} \tag{5.4.18}$$
$$\forall\, r \in C^3(\partial D_0) \quad \text{with} \quad \|r\|_{C^3(\partial D_0)} \le C_0,$$

i.e., as a quasi-solution. This defines a (possibly multi-valued) operator

$$Q : C(\partial D) \to C^3(\partial D) \tag{5.4.19}$$

which maps a function $\tilde{r}_{n+1}$ onto the quasi-solution with constraint C_0 of (5.4.18). As a first step towards a proof of convergence of the regularized Newton scheme (5.4.17) we calculate

$$
\begin{aligned}
(\tilde{r}_{n+1} - r_*) &= r_n - \left(u_* |_{\partial D(r_n)} \,/\, \frac{\partial u_n}{\partial \nu_0} \right) - r_* \\
&= \left(- u_* |_{\partial D(r_n)} - (r_* - r_n) \frac{\partial u_*}{\partial \nu_0} \Big|_{\partial D(r_n)} \right. \\
&\quad \left. - (r_* - r_n)\left(\frac{\partial u_n}{\partial \nu_0} - \frac{\partial u_*}{\partial \nu_0} \Big|_{\partial D(r_n)} \right) \right) \,/\, \frac{\partial u_n}{\partial \nu_0}.
\end{aligned}
\tag{5.4.20}
$$

To estimate the first term of (5.4.20) we remark that the field u_* is two times continuously differentiable. Taylor's expansion yields

$$\left\| u_* |_{D(r_*)} - u_* |_{\partial D(r_n)} - (r_* - r_n) \frac{\partial u_*}{\partial \nu_0} \Big|_{\partial D(r_n)} \right\|_{C(\partial D)}$$
$$= O\left(\|r_n - r_*\|_{C(\partial D)}^2 \right) \tag{5.4.21}$$

and since $u_* |_{D(r_*)} \equiv 0$ the first term of (5.4.20) is

$$O\left(\|r_n - r_*\|_{C(\partial D)}^2 \right).$$

Next, we choose the constant C_0 for the quasi-solution by some *a priori* information about the unknown boundary r_* such that $\|r_*\|_{C^3(\partial D_0)} \le C_0$. To estimate

$$\left\| \frac{\partial u_n}{\partial \nu_0} - \frac{\partial u_*}{\partial \nu_0} \Big|_{\partial D(r_n)} \right\|_{C(\partial D)}$$

by $\|r_n - r_*\|_{C(\partial D)}$ we use the fact that $\|r_n\|_{C^3(\partial D)} \le C_0$ and $\|r_*\|_{C^3(\partial D)} \le C_0$. In this case

$$\|r_n - r_*\|_{C(\partial D)} = \epsilon$$

yields

$$\|r_n - r_*\|_{C^2(\partial D)} \le c\epsilon^{1/4}$$

with some constant c. With $p := 5/4 > 1$ we thus obtain

$$\left\| (r_* - r_n) \left(\frac{\partial u_n}{\partial \nu_0} - \frac{\partial u_*}{\partial \nu_0} \Big|_{\partial D(r_n)} \right) \right\|_{C(\partial D)} = O\left(\|r_n - r_*\|^p_{C(\partial D)} \right). \quad (5.4.22)$$

This proves

$$\|\tilde{r}_{n+1} - r_*\|_{C(\partial D_0)} \leq c \|r_n - r_*\|^p_{C(\partial D_0)} \quad (5.4.23)$$

with constants $c > 0$ and $p > 1$. Since r_{n+1} is the quasi-solution (5.4.18) we obtain

$$\|r_{n+1} - \tilde{r}_{n+1}\|_{C(\partial D_0)} \leq \|r_* - \tilde{r}_{n+1}\|_{C(\partial D_0)} \quad (5.4.24)$$

and now (5.4.23) implies

$$\begin{aligned}\|r_{n+1} - r_*\|_{C(\partial D_0)} &\leq \|r_{n+1} - \tilde{r}_{n+1}\|_{C(\partial D_0)} + \|\tilde{r}_{n+1} - r_*\|_{C(\partial D_0)} \\ &\leq 2c \|r_n - r_*\|^p_{C(\partial D_0)}. \quad (5.4.25)\end{aligned}$$

This shows superlinear convergence of the regularized scheme given by (5.4.17), (5.4.18) for perfect data u_*^∞.

We now consider the more realistic case where the given far field pattern u_δ^∞ includes an error of size smaller or equal to $\delta > 0$ with respect to the perfect far field pattern of the domain $D(r_*)$, i.e., we have

$$\|u_\delta^\infty - u_*^\infty\| \leq \delta. \quad (5.4.26)$$

In this case we need to regularize the mapping $F'(r)^{-1}$ which maps the far field pattern of a solution to the exterior Dirichlet problem onto the boundary values of the field on $\partial D(r)$. We replace the unbounded operator $(\mathcal{F}'(r_n))^{-1}$ by a family of bounded operators R_α such that

$$R_\alpha (\mathcal{F}'(r_n)\varphi) \to \varphi, \ \alpha \to 0 \quad (5.4.27)$$

for each fixed $\varphi \in X$. For a regularization parameter $\alpha = \alpha(\delta)$ we define the fully regularized Newton scheme by

$$r_{n+1} = Q\left(r_n - R_{\alpha(\delta)} F(r_n) \right), \quad (5.4.28)$$

where Q is the quasi-solution defined in (5.4.19). We need to supplement the scheme by some stopping rule which specifies a function $N(\delta)$ such that $r_\delta := r_{N(\delta)}$ is considered as the solution of the inverse problem by the regularized Newton method (5.4.28). We are interested in convergence properties in the sense of the following definition:

DEFINITION 5.4.2 *The regularized Newton scheme is called locally convergent (or locally regular) if there is a neighbourhood U of the true solution such that any regularized solution r_δ obtained from starting values in U tends to the true solution r_* when the data error δ tends to zero.*

Our main ingredients to treat the ill-posedness of the inverse problem are the operator Q to obtain bounds for the C^2-norms of the iterates r_n and the operators R_α to reconstruct the field $u_*|_{\partial D(r_n)}$ from the measured version $\mathcal{F}(r_n) - u_\delta^\infty$ of the far field data

$$\mathcal{F}(r_n) - \mathcal{F}(r_*).$$

Usually (cf. [28], [8]), these two regularizations are implicit in a discretized regularization which is applied directly to the far field equation given by the first line of (5.4.11). We now consider four methods:

1. the point-source method or

2. minimum norm solutions or

3. the Tikhonov regularization or

4. a spectral cut-off

for the regularization of the inverse $\mathcal{F}^{-1}$ of $\mathcal{F}$.

For these operators we need to formulate a stopping rule to determine $N(\delta)$. We formulate this rule in terms of the error of the reconstruction in the boundary values of u_* on $\partial D(r_n)$ and the Neumann values $\frac{\partial u_n}{\partial \nu_n}$ of the solution to the exterior Dirichlet problem for $D(r_n)$ with boundary values $-u^i|_{\partial D(r_n)}$. We investigate convergence for boundaries ∂D_* of domains D_* where

$$\frac{\partial u_*}{\partial \nu}(x) \geq \frac{1}{\eta}, \quad x \in \partial D_*, \qquad (5.4.29)$$

for some $\eta > 0$. Then, in a neighbourhood U of ∂D_* we obtain the estimate

$$\frac{\partial u_n}{\partial \nu}(x) \geq \frac{1}{2\eta}, \quad x \in U, \qquad (5.4.30)$$

for the normal derivative of u_n. From Theorem 5.3.5 and (5.3.31) we derive that given $\tau > 0$ there is $\delta > 0$ and a regularization parameter $\alpha(\delta) > 0$ such that

$$|u_{\alpha(\delta)}(x) - u_*(x)| \leq \tau(\delta), \quad x \in \partial D, \qquad (5.4.31)$$

is satisfied in the neighbourhood U of D_*. Since δ depends monotonously on τ we may consider τ as some function $\tau(\delta)$ with

$$\tau(\delta) \to 0, \quad \delta \to 0. \qquad (5.4.32)$$

Given some *a priori* information about the domains we can estimate τ from the data error δ with the help of the techniques of the preceding sections, see (5.3.29). We are now prepared to formulate the

DEFINITION 5.4.3 (**A posteriori stopping rule.**) *For the reconstruction of an unknown boundary ∂D_* by a regularized Newton method from measured data u^∞_{meas} with data error (5.4.26) we stop the Newton iteration if for two successive approximations we observe*

$$\|r_{n+1}(x) - r_n(x)\|_{C(\partial D_0)} \leq \frac{2\tau(\delta)}{\eta}. \qquad (5.4.33)$$

We will now prove that the contribution of the data error in each Newton step is bounded by $\frac{\tau}{\eta}$ and that the regularized Newton scheme with the above stopping rule is locally convergent or regular, respectively.

THEOREM 5.4.4 *Let the unknown domain D_* have analytic boundary and assume that (5.4.15) is satisfied. Then the regularized Newton scheme (5.4.28) with a regularization scheme $R_{\alpha(\delta)}$ for $\mathcal{F}^{-1}$ which satisfies (5.4.31) and the stopping rule (5.4.33) is locally convergent, i.e., we have*

$$\|r_\delta - r_*\|_{\partial D} \to 0, \quad \delta \to 0. \qquad (5.4.34)$$

Proof. We need to modify the convergence analysis of the perfect scheme as follows. We calculate

$$
\begin{aligned}
(\tilde{r}_{n+1} - r_*) &= r_n - \left(u_\alpha|_{\partial D(r_n)} / \frac{\partial u_n}{\partial \nu_0}\right) - r_* \\
&= \left(u_\alpha|_{\partial D(r_n)} - (r_* - r_n)\frac{\partial u_n}{\partial \nu_0}\right) / \frac{\partial u_n}{\partial \nu_0} \\
&= \left(u_*|_{\partial D(r_n)} - (r_* - r_n)\frac{\partial u_*}{\partial \nu_n}\right. \qquad (5.4.35) \\
&\quad - (r_* - r_n)\left(\frac{\partial u_n}{\partial \nu_0} - \frac{\partial u_*}{\partial \nu_0}\Big|_{\partial D(r_n)}\right) \\
&\quad \left. + (u_\alpha|_{\partial D(r_n)} - u_*|_{\partial D(r_n)})\right) / \frac{\partial u_n}{\partial \nu_0}.
\end{aligned}
$$

We obtain

$$
\begin{aligned}
\|r_{n+1} - r_*\|_{C(\partial D_0)} &\leq 2\|\tilde{r}_{n+1} - r_*\|_{C(\partial D_0)} \qquad (5.4.36) \\
&\leq c\|r_n - r_*\|^p_{C(\partial D_0)} + \frac{1}{\eta}\|u_\alpha - u_*\|_{C(\partial D(r_n))}
\end{aligned}
$$

with some constant c. For calculations we abreviate $\|r_n - r_*\| = \beta$ and $\tau/\eta = \gamma$. The estimate

$$c\beta^p + \gamma \leq \frac{\beta}{2} \qquad (5.4.37)$$

is satisfied for

$$\gamma < \frac{\beta}{4}, \qquad c\beta^{p-1} < \frac{1}{4}. \qquad (5.4.38)$$

This shows that in the neighborhood $B(\partial D_*)$ of the true solution the estimate

$$\|r_{n+1} - r_*\|_{C(\partial D_0)} \leq \frac{\|r_n - r_*\|_{C(\partial D_0)}}{2} \qquad (5.4.39)$$

is satisfied as long as

$$4\gamma < \|r_n - r_*\|_{C(\partial D_0)}, \qquad \|r_n - r_*\|_{C(\partial D_0)} < \left(\frac{1}{4c}\right)^{1/(p-1)}, \qquad (5.4.40)$$

where c is a constant given by the first term on the right-hand side of (5.4.36). Under the conditions (5.4.40) we are now able to estimate the true error $\|r_n - r_*\|$ by the distance $\|r_{n+1} - r_n\|$ by

$$
\begin{aligned}
\|r_{n+1} - r_n\|_{C(\partial D_0)} &= \|r_{n+1} - r_* + r_* - r_n\|_{C(\partial D_0)} \\
&\geq \|r_n - r_*\|_{C(\partial D_0)} - \|r_{n+1} - r_*\|_{C(\partial D_0)} \\
&\geq \|r_n - r_*\|_{C(\partial D_0)} - \left(c\|r_n - r_*\|_{C(\partial D_0)}^p + \frac{\tau}{\eta}\right) \\
&= \|r_n - r_*\|_{C(\partial D_0)} - c\|r_n - r_*\|_{C(\partial D_0)}^p - \frac{\tau}{\eta} \\
&\geq \|r_n - r_*\|_{C(\partial D_0)} - \frac{1}{2}\|r_n - r_*\|_{C(\partial D_0)} \\
&\geq \frac{1}{2}\|r_n - r_*\|_{C(\partial D_0)}. \qquad (5.4.41)
\end{aligned}
$$

Thus for $\|r_{n+1} - r_n\|_{C(\partial D_0)} < \frac{2\tau}{\eta}$ we obtain $\|r_n - r_*\|_{C(\partial D_0)} < \frac{4\tau}{\eta}$ and the approximation r_n is no longer in the interval where we obtain sufficient improvements with further iterations. This substantiates the choice of the stopping rule. Since by the stopping rule we stop the iterations when

$$\|r_n - r_*\|_{C(\partial D_0)} < \frac{4\tau}{\eta} \qquad (5.4.42)$$

and for all regularization schemes under consideration by Theorem 5.3.5 we have the convergence $\tau = \tau(\delta) \to 0$ for $\delta \to 0$, we obtain the convergence (5.4.34). This ends the proof.   □

Remark: Equation (5.4.36) proves local superlinear convergence of the regularized scheme towards the regularized solution r_δ.

6 Singular sources and shape reconstruction

For the application of the methods discussed in the preceding Chapter 5 a knowledge of the boundary condition is required to reconstruct the unknown scatterer. But in many practical situations the physical properties of the scatterer are not known and these methods are not applicable. It is therefore of practical interest to develop reconstruction methods which do not require knowledge of the boundary condition or the physical properties of a scatterer.

We will develop a method of singular sources for the reconstruction of a scattering object when the physical properties of the scatterer are unknown. In a first step we will consider a sound-soft or sound-hard impenetrable scatterer and show in a second step how the results can be extended to the reconstruction of the shape of an unknown inhomogeneous medium scatterer.

6.1 Acoustic scattering

The main idea of the method of singular sources is the use of the field $\Phi^s(z, z)$ to reconstruct the shape of the scattering object. In a first part of this section we use the operator Q given by (3.1.35) for the reconstruction of $\Phi^s(z, z)$ for scattering by impenetrable scatterers. We estimate the error for the reconstruction of $\Phi^s(z, z)$ by $(Qu^\infty)(z, z)$ in the following theorem.

THEOREM 6.1.1 *Consider the far field patterns $u^\infty(\cdot, d)$ for scattering of plane waves $u^i(\cdot, d)$, $d \in \Omega$, by a sound-soft or sound-hard scatterer $D \in C$. Let the function $p : B \to \Omega$ be chosen such that (5.1.2) is satisfied. Given $\mu = 0, s = 1, \tau, \eta, \rho > 0$ and a measured far field pattern u_δ^∞ with data error*

$$\left\| u_\delta^\infty - u^\infty \right\|_{L^2(\Omega \times \Omega)} \le \delta, \tag{6.1.1}$$

the error for the approximation of $\Phi^s(x, z)$ by Qu_δ^∞ is estimated by

$$\left| \Phi^s(x, z) - (Qu_\delta^\infty)(x, z) \right| \le c \frac{\eta}{\rho^{m-1}} + C \, b_{\eta, \rho} \, \tau + \frac{1}{\gamma_m} b_{\eta, \rho} b_{\tau, \rho} \, \delta, \tag{6.1.2}$$

for $x, z \in B \setminus D_\rho$ with the constants $b_{\eta, \rho}$ defined by (3.1.33) and constants c, C uniformly for domains $D \in C$.

Proof. We use Theorem 3.1.6 and the Cauchy-Schwarz inequality applied to the first and second term of the right-hand side of the decomposition

$$\left| \Phi^s(x,z) - (Qu_\delta^\infty)(x,z) \right| \leq \left| \Phi^s(x,z) - (Qu^\infty)(x,z) \right| + \left| \left(Q\left(u^\infty - u_\delta^\infty \right) \right)(x,z) \right|$$

to obtain (6.1.2).   □

According to Theorem 6.1.1 we can use the operator Q to reconstruct Φ^s from the knowledge of the far field patterns $u^\infty(\cdot, d)$ of the scattered fields $u^s(\cdot, d)$ of incident plane waves for all directions of incidence $d \in \Omega$. Given the error δ in the measurements, the error for the reconstruction of Φ^s from u_δ^∞ is estimated by (6.1.2). The density functions $g_\eta(x, \cdot)$ and $g_\tau(z, \cdot)$ used to define Q can be computed according to some *a priori* knowledge on the unknown scatterer $\mathcal{D}$ as given by the class $\mathcal{C}$ of scatterers defined in Definition 2.1.5 and Definition 2.2.5.

The operator Q is strongly related to the operator A defined by (5.1.1) to set up the point-source method. We can use Lemma 5.1.3 to efficiently compute the densities g_τ and g_η using rotations and translations. The computation of g_τ for different $p \in \Omega$ is reduced to the computation of one density function g_τ of (5.1.9) with $p = p_0 \in \Omega$ and discrepancy τ and another density function g_η of (5.1.9) with $p = p_0 \in \Omega$ and discrepancy η.

Given an approximating function $a(z)$ for $\Phi^s(z,z), z \in B \setminus D_\rho$, we may follow Theorem 2.1.15 and search for the boundary of the unknown scatterer $\mathcal{D}$ as the set of points z where $a(z)$ is larger than a constant $C > 0$. We call this the method of singular sources, since the main reason for the behavior of Φ^s is the singularity of the source $\Phi(\cdot, z)$. The constant C plays the role of a regularization parameter and has to be chosen according to the other regularization parameters of the reconstruction, i.e., depending on ρ, τ and η.

The method of singular sources has some features, which are similar to the point-source method. The choice of the orientations $p(z)$ of the domain of approximation needs some knowledge about the unknown scatterer. The following multistep procedure describes a method to successively construct both the scatterer and an appropriate function $p(z)$.

- As a first step compute for a number of fixed orientations $p_\xi, \xi = 1, .., N$ approximations $a_\xi^{(1)}(z)$ for the field $\Phi^s(z, z)$ using the operator Q, where Q is depending on p_ξ via the densities $g_\tau(x, \cdot)$ and $g_\eta(z, \cdot)$.

Search an approximation D_1 for the shape D of the unknown scatterer $\mathcal{D}$ as the set

$$D_1 := \left\{ z \in B : |a_\xi^{(1)}(z)| > C \text{ for } \xi = 1, ..., N \right\}$$

of the points where all approximations $a_\xi^{(1)}(z)$ are larger than C.

- Adapt in each further step $n > 1$ the orientation $p_n(z)$ of the domain of approximation to the knowledge about the unknown scatterer of the previous steps $1, ..., n - 1$ and repeat the procedure to compute an approximation $a^{(n)}(z)$ to $\Phi^s(z, z)$ and to search an approximation D_n for D according to

$$D_n := \left\{ z \in B : |a^{(n)}(z)| > C \right\}.$$

An aspect of this procedure is the observation that the values of $a_\xi(z)$ in general are very large, if the unknown scatterer $\mathcal{D}$ is not contained in the domain of approximation G_{z, p_ξ, β_0}. Thus we obtain large values of $a_\xi(z)$, if z is in the interior of the unknown scatterer and we can replace the search for ∂D by the search for D.

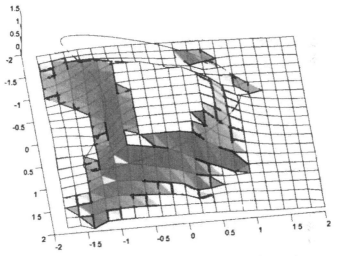

Figure 6.1: Illustration for the method of singular sources.

Figure 6.1 demonstrates the reconstruction of a kite-shaped domain using the method of singular sources. The plot shows $|a(z)| = |(Qu^\infty)(z, z)|$ on a rectangular grid for the second step of the method.

An obvious advantage of the method of singular sources is the fact that we do not need to take into account the boundary condition of the impenetrable scatterer. In contrast to the point-source method, for the method of singular sources the reconstruction algorithm is the same for the sound-soft and sound-hard boundary condition, i.e., the boundary condition does not need to be known for the algorism. This feature is one of the main reasons for the distinction of category II and III of reconstruction methods discussed in the introduction. The method of singular sources thus belongs to the third category.

One of the prices paid for the advantage of the method of singular sources is the amount of data necessary for reconstructions. The method of singular sources needs the far field pattern for a large number of incident plane waves, whereas with the point-source method we obtain reconstructions even for one or a small number of measured far field patterns.

Another price, which has to be paid to be independent of the boundary condition, is the ill-posedness of the reconstruction operator Q. The norm of Q is given by

$$\|g_\tau(z,\cdot)\|_{L^2(\Omega)} \|g_\eta(z,\cdot)\|_{L^2(\Omega)}$$

with the densities g_τ and g_η, which form the kernel of Q. For the point-source method the norm of the operator A is equal to $\|g_\tau(z,\cdot)\|_{L^2(\Omega)}$. This is only the square root of the norm of Q. We will observe below that the ill-posedness of the problem is becoming even worse, if we consider impenetrable and penetrable scatterers at the same time.

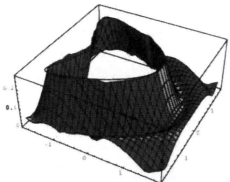

Figure 6.2[1]

Figure 6.2 shows a surface plot of the approximation $a(z)$ and the boundary ∂D of the unknown sound-soft scatterer for $\kappa = 2$ in the second step of the reconstruction algorithm. Here the orientation $p(z)$ of the domain of approximation is chosen as $p(z) = z/|z|$.

---

[1]Figures 6.2 to 6.8 reprinted from [70], with permission from Elsevier Science.

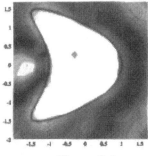
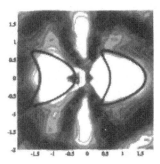

Figure 6.4   Figure 6.4

Figures 6.3 and 6.4 show contour plots of $a(z)$ in step 2 of the algorithm for one or two sound-soft domains, $\kappa = 3$ and $p(z) = z/|z|$. Here we used the same set of regularization parameters for the different data sets for one or two obstacles.

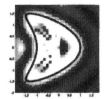
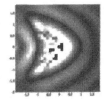
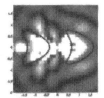

Figure 6.5   Figure 6.6   Figure 6.7   Figure 6.8

Figures 6.5 to 6.8 show density plots of $a(z)$ in the second step of the reconstruction algorithm for one or two sound-soft domains and one or two sound-hard domains, $\kappa = 4$ and $p(z) = z/|z|$. We used the same set of regularization parameters for all four images.

We now investigate the reconstruction of the shape of an object, which may be either an impenetrable or an inhomogeneous medium scatterer $D \in \mathcal{C}$. From the sections on uniqueness and stability we know that the shape of the scatterer is uniquely determined by the far field patterns for all incident plane waves and that the shape depends stably on the far field patterns with respect to the data error in $L^2(\Omega \times \Omega)$. Results on ϵ-uniqueness and ϵ-stability for a large finite number of measurements investigate the practical situation, where only a finite number of measurements of the far field pattern for a finite number of incident plane waves are possible.

For the reconstruction of the shape D of an inhomogeneous medium scatterer D we will use Theorem 2.2.12 and the operator Q defined by (3.1.35). We have to adapt the order μ of the multipole to the behavior of the refractive index n at the boundary of the unknown scatterer according to the situation S5 or S6 in (3.1) and Definition 2.2.5.

In the same way as in Theorem 6.1.1 we obtain

THEOREM 6.1.2 *Consider the far field patterns* $u^\infty(\cdot, d)$ *for scattering of plane waves* $u^i(\cdot, d)$, $d \in \Omega$, *by a sound-soft, sound-hard or inhomogeneous medium scatterer* $\mathcal{D} \in \mathcal{C}$. *For a measured far field pattern* u_δ^∞ *with data error*

$$\left\| u_\delta^\infty - u^\infty \right\|_{L^2(\Omega \times \Omega)} \leq \delta, \tag{6.1.3}$$

the error for the approximation of $\Phi_{\mu,q}^s(x,z)$ *by* $Q u_\delta^\infty$ *is estimated by*

$$\left| \Phi_{\mu,q}^s(x,z) - (Q u_\delta^\infty)(x,z) \right| \leq c \frac{\eta}{\rho^{m-1}} + C\, b_{\eta,\rho}\, \tau + \frac{1}{\gamma_m} b_{\eta,\rho} b_{\tau,\rho}\, \delta, \tag{6.1.4}$$

$x, z \in B \setminus D_\rho$, *with* $b_{\eta,\rho}$ *defined by (3.1.33) depending on* μ, $\rho > 0$ *and* $\eta > 0$ *and constants* c *and* C *uniformly for* $\mathcal{D} \in \mathcal{C}$.

Following Theorem 6.1.2 we can use the operator Q to reconstruct

$$\Phi_{\mu,-\nu(z_0)}^s(z,z) \tag{6.1.5}$$

from the knowledge of the far field patterns $u^\infty(\cdot, d)$ of the scattered fields $u^s(\cdot, d)$ for all directions of incidence $d \in \Omega$. Given the error δ in the measurements, the error for the reconstruction of (6.1.5) from u_δ^∞ is estimated by (6.1.4).

With these preparations we can formulate a method of singular sources for the reconstruction of the shape of an object, which may be either an impenetrable or an inhomogeneous medium scatterer. Due to Theorems 2.2.12 and 2.1.16 the singular behavior of (6.1.5) can be used to find the support of the unknown scatterer as the set of points where the approximation $a(z)$ for (6.1.5) is sufficiently large. A corresponding constant C has to be chosen according to the other parameters of the reconstruction, i.e., depending on ρ, τ and η. We have to use a strategy to adapt the orientation $p = p(z)$ of the domain of approximation and the polarization $q = q(z)$ of the multipole to the knowledge about the scatterer in each step of a multistep procedure.

At this point we have to realize a difference between the cases $\mu = 0$ and $\mu > 0$. For $\mu = 0$ the multipole has a rotational symmetry. In this case we can work with rotations and do not need to compute different densities for different vectors q or $\nu(z_0) \in \Omega$, respectively. For the cases $\mu > 0$ we need to compute different densities for the reconstruction of (6.1.5), if the angle between the orientation p of the domain of approximation and the polarization q of the approximated multipole changes. From a practical point of view this problem is not as serious as it first seems, since usually a small number of different angles is sufficient to obtain a reasonable reconstruction of the shape of the unknown scatterer.

We close with a remark on the ill-posedness of the method of singular sources. If it is not known whether the unknown scatterer is penetrable or impenetrable, we have to work with multipoles of order μ for reconstructions. Since the ill-posedness of the reconstruction of D is mainly influenced by the norm of the densities g_τ and g_η, and since these norms increase with μ, the ill-posedness for the general problem is considerably larger than the ill-posedness for the reconstruction of arbitrary impenetrable scatterers or the ill-posedness for the case of a given boundary condition of an impenetrable scatterer.

6.2   Electromagnetic scattering

In Theorems 2.3.12 and 2.4.9 of Chapter 2 the behavior of the scattered field $E_{edp}^s(z, z, p)$ for points near a scatterer $\mathcal{D}$ is estimated. The boundary ∂D of the scatterer $\mathcal{D}$ is the set of points, for which

$$\left| E_{edp}^s(z, z, \nu(z_0)) \right|$$

becomes infinite, where in a neighborhood of the boundary $z_0 \in \partial D$ is given by the unique representation

$$z = z_0 + h\nu(z_0).$$

In Chapter 4, Theorem 3.2.3, we constructed the operator Q for the reconstruction of E_{edp}^s from the far field patterns $E_{pl}^\infty(\cdot, \cdot, q)$, $q \in \Omega$. For a measured far field patterns E_δ^∞ the error for the reconstruction of E_{edp}^s by QE_δ^∞ is estimated in the following theorem.

THEOREM 6.2.1 *We consider the scattering of electromagnetic plane waves $E_{pl}^i(\cdot, d, p)$ for $d, p \in \Omega$ by a scatterer $\mathcal{D} \in C$. Given a measured far field pattern E_δ^∞ with data error*

$$\left| E_{pl}^\infty(\cdot, \cdot, p) - E_\delta^\infty(\cdot, \cdot, p) \right|_{L^2(\Omega \times \Omega)} \le \delta, \;\; p \in \Omega, \qquad (6.2.1)$$

the error for the approximation of $E_{edp}^s(x, z, p)$ by QE_δ^∞ is estimated by

$$\left| q \cdot E_{edp}^s(x, z, p) - (p \cdot QE_\delta^\infty(\cdot, \cdot, q))(x, z) \right| \qquad (6.2.2)$$

$$\le c\frac{\eta}{\rho^3} + C\|g_\eta\|_{L^2(\Omega)}\tau + \frac{1}{\gamma}\|g_\tau\|_{L^2(\Omega)}\|g_\eta\|_{L^2(\Omega)}\,\delta,$$

$x, z \in B \setminus D_\rho$, *with constants c and C uniformly for $\mathcal{D} \in C$ and functions g_τ and g_η depending on ρ, τ and η.*

Proof. The estimate can be obtained from (3.2.13) with the help of

$$
\left| q \cdot E_{edp}^s(x,z,p) - (p \cdot Q E_\delta^\infty(\cdot,\cdot,q))(x,z) \right|
$$

$$
\leq \left| q \cdot E_{edp}^s(x,z,p) - (p \cdot Q E_{pl}^\infty(\cdot,\cdot,q))(x,z) \right|
$$

$$
+ \left| \left\{ p \cdot Q \Big(E_{pl}^\infty(\cdot,\cdot,q) - E_\delta^\infty(\cdot,\cdot,q) \Big) \right\}(x,z) \right| \quad (6.2.3)
$$

and the Cauchy-Schwarz inequality.   □

A reconstruction of the boundary of a scatterer $\mathcal{D} \in \mathcal{C}$ with the help of the singular scattered field $E_{edp}^s(z,z,p)$ can be formulated in analogy to the acoustic case. As a difference to the acoustic case both for the reconstruction of a perfect conductor and the reconstruction of the shape of an inhomogeneous medium we can use the scattered field of electric dipoles. For inverse scattering from an acoustic inhomogeneous medium we had to use multipoles of higher order even for the simplest case of a medium with a jump at the boundary ∂D.

7 Linear sampling methods

The idea to use point-sources for the reconstruction of the unknown scatterer has been exploited by Colton and Kirsch and many others since 1995 to develop and investigate a class of inversion methods, known as linear sampling methods. We will introduce the original linear sampling method as proposed by Colton and Kirsch [6] and Colton, Monk, Piana and Potthast [10], [13] in our first Section 7.1. Section 7.2 will present the main ideas of a modified approach by Kirsch [41], [42], who used the spectral theory of the far field operator to study the linear sampling method and obtained characterizations of the unknown support of a scatterer by the expansion of the far field pattern of point-sources with respect to the eigenvalues and eigenvectors of the far field operator. We will conclude with two sections about anisotropic media: in Section 7.3 we use the linear sampling method to detect the support of an orthotropic medium from the knowledge of the far field operator; Section 7.4 treats the full three-dimensional anisotropic inverse electromagnetic scattering problem as proposed by Colton and Potthast [14].

7.1 The original linear sampling method

To describe the original linear sampling method consider the direct acoustic scattering problem as introduced in Definition 2.1.1 or the direct acoustic inhomogeneous medium scattering problem as given by Definition 2.2.1. In the last case we consider a piecewise smooth index of refraction n.

We are interested in determining D from the knowledge of the far field patterns for all incident plane waves $u^\infty(\hat{x}; d)$ for $\hat{x}, d \in \Omega$. In order to do this, we investigate the far field equation

$$(Fg)(\hat{x}) = \Phi^\infty_{true}(\hat{x}, z), \quad \hat{x} \in \Omega, \tag{7.1.1}$$

where the operator $F : L^2(\Omega) \to L^2(\Omega)$, defined by

$$(Fg)(\hat{x}) := \int_\Omega u^\infty(\hat{x}; d) g(d) \, ds(d), \quad \hat{x} \in \Omega, \tag{7.1.2}$$

is known as the far field operator and Φ^∞_{true} denotes the far field pattern of the fundamental solution (1.2.40). The far field operator F has smooth kernel, is thus compact in $L^2(\Omega)$ and in general equation (7.1.1) does not have a solution. We temporarily assume that we are able to solve (7.1.1) and in a second step show how to modify the approach to treat the general case.

Let $g(\cdot, z)$ be a solution of (7.1.1) with z in the interior of the scattering obstacle D. Then the incident field $u^i = v_g$ with the Herglotz wave function

$$v_g(x) := \int_\Omega u^i(x,d)g(d)\,ds(d),\quad x \in \mathbb{R}^m \tag{7.1.3}$$

has the scattered field

$$v_g^s(x) := \int_\Omega u^s(x,d)g(d)\,ds(d),\quad x \in \mathbb{R}^m \tag{7.1.4}$$

and far field pattern

$$\begin{aligned}v_g^\infty(\hat{x}) &= \int_\Omega u^\infty(\hat{x},d)g(d)\,ds(d),\\ &= (Fg)(\hat{x}).\end{aligned} \tag{7.1.5}$$

Since the far field patterns of v_g^s and $\Phi(\cdot,z)$ coincide, by Rellich's lemma we know that the scattered fields in the exterior of the scatterering domain D must be identical. We first consider the case of an impenetrable sound-soft scatterer, i.e., we consider the Dirichlet boundary condition (2.1.4). On the boundary ∂D of the scatterer D we have $v_g^s = -v_g$. Then, from the boundedness of the Herglotz wave operator

$$H : L^2(\Omega) \to C(\partial D),\quad g \mapsto v_g, \tag{7.1.6}$$

we obtain

$$\begin{aligned}\|g(\cdot,z)\|_{L^2(\Omega)} &\geq c\|v_g\|_{C(\partial D)}\\ &= c\|v_g^s\|_{C(\partial D)}\\ &= c\|\Phi(\cdot,z)\|_{C(\partial D)}\end{aligned} \tag{7.1.7}$$

with some constant c. Thus for $z \to \partial D$ the norm $\|g(\cdot,z)\|$ of the solution $g(\cdot,z)$ of (7.1.1) will blow up in the same way as the singularity of the fundamental solution Φ. Since for the calculation of g only the far field pattern $u^\infty(\hat{x},d)$ for $\hat{x},d \in \Omega$ is needed, we can formulate a method to find the unknown scatterer D:

DEFINITION 7.1.1 (**Linear Sampling Method.**) *For z on a grid which covers the region where the unknown scatterer D is supposed solve equation (7.1.1) with an appropriate regularization and calculate the norm $\|g(\cdot,z)\|$ of the solution. Then, the boundary of the unknown scatterer is found as a level curve where $\|g(\cdot,z)\|$ is large.*

Since for the determination of the unknown scatterer the main effort is the calculation of $g(\cdot,z)$ as a solution to a linear equation on a grid of sampling points, the method is called linear sampling method.

In general the equation (7.1.1) does not have a solution. But under appropriate assumptions on the scatterer we can show as in Lemma 3.1.2 that H has dense range in $C(\partial D)$. In this case given $\epsilon > 0$ for each $z \in D$ there is a density $\tilde{g}(\cdot, z)$ with

$$\|H\tilde{g}(\cdot, z) - (-\Phi(\cdot, z))\|_{C(\partial D)} \leq \frac{\epsilon}{\|\mathcal{F}\|}, \qquad (7.1.8)$$

and

$$\|F\tilde{g}(\cdot, z) - \Phi^{\infty}_{true}(\cdot, z)\|_{L^2(\Omega)} \leq \epsilon \qquad (7.1.9)$$

where $\mathcal{F} : C(\partial D) \to L^2(\Omega)$ denotes the mapping of a function f onto the far field pattern of the solution to the exterior Dirichlet problem with boundary values f. From (7.1.8) we obtain for $\tilde{g}$ or $v_{\tilde{g}(\cdot, z)}$, respectively, the behavior (7.1.7). Using a regularization method to solve the equation (7.1.1) may be considered as a way to numerically solve the equation

$$Fg(\cdot, z) = v^{\infty}_{\tilde{g}(\cdot, z)}(\cdot, z) \qquad (7.1.10)$$

with the right-hand side given by $\Phi^{\infty}(\cdot, z)$ which has a data error of size ϵ.

The case of a penetrable inhomogeneous medium. We now turn our attention to the inhomogeneous medium case. The index of refraction is assumed to be piecewise continuously differentiable with a jump discontinuity across the smooth boundary ∂D; compare Definition 2.2.5. Our aim is to show that the far field equation (7.1.1) has an approximate solution g such that $\|g\|_{L^2(\Omega)}$ and $\|v_g\|_{L^2(D)}$ become unbounded as z tends to the boundary ∂D of the support $\chi = 1 - n$. The proof of this result is more difficult than for impenetrable obstacles and involves examining a special interior transmission problem (cf. [8, 43]) instead of an interior Dirichlet or Neumann problem for the Helmholtz equation as in the case of obstacle scattering. Our analysis will be based on a projection theorem for Hilbert spaces where the inner product is replaced by a bounded sesquilinear form.

Let X be a Hilbert space with the scalar product $(\cdot, \cdot)$ and norm $\|\cdot\|$ induced by $(\cdot, \cdot)$. Let $\langle \cdot, \cdot \rangle$ be a bounded sesquilinear form on X such that

$$|\langle \phi, \phi \rangle| \geq \gamma \|\phi\|^2 \qquad (7.1.11)$$

for all $\phi \in X$ where γ is a positive constant. For a subspace $Z \subset X$ we define $Z^{\perp}$ to be the orthogonal complement of Z with respect to $(\cdot, \cdot)$ and $Z^{\perp_*}$ to be the orthogonal complement of Z with respect to $\langle \cdot, \cdot \rangle$. By the Lax-Milgram Theorem there exists a unique bounded linear operator $M : X \to X$ such that

$$\langle \phi, \psi \rangle = (M\phi, \psi) \qquad (7.1.12)$$

for all $\phi, \psi \in X$, M is bijective and the norm of M^{-1} is bounded by γ^{-1}.

LEMMA 7.1.2 *For every closed subspace $Z \subset X$ we have the decomposition*

$$X = Z^\perp + MZ$$

where $Z^\perp \cap MZ = \{0\}$.

Proof. Define $G := Z^\perp + MZ$ and let $\phi \in G^\perp$. Then $\phi \in Z \cup (MZ)^\perp$, i.e.,

$$\langle h, \phi \rangle = (Mh, \phi) = 0$$

for all $h \in Z$. Setting $h = \phi$ by (7.1.11) we obtain $\phi = 0$ and hence $X = Z^\perp + MZ$. Now assume that for $g \in X$ we have $g = \psi_1 + Mh_1 = \psi_2 + Mh_2$. Then for $\psi := \psi_1 - \psi_2$ and $h := h_1 - h_2$ we have $0 = \psi + Mh$ with $\psi \in Z^\perp$ and $h \in Z$. Therefore

$$0 = (\psi + Mh, h) = (Mh, h) = \langle h, h \rangle$$

and hence $h = 0$. This implies $\psi = 0$ and the proof is complete. $\square$

Now let P_0 be the orthogonal projection operator in X onto the space Z with respect to the scalar product $(\cdot, \cdot)$ and let P_M be the projection operator onto MZ as defined by Lemma 7.1.2. By the closed graph theorem, P_M is a bounded operator.

LEMMA 7.1.3 *For every closed subspace $Z \subset X$ we have*

$$M^{-1}Z^\perp = (M^*Z)^\perp = Z^{\perp_*}.$$

Proof. The first equality follows from the fact that $\phi \in (M^*Z)^\perp$ if and only if $(\phi, M^*h) = (M\phi, h) = 0$ for every $h \in Z$ and hence $M\phi \in Z^\perp$, i.e., $\phi \in M^{-1}Z^\perp$. The second equality follows from the fact that $(\phi, M^*h) = (M\phi, h) = \langle \phi, h \rangle$. $\square$

We are now in a position to show that every $\phi \in X$ can be uniquely written as a sum $\phi = v + w$ with $v \in Z^{\perp_*}$ and $w \in Z$, i.e., $X = Z^{\perp_*} \oplus_s Z$ where $\oplus_s$ is the orthogonal decomposition with respect to the sesquilinear form $\langle \cdot, \cdot \rangle$.

THEOREM 7.1.4 *For every closed subspace $Z \subset X$ we have the orthogonal decomposition*

$$X = Z^{\perp_*} \oplus_s Z.$$

The projection operator $P : X \to Z^{\perp_*}$ defined by this decomposition is bounded in X.

Proof. For $\phi \in X$ we define $\hat{\phi} := M\phi$. Then from Lemma 7.1.2 we have that

$$\hat{\phi} = (1 - P_M)\hat{\phi} + P_M\hat{\phi},$$

i.e.,

$$M\phi = (1 - P_M)M\phi + P_M M\phi.$$

Hence

$$\phi = M^{-1}(1 - P_M)M\phi + M^{-1}P_M M\phi = v + w$$

where

$$v := M^{-1}(1 - P_M)M\phi \in M^{-1}Z^\perp = Z^{\perp_*}$$

$$w := M^{-1}P_M M\phi \in Z.$$

We have thus shown that $X = Z^{\perp_*} + Z$. To show the uniqueness of this decomposition, suppose $v + w = 0$ with $v \in Z^{\perp_*}$ and $w \in Z$. Then

$$0 = |\langle v, w \rangle| = |\langle w, w \rangle| \geq \gamma \|w\|^2$$

which implies that $w = v = 0$. Finally, since from the above analysis we obtain $P = M^{-1}(1 - P_M)M$ and P_M is bounded, we have that P is bounded. $\square$

We will now turn our attention to the problem of showing the existence of a unique weak solution v, w of the interior transmission problem

$$\Delta w + \kappa^2 n(x)w = 0, \quad \Delta v + \kappa^2 v = 0 \quad \text{in} \ D \qquad (7.1.13)$$

$$w - v = \Phi(x, z), \quad \frac{\partial w}{\partial \nu} - \frac{\partial v}{\partial \nu} = \frac{\partial}{\partial \nu_x}\Phi(x, z) \quad \text{on} \ D \qquad (7.1.14)$$

where $n = 1 - \chi$ and D are as defined above (see also Definition 2.2.5), and $\Phi(\cdot, z)$ is given by (1.2.40) with $z \in D$. We will further assume that there exists a positive constant c such that

$$\text{Im} \ \chi(x) \geq c \qquad (7.1.15)$$

for $x \in D$. To motivate the following definition of a weak solution of (7.1.13), (7.1.14) we note that if a solution $v, w \in C^2(D) \cap C^1(\overline{D})$ to (7.1.13), (7.1.14) exists, then from Green's formula and (7.1.13) we have that

$$w(x) + \kappa^2 \int_D \Phi(x, y)\chi(y)w(y)dy = v(x), x \in D \qquad (7.1.16)$$

and (7.1.14) will be satisfied provided

$$-\kappa^2 \int_D \Phi(x, y)\chi(y)w(y)dy = \Phi(x, y), x \in \partial B \qquad (7.1.17)$$

where B is a ball centered at the origin with $\overline{D} \subset B$. This last statement follows from Rellich's lemma and the unique continuation principle.

DEFINITION 7.1.5 (**Weak interior transmission problem.**) *Let Z be the linear space*

$$Z := \{u \in C^2(\mathbb{R}^m) : \Delta u + \kappa^2 u = 0 \text{ in } \mathbb{R}^m\}$$

and $\overline{Z}$ the closure of Z in $L^2(D)$. With the volume potential $V\varphi$ defined by (2.2.2) a pair v, w with $v \in \overline{Z}$ and $w \in L^2(D)$ is said to be a weak solution of the interior transmission problem (7.1.13), (7.1.14) with the point source $z \in D$ if v and w satisfy the integral equation

$$(I + \kappa^2 V\chi)w(x) = v(x), \quad x \in D$$

and the boundary condition

$$-\kappa^2(V\chi w)(x) = \Phi(x, z), \quad x \in \partial B.$$

Before proceeding to establish the existence of a unique weak solution to the interior transmission problem (7.1.13), (7.1.14), we make a few preliminary observations. We first note that condition (7.1.15) implies that the multiplication operator

$$(M\phi)(x) := \chi(x)\phi(x)$$

for $\phi \in L^2(D)$ is continuously invertible in $L^2(D)$ and that the sesquilinear form

$$\langle \phi, \psi \rangle := \int_D \chi(y)\phi(y)\overline{\psi(y)}dy \qquad (7.1.18)$$

satisfies the assumption (7.1.11). We also note that from the Jacobi-Anger expansion we have that the space of Herglotz wave functions with kernel $g \in L^2(\Omega)$ is a dense subset of $\overline{Z}$. Finally, by the uniqueness of the solution to the exterior Dirichlet problem for the Helmholtz equation and the unique continuation principle, we see that if v, w is a weak solution of the interior transmission problem with point source $z \in D$ then $-\kappa^2(V\chi w)(x) = \Phi(x, z)$ for all $x \in \mathbb{R}^m \setminus D$.

THEOREM 7.1.6 *For every source point $z \in D$ there exists at most one weak solution of the interior transmission problem.*

Proof. Let w and v be the difference between two weak solutions of the interior transmission problem. Then from the boundary condition

$(V\chi w)(x) = 0$ for $x \in \partial B$ we have

$$\int_D \Phi(x,y)\chi(y)w(y)dy = 0, \quad x \in \partial B. \tag{7.1.19}$$

From the addition formula for Bessel functions we now see that

$$\langle w, h \rangle = \int_D \chi(y)w(y)\overline{h(y)}dy = 0 \tag{7.1.20}$$

for all functions h of the form $h(x) = u_n$ with u_n given by (1.2.38) in the three-dimensional and the corresponding expression in the two-dimensional case. By continuity (7.1.20) also holds for $h \in \overline{Z}$, i.e., $w \in Z^{\perp_*}$.

Now let $(v_j) \subset Z$ be a sequence with $v_j \to v$ as $j \to \infty$ in $L^2(D)$ and note that $(I + \kappa^2 V\chi)^{-1}$ exists and is bounded in $L^2(D)$. Hence, for $w_j := (I + \kappa^2 V\chi)^{-1}v_j$, we have that $w_j \to w \in L^2(D)$ as $j \to \infty$. If x does not belong to D we define w_j by

$$w_j(x) := v_j(x) - \kappa^2(V\chi w_j)(x), \quad x \in \mathbb{R}^m \setminus D.$$

The functions v_j and w_j satisfy

$$\Delta v_j + \kappa^2 v_j = 0 \tag{7.1.21}$$

and

$$\Delta w_j + \kappa^2 n(x)w_j = 0 \tag{7.1.22}$$

in both D and $B \setminus \overline{D}$. From (7.1.21) and (7.1.22) by Green's first and second theorems applied to D and $B \setminus \overline{D}$ we have

$$\int_{\partial B} \left(w_j \frac{\partial \overline{v}_j}{\partial \nu} - \overline{v}_j \frac{\partial w_j}{\partial \nu} \right) ds = -\kappa^2 \int_D \chi w_j \overline{v}_j dy \tag{7.1.23}$$

$$\text{Im} \int_{\partial B} v_j \frac{\partial \overline{v}_j}{\partial \nu} ds = 0 \tag{7.1.24}$$

and

$$\text{Im} \int_{\partial B} w_j \frac{\partial \overline{u}_j}{\partial \nu} ds = -\kappa^2 \int_D \text{Im}\chi |w_j|^2 dy. \tag{7.1.25}$$

We combine (7.1.23)-(7.1.25) to obtain

$$\text{Im} \int_{\partial B} (w_j - v_j) \frac{\partial}{\partial \nu} \overline{(w_j - v_j)} ds$$

$$= \text{Im} \int_{\partial B} \left(w_j \frac{\partial \overline{w}_j}{\partial \nu} + v_j \frac{\partial \overline{v}_j}{\partial \nu} - v_j \frac{\partial \overline{w}_j}{\partial \nu} - w_j \frac{\partial \overline{v}_j}{\partial \nu} \right) ds$$

$$= -\kappa^2 \int_D \text{Im}\chi |w_j|^2 dy + \kappa^2 \text{Im} \int_D \chi w_j \overline{v}_j dy. \tag{7.1.26}$$

We now note that from (7.1.19) and (7.1.20) we have by the Cauchy-Schwarz inequality that

$$w_j(x) - v_j(x) = -(\kappa^2 V \chi w_j)(x) \to -(\kappa^2 V \chi w)(x) = 0$$

uniformly for $x \in \partial B$ and

$$\int_D \chi(y) w_j(x) \overline{v_j(y)} dy \to \int_D \chi(y) w(y) \overline{v(y)} dy = 0$$

as $j \to \infty$. Hence, taking the limit $j \to \infty$ in (7.1.26) we obtain

$$\int_D \mathrm{Im}\chi |w|^2 dy = 0.$$

From (7.1.15) we can now conclude that $w(x) = 0$ for $x \in D$. Since

$$(I + \kappa^2 V \chi) w(x) = v(x) \quad \text{for} \quad x \in D$$

we can also conclude that $v(x) = 0$ for $x \in D$ and the proof is complete. □

THEOREM 7.1.7 *For every source point $y \in D$ there exists a weak solution to the interior transmission problem.*

Proof. Choosing an appropriate coordinate system, we can assume without loss of generality that $y = 0$. We consider the space

$$Z_1^0 := \mathrm{span}\{J_p(\kappa r)e^{\pm ip\phi}, \quad p = 1, 2, \ldots\}$$

in two dimensions and the corresponding space

$$Z_1^0 := \mathrm{span}\{j_p(\kappa r)Y_p^q(\hat{x}), \quad p = 1, 2, \ldots, \quad q = -p, \ldots, p\}$$

for the three-dimensional case. Let Z_1 be the closure of Z_1^0 in $L^2(D)$. Then the space $Z_1^{\perp_*} \cap \overline{Z}$ has dimension one (note that for sufficiently small ϵ a function $h \in Z_1$ satisfies

$$\int_{|x|<\epsilon} h dx = 0$$

but this is not true for $h = J_0$ or $h = j_0$). Let ψ be a unit vector in $Z_1^{\perp_*} \cap \overline{Z}$ and P the projection operator from $L^2(D)$ onto $Z^{\perp_*}$ as defined by Theorem 7.1.4. We first consider the integral equation

$$(I + P\kappa^2 V \chi)u = P\kappa^2 V \chi \psi \tag{7.1.27}$$

in $L^2(D)$. Since $\kappa^2 V\chi$ is compact and P is bounded, the operator $P\kappa^2 V\chi$ is compact in $L^2(D)$. In order to apply the Riesz theory, we will prove uniqueness for the homogeneous equation. To this end, assume that $\hat{u} \in L^2(D)$ satisfies

$$(I + P\kappa^2 V\chi)\hat{u} = 0.$$

Then $\hat{u} \in Z^{\perp_*}$ and

$$\begin{aligned} (I + \kappa^2 V\chi)\hat{u} &= (I + P\kappa^2 V\chi)\hat{u} + (I - P)\kappa^2 V\chi\hat{u} \\ &= (I - P)\kappa^2 V\chi\hat{u} \\ &=: \hat{v} \in \overline{Z}. \end{aligned}$$

Since $\hat{v} \in \overline{Z}$ and $\hat{u} \in Z^{\perp_*}$, from Theorem 7.1.6 we have that $\hat{v} = \hat{u} = 0$. By the Riesz theory we now obtain the continuous invertibility of $I + P\kappa^2 V\chi$ in $L^2(D)$.

Now let u_0 be the solution of (7.1.27) and note that $u_0 \in Z^{\perp_*}$. We define the constant c and function $u \in L^2(D)$ by

$$c := 1/(-\kappa^2\langle J_0, \psi\rangle)$$

$$u := c(u_0 - \psi)$$

and the same expressions with J_0 replaced by j_0 for the three-dimensional case. Then we compute

$$\begin{aligned} (I + P\kappa^2 V\chi)u &= c(I + P\kappa^2 V\chi)u_0 - c(I + P\kappa^2 V\chi)\psi \\ &= -c\psi \end{aligned}$$

and

$$\begin{aligned} (I + \kappa^2 V\chi)u &= (I + P\kappa^2 V\chi)u + (I - P)\kappa^2 V\chi u \\ &= -c\psi + (I - P)\kappa^2 V\chi u \\ &=: v \in \overline{Z}. \end{aligned}$$

Since

$$\begin{aligned} \langle h, u\rangle &= c\langle h, u_0 - \psi\rangle \\ &= c\langle h, u_o\rangle - c\langle h, \psi\rangle \\ &= 0 \end{aligned}$$

for all $h \in Z_1$ and

$$\begin{aligned} \langle J_0, u\rangle &= c\langle J_0, u_0 - \psi\rangle \\ &= c\langle J_0, \psi\rangle \\ &= -1/\kappa^2 \end{aligned}$$

we have from the multipole-expansions (1.2.51) or (1.2.41), respectively, that

$$-(\kappa^2 V \chi u)(x) \;=\; \frac{i}{4} H_0^{(1)}(kr)$$
$$\;=\; \Phi(x,0), \quad x \in \partial B$$

for the two-dimensional case and a corresponding equation with H_0^1 replaced by $h_0^{(1)}$ for three dimensions, and the proof is complete. □

We will now show how the support $\overline{D}$ of χ can be determined from the far field pattern u^∞ corresponding to the inhomogeneous medium scattering problem. Following [13] we will solve the inverse problem of determining ∂D from u^∞ by calculating special approximate solutions of the equation

$$Fg = \Phi^\infty_{true}(\cdot, z), \quad z \in \mathbb{R}^m. \tag{7.1.28}$$

Note that (7.1.28) has a solution if and only if the interior transmission problem (7.1.13), (7.1.14) with source point $z \in D$ has a solution $v, w \in C^2(D) \cap C^1(\overline{D})$ such that $v = v_g$ is a Herglotz wave function with kernel g (cf. Theorem 5.16 of [43]). This is true only in very special cases. However, we will show that we can always find an approximate solution of the far field equation which has the crucial property

$$\lim_{z \to \partial D} \|g(\cdot, z)\|_{L^2(\Omega)} = \infty \quad \text{and} \quad \lim_{z \to \partial D} \|v_g\|_{L^2(D)} = \infty$$

which suffices to determine ∂D. The existence of such functions is shown in the following theorem.

THEOREM 7.1.8 *For every $\epsilon > 0$ and $z \in D$ there is a solution $g = g(\cdot, z) \in L^2(\Omega)$ of the inequality*

$$\|Fg - \Phi^\infty_{true}(\cdot, z)\|_{L^2(\Omega)} \le \epsilon \tag{7.1.29}$$

such that

$$\lim_{z \to \partial D} \|g\|_{L^2(\Omega)} = \infty \quad \text{and} \quad \lim_{z \to \partial D} \|v_g\|_{L^2(D)} = \infty \tag{7.1.30}$$

where v_g is the Herglotz wave function with kernel g.

Proof. Let $v(\cdot, z)$, $w(\cdot, z)$ be the unique weak solution to the interior transmission problem with source point z. From the remarks preceding Theorem 7.1.7 we can approximate $v(\cdot, z) \in \overline{Z}$ by a Herglotz wave function

v_g with kernel $g = g(\cdot, z)$, i.e., for every $\tilde{\epsilon} > 0$ and $y \in D$ there exists $g \in L^2(\Omega)$ such that

$$\|v(\cdot, z) - v_g\|_{L^2(D)} \leq \tilde{\epsilon}. \qquad (7.1.31)$$

Then by the continuity of the operator $(I + \kappa^2 V \chi)^{-1}$ we have for $u_g := (I + \kappa^2 V \chi)^{-1} v_g$ that

$$\|w(\cdot, z) - u_g\|_{L^2(D)} \leq c\tilde{\epsilon} \qquad (7.1.32)$$

for some positive constant c and by the continuity of $\kappa^2 V \chi : L^2(D) \to C(\partial B)$ we have that

$$\|\Phi(\cdot, z) - (-\kappa^2 V \chi u_g)\|_{C(\partial B)} \leq c' \tilde{\epsilon} \qquad (7.1.33)$$

for some positive constant c'. By the continuous dependence of the solution of the exterior Dirichlet problem with respect to the boundary data, we obtain from (7.1.33) the estimate

$$\|\Phi^\infty_{true}(\cdot, z) - (-\kappa^2 V^\infty \chi u_g)\| \leq c'' \tilde{\epsilon}$$

for a positive constant c''. We now note that the far field pattern $-\kappa^2 V^\infty \chi u_g$ of $-\kappa^2 V \chi u_g$ is given by

$$-(\kappa^2 V^\infty \chi u_g)(\hat{x}) = \int_\Omega u^\infty(\hat{x}; d) g(d) ds(d), \quad \hat{x} \in \Omega.$$

Choosing $\tilde{\epsilon} = \epsilon / c''$ we get a solution of the inequality (7.1.29).

We now need to verify (7.1.30). In the following we let c denote a generic constant. We begin by noting that since the Sobolev space $H^{3/2}(\partial D)$ is continuously imbedded in $C(\partial D)$ we have that

$$\|\Phi(\cdot, z)\|_{C(\partial D)} \leq c\|\Phi(\cdot, z)\|_{H^{3/2}(\partial D)} = c\|\kappa^2 V \chi w(\cdot, z)\|_{H^{3/2}(\partial D)}. \qquad (7.1.34)$$

Using the trace theorem and the boundedness of the operators

$$(I + \kappa^2 V \chi)^{-1} : L^2(D) \to L^2(D)$$

and $\kappa^2 V \chi : L^2(D) \to H^2(D)$ we have the estimates

$$\|\kappa^2 V \chi w(\cdot, z)\|_{H^{3/2}(\partial D)} \leq c\|\kappa^2 V \chi w(\cdot, z)\|_{H^2(D)} \leq c\|w(\cdot, z)\|_{L^2(D)} \qquad (7.1.35)$$

and

$$\|w(\cdot, z)\|_{L^2(D)} \leq c\|v(\cdot, z)\|_{L^2(D)}. \qquad (7.1.36)$$

From (7.1.31) we have that

$$\|v(\cdot, z)\|_{L^2(D)} = \|v_g + (v(\cdot, z) - v_g)\|_{L^2(D)} \leq \|v_g\|_{L^2(d)} + \tilde{\epsilon}. \qquad (7.1.37)$$

If we now combine (7.1.34)-(7.1.37) together we get

$$\|\Phi(\cdot, z)\|_{C(\partial D)} \le c(\|v_g\|_{L^2(D)} + \tilde{\epsilon}). \qquad (7.1.38)$$

Since $\Phi(x, z)$ has at least a logarithmic singularity as z tends to x we obtain the second equation of (7.1.30) from (7.1.38). The first equation of (7.1.30) is now a consequence of the observation that a Herglotz wave function with bounded kernel $g \in L^2(\Omega)$ will be bounded in $L^2(D)$. The proof is now complete.   □

Regularization and numerical applications. As a consequence of the compactness of the far field operator, the linear inverse problem of determining the function $g(\cdot; z)$ from the far field equation is ill-posed in the sense of Hadamard. Thus we need to use an appropriate regularization algorithm.

The main problems in choosing a regularization method for solving the far field equation (7.1.1) are the following:

1. Whatever regularization method we use to solve the far field equation we must allow the regularization parameter α to depend on y. This is due to the fact that according to Theorem 7.1.8 we are looking for a solution $g = g(\cdot, z)$ that blows up as z approaches ∂D. In particular, if Tikhonov's method in $L^2(\Omega)$ is adopted, the regularized solution is determined from the minimization problem

$$\|Fg - \Phi^\infty_{true}(\cdot, z)\|^2_{L^2(\Omega)} + \alpha\|g(\cdot, z)\|^2_{L^2(\Omega)} = \text{minimum} \qquad (7.1.39)$$

where the solution $\|g(\cdot, z)\|^2_{L^2(\Omega)}$ becomes unbounded as y approaches ∂D and hence the choice of α depends on z.

2. The far field equation is different than the usual linear inverse problem

$$A\varphi = f \qquad (7.1.40)$$

where the data f on the right-hand side is affected by noise and the linear operator A acting at the left hand side is known exactly. On the contrary, in our case the noisy data is the far field pattern (which is the kernel of the operator A in (7.1.40)) while the exponential function at the right-hand side of equation (7.1.1) (which is the f in (7.1.40)) is known exactly.

We will follow [13] and regularize the far field equation by the Tikhonov-Morozov regularization, i.e., we calculate the solution $g(\cdot, z)$ of

$$(\alpha I + F^* F)g_\alpha = F^* \Phi^\infty_{true}(\cdot, z), \qquad (7.1.41)$$

where α is chosen such that

$$\|Fg_\alpha - \Phi^\infty_{true}(\cdot, z)\| = \delta\|g_\alpha\|. \qquad (7.1.42)$$

The following image (Figure 7.1) shows a plot of $1/\|g(\cdot, z)\|$ on a rectangular grid containing the unknown scatterer which consists of a single kite of the form

$$x(t) = (1.5\,\sin(t), \cos(t) + 0.65\,\cos(2t) - 0.65), \quad t \in [0, 2\pi].$$

Here, we worked with $\kappa = \sqrt{2}$ and $\delta = 0.0001$.

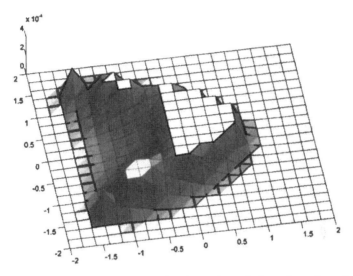

Figure 7.1: Reconstruction of a kite-shaped domain using the linear sampling method. The image shows $1/\|g(\cdot, z)\|$.

7.2   Spectral theory and a modified linear sampling method

To analyse the convergence properties of the linear sampling method Kirsch [41], [42] investigated the spectral theory of the far field operator

$$Fg(\hat{x}) := \int_\Omega u^\infty(\hat{x}, d)g(d)ds(ds), \quad \hat{x} \in \Omega \qquad (7.2.1)$$

and proposed a modified linear sampling method where instead of F^*F the operator $(F^*F)^{1/4}$ is used for the reconstructions. For this operator Kirsch proved the convergence of the method and derived a characterization of the support of a scatterer by the behaviour of the expansion coefficients of the far field

$$\Phi_{true}^{\infty}(\cdot, z), \quad z \in I\!\!R^m,$$

of point-sources with respect to the singular system of the far field operator F. We will introduce the main results and relate them to the original linear sampling method, the point-source method and the method of singular sources.

Factorizations of the far field operator. The far field operator F maps a density g onto the far field pattern of the scattered field for the incident field $u^i = v_g$ with the Herglotz wave operator H given by (3.1.3). If we consider domains for which the homogeneous interior Dirichlet problem has only the trivial solution, as used in (5.3.19) we can solve the scattering problem by a single-layer approach $u^s = S\varphi$ in an appropriate space. Then, the density φ is given by $\varphi = -S^{-1}u^i$ and the far field pattern of the solution is given by $S^\infty\varphi$ where the operator S^∞ is the L^2-adjoint of the operator $\gamma_m H$. Thus the operator $\mathcal{F}$, which maps the boundary values of the incident field onto the far field pattern of the scattering problem, satisfies $\overline{\gamma_m}H^*S^{-1} = \mathcal{F}$. We obtain the representation

$$F = -\overline{\gamma_m}H^*S^{-1}H = \mathcal{F}H. \qquad (7.2.2)$$

We take the L^2-adjoint of the equality $\overline{\gamma_m}H^* = \mathcal{F}S$ to derive $\gamma_m H = S^*\mathcal{F}^*$ and use it to transform (7.2.2) into

$$Fg = -\frac{1}{\gamma_m}\mathcal{F}S^*\mathcal{F}^*, \qquad (7.2.3)$$

where S is considered as operator on $L^2(\partial D)$ and $\mathcal{F}$ as operator $\mathcal{F}$: $L^2(\partial D) \to L^2(\Omega)$.

Spectral theory and the ranges of $\mathcal{F}$ and F. It has been shown in [9] for scattering by impenetrable obstacles, that for domains D for which the homogeneous interior Dirichlet problem has only the trivial solution the operator F is injective, normal, its eigenvalues $\lambda_j \in \mathbb{C}$ lie on a circle of radius $\frac{2\pi}{\kappa}$ with center $\frac{2\pi}{\kappa}i$ and that the corresponding eigenfunctions $\{\psi_j : j \in I\!\!N\}$ form a complete orthonormal set in $L^2(\Omega)$.

Consider a singular system $(\mu_j, \psi_j, \tilde{\psi}_j)$ of the operator F in $L^2(\Omega)$ as defined in Theorem 1.2.8. For normal F we have $\mu_j = |\lambda_j|$ and $\tilde{\psi} = \text{sign}(\lambda_j)\psi$. We will now follow [42] and use this special form together with the factorization (7.2.3) to establish a relationship between the singular system of the operator F and a special Riesz basis of the space $H^{-1/2}(\partial D)$ on the boundary ∂D of the scatterer. It will allow us in a second step to express the range of the scattering operator $\mathcal{F} : H^{1/2}(\partial D) \to L^2(\Omega)$ in terms of the decay

$$\sum_{j=1}^{\infty} \frac{|\rho_j|^2}{\mu_j} < \infty \qquad (7.2.4)$$

of the coefficients ρ_j, $j \in I\!N$, in the expansion

$$\psi = \sum_{j=1}^{\infty} \rho_j \psi_j \qquad (7.2.5)$$

of elements $\psi \in L^2(\Omega)$ with respect to the singular system $(\mu_j, \psi_j, \tilde{\psi}_j)$. Since the far field patterns $\Phi_{true}^{\infty}(\cdot, z)$ of point-sources $\Phi(\cdot, z)$ will be in the range of $\mathcal{F}$ if and only if $z \in D$, the characterization (7.2.4) directly leads to a characterization of the scattering domain D by F.

We define

$$\varphi_j := \frac{1}{i\sqrt{\lambda_j}} \mathcal{F}^* \psi_j, \quad j \in I\!N, \qquad (7.2.6)$$

where the branch of the root $\sqrt{\lambda_j}$ is chosen such that $\mathrm{Im}(\sqrt{\lambda_j}) > 0$. Then we obtain the following central theorem.

THEOREM 7.2.1 (Kirsch.) *We assume that the homogeneous interior Dirichlet problem for the domain D has only the trivial solution. Then the functions $\{\varphi_j : j \in I\!N\}$ form a Riesz basis of $H^{-1/2}(\partial D)$, i.e., the space $H^{-1/2}(\partial D)$ consists exactly of functions φ of the form*

$$\varphi = \sum_{j=1}^{\infty} \alpha_j \varphi_j \quad \text{with} \quad \sum_{j=1}^{\infty} |\alpha_j|^2 < \infty. \qquad (7.2.7)$$

Before we present the proof of Theorem 7.2.1 we will collect some well-known properties of the single-layer operator.

LEMMA 7.2.2 *We assume that the homogeneous interior Dirichlet problem for the domain D has only the trivial solution. Then the operator S and its L^2-adjoint S^* are isomorphisms from $H^{-1/2}(\partial D)$ onto $H^{1/2}(\partial D)$. $\mathrm{Im}(\varphi, S\varphi) = 0$ for some $\varphi \in H^{-1/2}(\partial D)$ implies that $\varphi = 0$, where $(\cdot, \cdot)$ denotes the duality pairing in $(H^{-1/2}(\partial D), H^{1/2}(\partial D))$. Let S_i be the single-layer potential corresponding to the wave number $\kappa = i$. The operator S_i is compact, self adjoint, and positive definite (i.e., the eigenvalues are positive) in $L^2(\partial D)$. S_i is coercive as an operator from $H^{-1/2}(\partial D)$ onto $H^{1/2}(\partial D)$, i.e.,*

$$(\varphi, S_i \varphi) \geq c_0 \|\varphi\|^2_{H^{-1/2}(\partial D)}, \quad \varphi \in H^{-1/2}(\partial D). \qquad (7.2.8)$$

Thus there exists a self adjoint and positive definite square root $S_i^{1/2}$ *of* S_i, *and* $S_i^{1/2}$ *is an isomorphism from* $H^{-1/2}(\partial D)$ *onto* $L^2(\partial D)$ *and from* $L^2(\partial D)$ *onto* $H^{1/2}(\partial D)$. *The difference* $S - S_i$ *is compact from* $H^{-1/2}(\partial D)$ *into* $H^{1/2}(\partial D)$.

Proof. See [42]. $\square$

Proof of Theorem 7.2.1. The central part of the proof of Theorem 7.2.1 will be a special orthogonality relation for the elements φ_j from which with the help of Theorem 1.2.9 we will establish that $\{\varphi_j : j \in \mathbb{N}\}$ forms a Riesz basis. We calculate

$$
\begin{aligned}
(S\varphi_j, \varphi_l)_{L^2(\partial D)} &= (\varphi_j, S^*\varphi_l) \\
&= \frac{1}{\sqrt{\lambda_j}\,\sqrt{\lambda_l}} (\mathcal{F}^*\psi_j, S^*\mathcal{F}^*\psi_l)_{L^2(\partial D)} \\
&= \frac{1}{\sqrt{\lambda_j}\,\sqrt{\lambda_l}} (\psi_j, \mathcal{F}S^*\mathcal{F}^*\psi_l)_{L^2(\partial D)} \\
&= \frac{\overline{\lambda_l}}{\sqrt{\lambda_j}\,\sqrt{\lambda_l}} (\psi_j, \psi_l)_{L^2(\partial D)} \\
&= \frac{s_j}{4\pi}\delta_{jl}, \quad j, l \in \mathbb{N},
\end{aligned}
\tag{7.2.9}
$$

where we define

$$
s_j := -\frac{\overline{\lambda_j}}{|\lambda_j|}.
\tag{7.2.10}
$$

From the above remarks about the behaviour of the eigenvalues λ_j we obtain that $|s_j| = 1$, $\mathrm{Im}(s_j) > 0$, $j \in \mathbb{N}$ and $\mathrm{Im}(s_j) \to 0$ for $j \to \infty$. We cannot directly take the square root of S, but for S_i this is possible and thus with K defined by $K := S_i^{-1/2}(S - S_i)S_i^{-1/2}$ we decompose S into $S = S_i^{1/2}(I+K)S_i^{1/2}$. According to Lemma 7.2.2 K is compact in $L^2(\partial D)$. We now apply Theorem 1.2.9 with $X = L^2(\partial D)$ to the operator $I + K$ and the functions

$$
\tilde{\varphi}_j := S_i^{1/2}\varphi_j, \quad j \in \mathbb{N}.
\tag{7.2.11}
$$

Because of the injectivity of $\mathcal{F}$, $\mathcal{F}^*$ and $S_i^{1/2}$ the functions $\tilde{\varphi}_j$, $j \in \mathbb{N}$ are linearly independent and complete in $L^2(\partial D)$. To prove that

$$
\mathrm{Im}(\varphi, K\varphi)_{L^2(\partial D)} \neq 0
$$

for all $\varphi \neq 0$ let φ be an element in $L^2(\partial D)$ with $\mathrm{Im}(\varphi, K\varphi)_{L^2(\partial D)} = 0$. We define $\psi := S_i^{-1/2}\varphi$ and obtain

$$
\mathrm{Im}(\psi, S\psi) = \mathrm{Im}(\psi, (S - S_i)\psi) = 0,
\tag{7.2.12}
$$

where we used the fact that $(\psi, S_i \psi)$ is real valued and where $(\cdot, \cdot)$ denotes the duality pairing in $(H^{-1/2}(\partial D), H^{1/2}(\partial D))$. Equation (7.2.12) by Lemma 7.2.2 implies $\varphi = 0$.

Now, by Theorem 1.2.9 we conclude that $\tilde{\varphi}_j$, $j \in I\!N$ is a Riesz basis of $L^2(\partial D)$ and by the isomorphism S_i the set $\{\varphi_j : j \in I\!N\}$ is a Riesz basis of $H^{-1/2}(\partial D)$. This completes the proof.

$\square$

As a corollary we now obtain a characterization of the range of $\mathcal{F}$ by the expansion of functions with respect to the singular system of F and a characterization of the scattering domain D by the far field pattern of point-sources.

COROLLARY 7.2.3 *Given a singular system* $(\mu_j, \psi_j, \tilde{\psi}_j)$ *of* F *the range* $R(\mathcal{F})$ *of* $\mathcal{F} : H^{1/2}(\partial D) \to L^2(\Omega)$ *is given by*

$$R(\mathcal{F}) = \left\{ \sum_{j=1}^{\infty} \rho_j \psi_j : \sum_{j=1}^{\infty} \frac{|\rho_j|^2}{\mu_j} < \infty \right\}. \tag{7.2.13}$$

The function $\Phi_{true}^{\infty}(\cdot, z)$ *is in* $R(\mathcal{F})$ *if and only if* $z \in D$.

Proof. An element φ is in $H^{-1/2}(\partial D)$ if and only if the expansion (7.2.7) is satisfied. By an application of $\mathcal{F}S^*$ we obtain

$$\begin{aligned}
\mathcal{F}S^*\varphi &= \mathcal{F}S^* \sum_{j=1}^{\infty} \alpha_j \varphi_j \\
&= \frac{i}{4\pi} \sum_{j=1}^{\infty} \alpha_j \sqrt{\lambda_j}\, \psi_j \\
&= \sum_{j=1}^{\infty} \rho_j \psi_j
\end{aligned} \tag{7.2.14}$$

with $\rho_j = i\alpha_j \sqrt{\lambda_j}/4\pi$. Since S^* is an isomorphism $H^{-1/2}(\partial D) \to H^{1/2}(\partial D)$ an element is in $R(\mathcal{F})$ if and only if its expansion coefficients ρ_j with respect to the elements ψ_j of the singular system of F satisfy

$$\sum_{j=1}^{\infty} \frac{|\rho_j|}{\mu_j} = \frac{1}{4\pi} \sum_{j=1}^{\infty} |\alpha_j|^2 < \infty. \tag{7.2.15}$$

Clearly, the far field pattern $\Phi^\infty_{true}(\cdot, z)$ can be in the range of the scattering operator if and only if the corresponding unique scattered field $\Phi(\cdot, z)$ is well defined as a solution of the Helmholtz equation in $\mathbb{R}^m \setminus \overline{D}$ with boundary values in $H^{1/2}(\partial D)$. This is true if and only if $z \in D$.

$\square$

Analogous statements have been derived by Kirsch for the Neumann- and Impedance boundary condition in [42], for the inhomogeneous medium in [41] and for the open arc by Kirsch and Ritter [49].

We close with a description of the reconstruction algorithm which is obtained in analogy to the original linear sampling method. We first remark that we might express $R(\mathcal{F})$ in terms of F by

$$R(\mathcal{F}) = R\left((F^*F)^{1/4}\right) \qquad (7.2.16)$$

with the operator $(F^*F)^\delta$ defined by

$$(F^*F)^\delta \psi := \sum_{j=1}^\infty \mu_j^{2\delta}(\psi, \psi_j)_{L^2(\Omega)} \psi_j, \quad \psi \in L^2(\Omega), \qquad (7.2.17)$$

since the power $1/4$ exactly mirrors the decay condition (7.2.13) on the coefficients ρ_j with respect to the singular system $(\mu_j, \psi_j, \tilde\psi_j)$.

Consider the equation

$$(F^*F)^{1/4}\varphi = \Phi^\infty_{true}(\cdot, z), \quad z \in \mathbb{R}^m. \qquad (7.2.18)$$

The equation is solvable if and only if $z \in D$. Since the operator F (not the right-hand side) contains the measured far field patterns for scattering of plane waves with error of size δ, we need to apply the Tikhonov-Morozov regularization as given by Theorem 1.2.14, i.e., we numerically solve the equation

$$\alpha\varphi_\alpha + (F^*F)^{1/2}\varphi_\alpha = (F^*F)^{1/4}\Phi^\infty_{true}(\cdot, z) \qquad (7.2.19)$$

with $\alpha = \alpha(\delta)$ determined by the zero of

$$G(\alpha) := \delta\|\varphi_\alpha\| - \|(F^*_\delta F_\delta)^{1/2}\varphi_\alpha - \Phi^\infty_{true}(\cdot, z)\|, \quad \alpha > 0, \qquad (7.2.20)$$

to obtain a regularized solution φ_α of (7.2.18). Due to Theorem 1.2.14 we know that for $\delta \to 0$ the solution $\varphi_{\alpha(\delta)}$ will be bounded if and only if $z \in D$. For the following pictures we chose $\kappa = \sqrt{2}$ and $\delta = 0.0001$ and $\delta = 10^{-8}$.

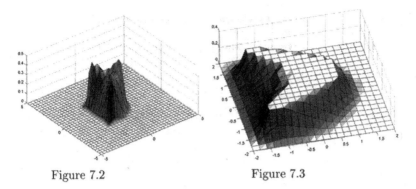

Figure 7.2   Figure 7.3

The Figures 7.2 and 7.3 above show reconstructions of a kite shaped obstacle by the modified linear sampling method, where we plot $1/\|g_\alpha(\delta)\|$ on a rectangular grid.

Finally, we would like to point out the fundamental conceptional differences between the point-source method (reconstructing the scattered field from the measurement of one far field pattern), the method of singular sources (which uses the far field patterns for many incident plane waves and detects the scatterer from the reconstructed scattered field for singular sources), and the linear sampling methods (which also need many incident plane waves and reconstruct the scatterer from the behaviour of the solution of an integral equation of the first kind).

7.3   Shape reconstruction for orthotropic media

After introducing two main branches of linear sampling methods in the preceding sections we now study the extension of these methods to electromagnetic scattering from anisotropic media. Anisotropic media are of particular interest in medical imaging, where parts of the body such as nerves are strongly anisotropic. Since the number of unknowns for anisotropic media is considerably larger than for isotropic scatterers, the reconstruction of the full density matrix is particularly difficult. Thus it is a natural step to use methods which avoid the reconstruction of the full anisotropic densities but focus on the reconstruction of special features or components of the scatterers, for example the support of an inhomogeneity. We will solve this inverse problem for an orthotropic medium (i.e., an anisotropic medium with symmetry along the third axis) in this section and treat the full three-dimensional anisotropic case in Section 7.4.

Our plan for the current section is as follows. First, we introduce and solve an interior transmission problem for an orthotropic medium. It will be used to show that there is a density $g(\cdot, z) \in L^2(\Omega)$ which approximately solves the far field equation

$$Fg = \Phi^\infty_{true}(\cdot, z), \quad z \in \mathbb{R}^2 \qquad (7.3.1)$$

and together with the Herglotz wave function v_g blows up when $z \in D$ tends to the boundary ∂D of the support $\overline{D}$ of the orthotropic medium. The results will be used to derive a linear sampling method for solving the inverse scattering problem for orthotropic media.

The interior transmission problem for an orthotropic medium. As a first step we formulate and solve a weak interior transmission problem for an orthotropic medium. Since for the solution of the interior transmission problem we will work in special weighted L^2-spaces we first spend some time to describe necessary assumptions on the matrix $M = 1 - N$ for these spaces to be suited to our needs.

Using the strongly singular integral operator T as defined by (2.5.28), from Section 2.5 we know that for electromagnetic scattering by an orthotropic medium the scattering problem can be reduced to the integral equation

$$\left(I + TM\right)\hat{u} = \nabla u^i \qquad (7.3.2)$$

in $L^2(D)$. Recall that by T_0 we denote the operator T with $\Phi(x, y)$ replaced by $\frac{1}{2\pi} \ln(|x - y|)$, i.e., the fundamental solution to the Helmholtz equation in $\mathbb{R}^2$ for wave number $\kappa = 0$. The scattered field u^s in the exterior $\mathbb{R}^2 \setminus \overline{D}$ is given by $WM\hat{u}$ with

$$(Wv)(x) := \int_D \nabla_y \Phi(x, y) \cdot v(y) dy, \quad x \in \mathbb{R}^2 \setminus D. \qquad (7.3.3)$$

The operator T_0 is bounded in $L^2(D)$ and so is the operator $T - T_0$. We follow Section 2.5 and assume that $M(x)$ can be pointwise diagonalized with a unitary complex matrix $U(x)$, i.e., we have $M(x) = U^*(x)M_D(x)U(x)$ with a diagonal matrix

$$M_D(x) = \begin{pmatrix} \eta_1 + i\eta_2 & 0 \\ 0 & \xi_1 + i\xi_2 \end{pmatrix}$$

and $U^*(x)U(x) = I$ for all $x \in \mathbb{R}^2$. We further assume that $I - M_D$ has a positive definite real part and a negative definite imaginary part. This implies that for $x \in D$ the matrix M is coercive, i.e.,

$$\mathrm{Im}(a\overline{M}a) \geq \gamma(x)|a|^2$$

for every $a \in \mathbb{C}^2$ where $\gamma(x) > 0$ for $x \in D$. It is shown in Section 2.5 that under these assumptions the operator $I + T_0 \mathcal{M}$ is continuously invertible in $L^2(D)$ and

$$I + T\mathcal{M} = I + T_0 \mathcal{M} + (T - T_0)\mathcal{M}$$

is injective. We now consider the integral equation (7.3.2) in the weighted space $L^2_\Gamma(D)$ defined by the scalar product

$$(a, b)_\Gamma := \int_D a(x) \cdot \overline{\Gamma(x)b(x)}dx \qquad (7.3.4)$$

with the positive definite Hermitean matrix function

$$\Gamma(x) := U^*(x)\mathrm{Im}(M_D)(x)U(x).$$

Then for $\tilde{\mathcal{M}} := \Gamma^{-1}\mathcal{M}$ we compute

$$\tilde{\mathcal{M}}(x) = U^*(x) \begin{pmatrix} \tilde{\eta}(x) & 0 \\ 0 & \tilde{\xi}(x) \end{pmatrix} U(x) \qquad (7.3.5)$$

for $x \in D$ with

$$\tilde{\eta} = \frac{\frac{\eta_1}{\eta_2}(1 - \frac{1}{2}\eta_1) - \frac{1}{2}\eta_2}{(1 - \frac{1}{2}\eta_1)^2 + (\frac{1}{2}\eta_2)^2} + i\frac{1}{(1 - \frac{1}{2}\eta_1)^2 + (\frac{1}{2}\eta_2)^2} \qquad (7.3.6)$$

and a corresponding expression for $\tilde{\xi}$. From this we obtain the estimate

$$\mathrm{Im}(a \cdot \tilde{\mathcal{M}}(x)\bar{a}) \geq c_1|a|^2 \qquad (7.3.7)$$

for all $x \in D$ and $a \in \mathbb{C}^3$ with

$$c_1 := \min \left\{ \inf_{x \in D} \frac{1}{(1 - \frac{1}{2}\eta_1(x))^2 + (\frac{1}{2}\eta_2(x))^2}, \qquad (7.3.8) \right.$$
$$\left. \inf_{x \in D} \frac{1}{(1 - \frac{1}{2}\xi_1(x))^2 + (\frac{1}{2}\xi_2(x))^2} \right\} > 0,$$

i.e., we have that $\tilde{\mathcal{M}}$ is coercive. With the additional assumption

$$\sup_{x \in D} \max \left\{ |\frac{\eta_1(x)}{\eta_2(x)}|, |\frac{\xi_1(x)}{\xi_2(x)}| \right\} \leq c_2' < \infty \qquad (7.3.9)$$

the matrix $\tilde{\mathcal{M}}$ satisfies the condition

$$\|\tilde{\mathcal{M}}a\|_{L^2(D)} \leq c_2\|a\|_{L^2(D)}, \qquad (7.3.10)$$

with some constant c_2 for all $a \in L^2(D)$. We also note that the square root of Γ exists and that Γ and $\mathcal{M}$ commute. We now observe from

$$
\begin{aligned}
\|\tilde{\mathcal{M}}a\|_\Gamma &= \left| \int_D \Gamma(y)\tilde{\mathcal{M}}(y)a(y) \cdot \overline{\tilde{\mathcal{M}}(y)a(y)} \, dy \right|^{1/2} \\
&= \left| \int_D \tilde{\mathcal{M}}(y)\Gamma^{1/2}(y)a(y) \cdot \overline{\tilde{\mathcal{M}}(y)\Gamma^{1/2}(y)a(y)} \, dy \right|^{1/2} \\
&\leq c_2\|a\|_\Gamma \quad\quad\quad\quad\quad\quad\quad\quad\quad\quad\quad (7.3.11)
\end{aligned}
$$

that multiplication by $\tilde{\mathcal{M}}$ is a bounded operation in $L_\Gamma^2(D)$. In the same way we compute the bound

$$
\|\tilde{\mathcal{M}}^{-1}a\|_\Gamma \leq \frac{1}{c_1}\|a\|_\Gamma, \quad\quad\quad\quad\quad\quad (7.3.12)
$$

i.e., the fact that c_1 is positive assures that $\tilde{\mathcal{M}}$ is a continuously invertible multiplication operator in $L_\Gamma^2(D)$. We also estimate

$$
\begin{aligned}
\left| \int_D a(y) \cdot \overline{\mathcal{M}(y)b(y)} \, dy \right| &= \left| \int_D a(y) \cdot \overline{\Gamma(y)\tilde{\mathcal{M}}(y)b(y)} \, dy \right| \\
&= \left| \int_D \Gamma^{1/2}(y)a(y) \cdot \overline{\Gamma^{1/2}(y)\tilde{\mathcal{M}}(y)b(y)} \, dy \right| \\
&\leq c_2\|a\|_\Gamma\|b\|_\Gamma \quad\quad\quad\quad\quad\quad (7.3.13)
\end{aligned}
$$

and obtain that the sesquilinear form defined by

$$
\langle a, b \rangle := \int_D a(x) \cdot \overline{\mathcal{M}(x)b(x)} \, dx \quad\quad\quad\quad (7.3.14)
$$

is bounded in $L_\Gamma^2(D)$.

We will now turn our attention to the problem of showing the existence of a unique weak solution $\tilde{u}, \tilde{v}$ of the two-dimensional interior transmission problem

$$
\nabla \cdot N\nabla\tilde{u} + \kappa^2\tilde{u} = 0, \quad \Delta_2\tilde{v} + \kappa^2\tilde{v} = 0 \quad \text{in } D \quad\quad (7.3.15)
$$

$$
\tilde{u} - \tilde{v} = \Phi(., x_0), \quad \frac{\partial\tilde{u}}{\partial\nu} - \frac{\partial\tilde{v}}{\partial\nu} = \frac{\partial}{\partial\nu}\Phi(., x_0) \quad \text{on } \partial D \quad\quad (7.3.16)
$$

where N is the refractive index of the orthotropic medium defined in Section 2.5. Later we will solve the inverse scattering problem with the help of a linear integral equation which has a solution if and only if the interior transmission problem (7.3.15), (7.3.16) has a solution which is a Herglotz wave function. This is true only in very special cases. Thus we need to consider appropriate approximate solutions of this integral equation. We will show the existence of such approximate solutions using a weak form of the interior transmission problem (7.3.15), (7.3.16). To motivate the following definition, we note that if a solution $u, v \in C^2(D) \cap C^1(\overline{D})$ to (7.3.15), (7.3.16) exists, then from Green's formula we have that

$$\tilde{u}(x) + \int_D \nabla_y \Phi(x, y) \cdot M(y) \nabla \tilde{u}(y) \, dy = \tilde{v}(x), \quad x \in \mathbb{R}^2 \qquad (7.3.17)$$

which is equivalent to the system of strongly singular integral equations

$$w(x) + \int_D \nabla_x \left(\nabla_x \cdot \{ \Phi(x, y) M(x) w(y) \} \right) dy = v(x), \quad x \in \mathbb{R}^2 \quad (7.3.18)$$

with $v := \nabla \tilde{v}$ and $w := (1 - \frac{1}{2}M)\nabla \tilde{u}$, i.e., using the operator T defined by (2.5.28),

$$(I + TM)w = v.$$

This motivates the following definition of a weak solution to (7.3.15) and (7.3.16) where Z^0 is the space of gradients $\nabla \tilde{v}$ of solutions $\tilde{v}$ to the Helmholtz equation on $\mathbb{R}^2$, Z is the closure of Z^0 in $L^2_\Gamma(D)$ and B is a disk with $\overline{D} \subset B$.

DEFINITION 7.3.1 *A pair w, v with $w \in L^2_\Gamma(D)$ and $v \in Z$ is said to be a solution of the weak interior transmission problem with source point $x_0 \in D$ if w, v satisfy the integral equation*

$$(I + TM)w = v \qquad (7.3.19)$$

and the boundary condition

$$(WMw)(x) = \Phi(x, x_0), \quad x \in \partial B. \qquad (7.3.20)$$

Remark. The homogeneous weak interior transmission problem is to find a solution $v \in Z, w \in L^2_\Gamma(D)$ of (7.3.19), (7.3.20) with Φ set equal to zero.

First we need to investigate the properties of the integral equation (7.3.19) in the space $L^2_\Gamma(D)$.

LEMMA 7.3.2 *The operator $I + TM$ is injective and continuously invertible in $L^2_\Gamma(D)$.*

Proof. Using (7.3.11) we obtain the estimate

$$\begin{aligned}
\|Ma\|^2_{L^2(D)} &= \int_D M(y)a(y) \cdot \overline{M(y)a(y)} \, dy \\
&= \int_D \Gamma(y)\tilde{M}(y)a(y) \cdot \overline{\Gamma(y)\tilde{M}(y)a(y)} \, dy \\
&\leq \bar{c}\|\tilde{M}a\|^2_\Gamma \\
&\leq c\|a\|^2_\Gamma
\end{aligned}$$

for some constant c and hence that multiplication by $\mathcal{M}$ maps $L^2_\Gamma(D)$ continuously into $L^2(D)$. Clearly the imbedding from $L^2(D)$ into $L^2_\Gamma(D)$ is bounded. Since every solution $w \in L^2_\Gamma(D)$ of

$$(I + T\mathcal{M})w = 0$$

is in $L^2(D)$, the injectivity of the operator $I + T\mathcal{M}$ in $L^2_\Gamma(D)$ is an immediate consequence of its injectivity in $L^2(D)$ (proven in [72]). From

$$
\begin{aligned}
(I + T\mathcal{M})^{-1} & \\
= \ & (I + T\mathcal{M})^{-1}\Big(I + T\mathcal{M} - T\mathcal{M}\Big) \\
= \ & I - (I + T\mathcal{M})^{-1}T\mathcal{M}
\end{aligned}
\tag{7.3.21}
$$

in $L^2(D)$ we see that $(I + T\mathcal{M})^{-1}$ can be extended to a bounded operator on $L^2_\Gamma(D)$ and the proof is complete. $\square$

THEOREM 7.3.3 *Let M be coercive. Then for every source point $x_0 \in D$ there exists at most one solution of the weak interior transmission problem.*

Proof. Let w, v be the difference between two solutions of the weak interior transmission problem, i.e., w, v satisfy the homogeneous weak interior transmission problem. Then from the boundary condition (7.3.20) we have for $\hat{u} := \mathcal{M}w$ that

$$\int_D \nabla_y \Phi(x, y)\hat{u}(y)dy = 0, \quad x \in \partial B. \tag{7.3.22}$$

From the addition formula for Bessel functions we now see that

$$(h, \hat{u}) = \int_D \overline{h(\xi)} \cdot \hat{u}(\xi) \, d\xi = 0 \tag{7.3.23}$$

for all functions h of the form $h(x) = \nabla_x(J_p(\kappa r)e^{ip\theta})$, $p = 0, \pm 1, \pm 2, \ldots$ where $x = (r\cos\theta, r\sin\theta)$. Note that the span of the functions

$$\nabla_x(J_p(\kappa r)e^{ip\theta}), \quad p = 0, \pm 1, \pm 2, \ldots$$

is dense in Z which can be shown by the expansion of solutions to the Helmholtz equation with respect to the basis

$$\big\{ J_p(\kappa r)e^{ip\theta} : p = 0, \pm 1, \pm 2, \ldots \big\}.$$

By continuity (7.3.23) also holds for $h \in \overline{Z}$, i.e., $\hat{u} \in \overline{Z}^\perp$.

Now let $(v_j) \subset Z$ be a sequence with $v_j \to v$ as $j \to \infty$ in $L^2_\Gamma(D)$. Then for $w_j := (I + T\mathcal{M})^{-1}v_j$ we have that $w_j \to w$, $j \to \infty$, in $L^2_\Gamma(D)$. On $\mathbb{R}^2 \setminus D$ we define w_j by

$$w_j(x) := v_j(x) - \Big(T\mathcal{M}w_j\Big)(x), \quad x \in \mathbb{R}^2 \setminus D.$$

Note that we have $v_j = \nabla u_j^i$ where u_j^i is a solution to the Helmholtz equation

$$\Delta u_j^i + \kappa^2 u_j^i = 0. \tag{7.3.24}$$

To w_j there correspond the functions

$$u_j(x) := -\nabla \cdot \int_D \Phi(x,y)(\mathcal{M}w_j)(y)\,dy \;+\; u_j^i(x), \quad x \in \mathbb{R}^2, \tag{7.3.25}$$

such that $w_j = \mathcal{M}^{-1}M\nabla u_j$ and

$$\nabla \cdot (I - M)\nabla u_j + \kappa^2 u_j = 0 \tag{7.3.26}$$

in $\mathbb{R}^2$. This follows from the definition of w_j and v_j. We apply Green's first and second theorems to obtain

$$\int_{\partial D}\left(u_j \frac{\partial \overline{u_j^i}}{\partial \nu} - \overline{u_j^i}\frac{\partial u_j}{\partial \nu}\right)ds \;=\; \int_D \overline{u_j^i}\,\nabla \cdot M\nabla u_j\,dy$$
$$=\; -\int_D \overline{\nabla u_j^i} \cdot M\nabla u_j\,dy \tag{7.3.27}$$

$$\mathrm{Im}\int_{\partial B} u_j^i \frac{\partial \overline{u_j^i}}{\partial \nu}ds \;=\; 0 \tag{7.3.28}$$

and

$$\mathrm{Im}\int_{\partial B} u_j \frac{\partial \overline{u_j}}{\partial \nu}ds \;=\; -\int_D \nabla u_j \cdot \overline{M\nabla u_j}\,dy. \tag{7.3.29}$$

Combining (7.3.27) to (7.3.29), we obtain

$$\mathrm{Im}\int_{\partial B}(u_j - u_j^i)\frac{\partial}{\partial \nu}\overline{(u_j - u_j^i)}ds$$
$$= \; \mathrm{Im}\int_{\partial B}\left(u_j^i \frac{\partial \overline{u_j^i}}{\partial \nu} + u_j \frac{\partial \overline{u_j}}{\partial \nu} - u_j \frac{\partial \overline{u_j^i}}{\partial \nu} - u_j^i \frac{\partial \overline{u_j}}{\partial \nu}\right)ds \tag{7.3.30}$$
$$= \; \mathrm{Im}\int_D \overline{\nabla u_j} \cdot M\nabla u_j\,dy + \mathrm{Im}\int_D \overline{\nabla u_j^i} \cdot M\nabla u_j\,dy.$$

From (7.3.22) and (7.3.23) with the help of the Cauchy-Schwarz inequality we get that

$$u_j^i(x) - u_j(x) = (WMw_j)(x) \to (WMw)(x) = 0 \tag{7.3.31}$$

uniformly for $x \in \partial B$ and an analogous statement for the gradient of (7.3.31). For $\hat{u} = Mw \in L^2(D)$ we obtain from (7.3.23) that

$$\int_D \overline{\nabla u_j^i} \cdot M\nabla u_j\,dy \to \int_D \overline{v} \cdot \hat{u}\,dy = 0 \tag{7.3.32}$$

as $j \to \infty$. Define ∇u by taking the gradient of (7.3.25) with w_j replaced by w and ∇u_j^i replaced by v noting that by the mapping properties of volume potentials (cf. [8]) this is well defined. Then in the limit $j \to \infty$ from (7.3.30) we obtain

$$\text{Im} \int_D \overline{\nabla u} \cdot M \nabla u \, dy = 0. \tag{7.3.33}$$

From the coercivity of M we can now conclude that $\nabla u = 0$ in D and thus that $w = \mathcal{M}^{-1} M \nabla u = 0$ and $v = (I + T\mathcal{M})w = 0$. This ends the proof. $\square$

For any subspace Z of $L^2_\Gamma(D)$ we define the spaces

$$Z^{\perp_*} := \left\{ a \in L^2_\Gamma(D) : \langle a, h \rangle = 0 \text{ for all } h \in Z \right\}$$

and

$$Z^{\perp_\Gamma} := \left\{ a \in L^2_\Gamma(D) : (a, h)_\Gamma = 0 \text{ for all } h \in Z \right\}.$$

THEOREM 7.3.4 *Let Z be a closed subspace of $L^2_\Gamma(D)$. Then*

$$L^2_\Gamma(D) = Z^{\perp_*} \oplus Z \tag{7.3.34}$$

and the projector P onto $Z^{\perp_*}$ defined by (7.3.34) is bounded as an operator on $L^2_\Gamma(D)$ by

$$\|P\|_\Gamma \leq \left(1 + \frac{c_2}{c_1}\right) \frac{c_2}{c_1}, \tag{7.3.35}$$

i.e., the bound holds uniformly for all matrices M which satisfy the conditions (7.3.7) and (7.3.10) with constants c_1 and c_2.

Proof. The first statement is a consequence of Theorem 7.1.4 applied to the space $X = L^2_\Gamma(D)$ provided

$$|\langle \phi, \phi \rangle| \geq c \|\phi\|_\Gamma^2 \tag{7.3.36}$$

for all $\phi \in X$ with a positive constant c. We see that (7.3.36) is true by using (7.3.6) - (7.3.8) to estimate

$$
\begin{aligned}
|\langle \phi, \phi \rangle| &= \left| \int_D \phi(y) \cdot M(y) \overline{\phi(y)} \, dy \right| \\
&= \left| \int_D \phi(y) \cdot \Gamma(y) \tilde{M}(y) \overline{\phi(y)} \, dy \right| \\
&= \left| \int_D \Gamma(y)^{1/2} \phi(y) \cdot \tilde{M}(y) \overline{\Gamma(y)^{1/2} \phi(y)} \, dy \right| \\
&\geq c_1 \int_D |\Gamma(y)^{1/2} \phi(y) \cdot \overline{\Gamma(y)^{1/2} \phi(y)}| \, dy \\
&= c_1 \|\phi\|_\Gamma^2
\end{aligned}
\tag{7.3.37}
$$

and hence (7.3.34) follows. To compute the bound for P we investigate the projection operator $P_{\tilde{\mathcal{M}}}$ onto $Z^{\perp_\Gamma}$ defined by

$$X = Z^{\perp_\Gamma} \oplus \tilde{\mathcal{M}} Z$$

where $\tilde{\mathcal{M}}$ is defined by (7.3.5). Let P_0 be the orthogonal projection operator onto Z in X with respect to the scalar product $(.,.)_\Gamma$ of X. Note that we have $P_0 \psi = P_0 P_{\tilde{\mathcal{M}}} \psi$ for all $\psi \in X$. For $g \in X \setminus Z^{\perp_\Gamma}$ we define

$$\psi := \frac{g}{\|P_{\tilde{\mathcal{M}}} g\|_\Gamma}, \quad w := P_{\tilde{\mathcal{M}}} \psi, \quad h := \tilde{\mathcal{M}}^{-1} w.$$

Then we have $\|w\|_\Gamma = \|\tilde{\mathcal{M}} h\| = 1$ and $h \in Z$. We estimate

$$
\begin{aligned}
\|\psi\|_\Gamma^2 &= \|(I - P_0)\psi\|_\Gamma^2 + \|P_0\psi\|_\Gamma^2 \geq \|P_0\psi\|_\Gamma^2 = \|P_0 w\|_\Gamma^2 \\
&= \sup_{g \in Z, \|g\|_\Gamma = 1} |(P_0 \tilde{\mathcal{M}} h, g)_\Gamma|^2 \geq |(\tilde{\mathcal{M}} h, \frac{h}{\|h\|})_\Gamma|^2 = \frac{1}{\|h\|_\Gamma^2} |\langle h, h \rangle|^2 \\
&\geq c_1 \|h\|_\Gamma^2 = c_1^2 \|\tilde{\mathcal{M}}^{-1} w\|_\Gamma^2 \geq \frac{c_1^2}{c_2^2} \|w\|_\Gamma^2.
\end{aligned}
\tag{7.3.38}
$$

Since we have $\|w\|_\Gamma = 1$, this yields

$$\|P_{\tilde{\mathcal{M}}} g\|_\Gamma \leq \frac{c_2}{c_1} \|g\|_\Gamma. \tag{7.3.39}$$

The estimate (7.3.35) now follows from the identity $P = \tilde{\mathcal{M}}^{-1}(I - P_{\tilde{\mathcal{M}}})\tilde{\mathcal{M}}$ together with (7.3.11) and (7.3.12). $\qquad\square$

THEOREM 7.3.5 *Let $\|\Gamma(x)\|_2 \leq \gamma$ for all $x \in D$ with a sufficiently small constant γ where $\|.\|_2$ denotes the Euclidean matrix norm. Then for every source point $x_0 \in D$ there exists a solution to the weak interior transmission problem.*

Proof. Choosing an appropriate coordinate system, we can assume without loss of generality that $x_0 = 0$. We consider the space

$$Z_1^0 := \text{span}\left\{ \nabla_x J_p(\kappa r) e^{\pm i p \theta}, \ p = 1, 2, ... \right\}$$

where $x = (r \cos\theta, r \sin\theta)$ and let Z_1 denote the closure of Z_1^0 in $L_\Gamma^2(D)$. The space $Z_1^{\perp_\bullet} \cap \overline{Z}$ is not empty. Let ψ be a unit vector in $Z_1^{\perp_\bullet} \cap \overline{Z}$ and define c_ψ by

$$c_\psi := \frac{-4i}{\langle \nabla J_0, \psi \rangle}.$$

Then for every element $w_{\perp_a} \in Z^{\perp_a}$, using the expansion (1.2.51) we obtain

$$\left(W\mathcal{M}c_\psi(w_{\perp_a} - \psi)\right)(x)$$

$$= c_\psi \int_D \nabla_y \Phi(x,y) \cdot \mathcal{M}(y)(w_{\perp_a} - \psi)(y)\, dy$$

$$= \frac{i}{4} H_0^{(1)}(\kappa|x|)\, c_\psi \int_D \nabla_y J_0(\kappa|y|) \cdot \mathcal{M}(y)\psi(y)\, dy$$

$$= \Phi(x,0). \tag{7.3.40}$$

Let P be the projection operator onto $Z^{\perp_a}$ defined in Theorem 7.3.4 and let us consider the integral equation

$$(I + PT_0\mathcal{M} + P(T - T_0)\mathcal{M})w_{\perp_a} = PT\mathcal{M}\psi \tag{7.3.41}$$

in $L^2_\Gamma(D)$. It is easily verified that the operator $P(T - T_0)\mathcal{M}$ is compact in $L^2_\Gamma(D)$ and the operators

$$\mathcal{M} : \quad L^2_\Gamma(D) \to L^2(D),$$
$$T_0 : \quad L^2(D) \to L^2(D),$$
$$P : \quad L^2_\Gamma(D) \to L^2_\Gamma(D)$$

are bounded uniformly for all matrices M for which (7.3.8) and (7.3.9) holds with fixed constants c_1 and c_2. Estimating

$$\|\psi\|^2_\Gamma = \int_D \psi(y) \cdot \overline{\Gamma(y)\psi(y)}\, dy$$

$$\leq \gamma \int_D |\psi(y)|^2\, dy = \gamma\|\psi\|_{L^2(D)} \tag{7.3.42}$$

we obtain that the norm of the imbedding operator $I_\Gamma : L^2(D) \to L^2_\Gamma(D)$ is bounded by γ and thus that for sufficiently small γ we have for $PT_0\mathcal{M} = PI_\Gamma T_0\mathcal{M}$ the estimate

$$\|PT_0\mathcal{M}\|_{L^2_\Gamma(D)} < 1$$

and hence $I + PT_0\mathcal{M}$ is continuously invertible in $L^2_\Gamma(D)$. We will now show the injectivity of $I + PT_0\mathcal{M} + P(T - T_0)\mathcal{M}$ and use the Riesz-Fredholm theory to show the existence of a solution to (7.3.41). Assume that for $\hat{w} \in L^2_\Gamma(D)$ we have

$$(I + PT\mathcal{M})\hat{w} = (I + PT_0\mathcal{M} + P(T - T_0)\mathcal{M})\hat{w} = 0.$$

Then $\hat{w} \in Z^{\perp_a}$ and

$$(I + T\mathcal{M})\hat{w} = (I + PT\mathcal{M})\hat{w} + (I - P)T\mathcal{M}\hat{w}$$
$$= (I - P)T\mathcal{M}\hat{w}$$
$$=: \hat{v} \in Z. \tag{7.3.43}$$

Now from Theorem 7.3.3 we have that $\hat{v} = \hat{w} = 0$. Applying $(I + PT_0\mathcal{M})^{-1}$ to both sides of (7.3.41), by the Riesz-Fredholm theory we now obtain the continuous invertibility of $I + PT\mathcal{M}$ in $L_{\Gamma}^2(D)$.

Now let $w_{\perp_*} \in Z^{\perp_*}$ be the solution of (7.3.41) and define

$$w := c_{\psi}(w_{\perp_*} - \psi).$$

We compute

$$
\begin{aligned}
(I + PT\mathcal{M})w &= c_{\psi}(I + PT\mathcal{M})w_{\perp_*} - c_{\psi}(I + PT\mathcal{M})\psi \\
&= -c_{\psi}\psi && (7.3.44)
\end{aligned}
$$

and

$$
\begin{aligned}
(I + T\mathcal{M})w &= (I + PT\mathcal{M})w + (I - P)T\mathcal{M}w \\
&= -c_{\psi}\psi + (I - P)T\mathcal{M}w \\
&=: v \in Z. && (7.3.45)
\end{aligned}
$$

Thus the pair w, v solves the weak interior transmission problem and the proof is finally complete. □

The inverse scattering problem. We are now prepared to use the results we have obtained for the interior transmission problem to derive a method for solving the inverse scattering problem of determining the support of a penetrable inhomogeneous orthotropic medium from a knowledge of the far field data.

Assume that the far field patterns $u^{\infty}(\hat{x}, d), \hat{x} \in \Omega$ of the scattered fields $u^s(., d)$ for incident plane waves for all directions of incidence $d \in \Omega$ are given. Recall that by $\Phi_{true}^{\infty}(\hat{x}, z), \hat{x} \in \Omega$, we denote the far field pattern of the point-source $\Phi(., z)$ defined in (1.2.40) and (1.2.50) and that a Herglotz wave function with kernel g is a solution of the Helmholtz equation of the form (1.1.39); see also (3.1.3).

To solve the inverse scattering problem of determining the support of the inhomogeneity from the far field data, we look for special approximate solutions of the equation

$$(Fg)(\hat{x}) = \Phi_{true}^{\infty}(\hat{x}, z), \quad \hat{x} \in \Omega \qquad (7.3.46)$$

where the far field operator F is defined as in (7.1.2). Note that (7.3.46) has a solution if and only if the interior transmission problem (7.3.15), (7.3.16) with source point $z \in D$ has a solution $\{w, v\}$ with a Herglotz wave function $v = v_g$. Such solutions exist only in very special cases. However, we will show that we can always find an approximate solution of (7.3.46) which has the crucial property

$$\lim_{z \to \partial D} \|g(.,z)\|_{L^2(\Omega)} = \infty, \qquad \lim_{z \to \partial D} \|v_g(.,z)\|_{L^2(D)} = \infty$$

which suffices to determine ∂D. The existence of such functions is shown in the following theorem.

THEOREM 7.3.6 *For every $\epsilon > 0$ and every $z \in D$ there is a solution $g(.,z) \in L^2(\Omega)$ of the inequality*

$$\| \int_{\Omega} u^{\infty}(\hat{x}, d) g(d) ds(d) - \Phi_{true}^{\infty}(\hat{x}, z) \|_{L^2(\Omega)} \leq \epsilon, \qquad (7.3.47)$$

such that

$$\lim_{z \to \partial D} \|g(.,z)\|_{L^2(\Omega)} = \infty \qquad and \qquad \lim_{z \to \partial D} \|v_g(.,z)\|_{L^2(D)} = \infty, \qquad (7.3.48)$$

where $v_g(.,z)$ is the Herglotz wave function with density $g(.,z)$.

Remark. The second condition in (7.3.48) ensures that the blowing up of g as $z \to \partial D$ is not only due to wild oscillations of the high frequency components of g.

Proof. Let $w(.,z)$, $v(.,z)$ be the weak solution to the interior transmission problem with source point z. Using the denseness of

$$\{ \nabla_x (J_p(\kappa r) e^{\pm i p \theta}) : \quad p = 0, 1, 2, ... \}$$

in Z and the Funk-Hecke formula it can easily be seen that we can approximate $v(.,z) \in Z$ by a Herglotz wave function $v_g(.,z)$ with density $g(.,z) \in L^2(\Omega)$, i.e., for any given $\tilde{\epsilon} > 0$ we can find $g \in L^2(\Omega)$ such that

$$\|v(.,z) - v_g(.,z)\|_{L^2(D)} \leq \tilde{\epsilon}. \qquad (7.3.49)$$

Then by continuity of the operator $(I + T\mathcal{M})^{-1}$ we have for

$$w_g := (I + T\mathcal{M})^{-1} v_g$$

that $w_g = w_g(.,z)$ satisfies

$$\|w(.,z) - w_g(.,z)\|_{L_\Gamma^2(D)} \leq c\,\tilde{\epsilon},$$

for some positive constant c and by continuity of $\mathcal{M} : L_\Gamma^2(D) \to L^2(D)$ and $W : L^2(D) \to C(\partial B)$ we have

$$\|W\mathcal{M}w_g(.,z)(.) - \Phi(.,z)\|_{C(\partial B)} \leq c'\,\tilde{\epsilon} \qquad (7.3.50)$$

for some positive constant c'. Note that the far field pattern $W^\infty \mathcal{M} w_g(., z)$ of the function $W \mathcal{M} w_g(., z)$ is given by

$$(W^\infty \mathcal{M} w_g(., z))(\hat{x}) = \int_\Omega u^\infty(\hat{x}, d) g(d) ds(d), \quad \hat{x} \in \Omega. \tag{7.3.51}$$

By the continuous dependence of the solution of the exterior Dirichlet problem with respect to the boundary data we obtain from (7.3.50) the estimate

$$\|(W^\infty \mathcal{M} w_g(., z)) - \Phi^\infty(., z)\|_{L^2(\Omega)} \leq c'' \tilde{\epsilon}$$

for a positive constant c''. Choosing $\tilde{\epsilon} = \epsilon/c''$ we get a solution of the inequality (7.3.47).

We now have to verify (7.3.48). In the following let c denote a generic constant. For a solution $w(., z), v(., z)$ of the weak interior transmission problem with source point z we have

$$\|\Phi(., z)\|_{H^{1/2}(\partial D)} = \|W \mathcal{M} w(., z)\|_{H^{1/2}(\partial D)}. \tag{7.3.52}$$

Using the trace theorem and the boundedness of the operators $S : L^2(D) \to H^1(D)$ and $(I + T\mathcal{M})^{-1} : L^2_\Gamma(D) \to L^2_\Gamma(D)$ we estimate

$$\|W \mathcal{M} w(., z)\|_{H^{1/2}(\partial D)} \leq c \|W \mathcal{M} w(., z)\|_{H^1(D)} \leq c \|w(., z)\|_{L^2_\Gamma(D)} \tag{7.3.53}$$

and

$$\|w(., z)\|_{L^2_\Gamma(D)} \leq c \|v(., z)\|_{L^2_\Gamma(D)}. \tag{7.3.54}$$

From (7.3.49) we have that

$$\|v(., z)\|_{L^2_\Gamma(D)} \tag{7.3.55}$$
$$= \|v_g(., z) + (v(., z) - v_g(., z))\|_{L^2_\Gamma(D)}$$
$$\leq \|v_g(., z)\|_{L^2_\Gamma(D)} + \tilde{\epsilon}.$$

Now combining (7.3.52) - (7.3.55) and $\|.\|_{L^2_\Gamma(D)} \leq c \|.\|_{L^2(D)}$ we get

$$\|\Phi(., z)\|_{H^{1/2}(\partial D)} \leq c \left(\|v_g(., z)\|_{L^2(D)} + \tilde{\epsilon} \right). \tag{7.3.56}$$

We complete the proof by showing that the norm $\|\Phi(., z)\|_{H^{1/2}(\partial D)}$ blows up for $z \to \partial D$. Then we obtain the second equation of (7.3.48) and the first equation of (7.3.48) follows from this.

To show that $\|\Phi(., z)\|_{H^{1/2}(\partial D)} \to \infty$ for $z \to \partial D$ let us consider the case of a unit circle Ω. Then using the integral

$$\frac{1}{2\pi} \int_0^{2\pi} \ln 4 \sin^2 \frac{\tau}{2} e^{im\tau} d\tau = \begin{cases} 0, & m = 0 \\ -\frac{1}{m}, & m = 1, 2, \dots \end{cases} \tag{7.3.57}$$

(cf. (12.16) of [51]) the Fourier coefficients a_m of the leading term $\log(|x-y|)$ of $\Phi(x,z)$ for $z \in \Omega$ can be shown to be proportional to $1/m$ and therefore $\log(|x-z|)$ is not a function in $H^{1/2}(\Omega)$. The general case can be reduced to the case of a unit circle by a simple coordinate transformation. This ends the proof.   □

7.4   Anisotropic electromagnetic media

We will now outline the proof for the case of an anisotropic medium. We use the strongly singular operator T defined by (2.6.11) and the volume potential V given by (2.2.2) for $m = 3$. Given the set $Y_n^m, m = -n, ..., n, n = 1, 2, ...$ of orthonormal spherical harmonics we define (compare equations (1.2.38) and (1.2.39))

$$u_n^m(x) := j_n(\kappa|x|)Y_n^m(\hat{x}), \qquad v_n^m(x) := h_n^{(1)}(\kappa|x|)Y_n^m(\hat{x}) \qquad (7.4.1)$$

and

$$M_n^m(x) := \text{curl}\,\{xu_n^m(x)\}, \qquad N_n^m(x) := \text{curl}\,\{xv_n^m(x)\}, \qquad (7.4.2)$$

where j_n and $h_n^{(1)}$ denote the spherical Bessel function and the spherical Hankel function of the first kind of order n. The analogue of the addition formula used in Theorem 7.3.5 for spherical vector wave functions is given by (cf. [8], Theorem 6.27)

$$
\begin{aligned}
\Phi(x,y)p &= i\kappa \sum_{n=1}^{\infty} \frac{1}{n(n+1)} \sum_{m=-n}^{n} N_n^m(x)\,\overline{M_n^m(y)}\cdot p \\
&+ \frac{i}{\kappa} \sum_{n=1}^{\infty} \frac{1}{n(n+1)} \sum_{m=-n}^{n} \text{curl}\,N_n^m(x)\,\text{curl}\,\overline{M_n^m(y)}\cdot p \\
&+ \frac{i}{\kappa} \sum_{n=1}^{\infty} \sum_{m=-n}^{n} \text{grad}\,v_n^m(x)\,\text{grad}\,\overline{u_n^m(y)}\cdot p \qquad (7.4.3)
\end{aligned}
$$

for an arbitrary vector $p \in \mathbb{R}^3$.

Following the orthotropic case, we assume that M can be diagonalized by a complex unitary matrix U, i.e., $M(x) = U(x)M_D(x)U^*(x)$ where $M_D(x)$ is a diagonal matrix and assume that $I-M_D(x)$, $x \in D$, has positive definite real part and negative definite imaginary part. This implies that, for $x \in D$, $M(x)$ is coercive. We again define the scalar product $(.,.)_\Gamma$ by (7.3.4) (where now of course $D \subset \mathbb{R}^3$). The diagonal matrix M_D gets a third diagonal element $\zeta = \zeta_1 + i\zeta_2$ and the diagonal matrix in (7.3.5) is replaced by

$$
\begin{pmatrix}
\tilde{\eta}(x) & 0 & 0 \\
0 & \tilde{\xi}(x) & 0 \\
0 & 0 & \tilde{\zeta}(x)
\end{pmatrix}
\tag{7.4.4}
$$

where in (7.3.6) the factor $\frac{1}{2}$ is replaced by $\frac{1}{3}$. We then define $\tilde{\mathcal{M}}$ as in (7.3.5). The estimates (7.3.7) to (7.3.13) now follow with the factors $\frac{1}{2}$ in (7.3.8) replaced by $\frac{1}{3}$ and

$$
|\frac{\zeta_1(x)}{\zeta_2(x)}|
$$

added to (7.3.9). Finally, the sesquilinear form $\langle .,. \rangle$ is defined by (7.3.14) and is bounded in $L_{\Gamma}^2(D)$.

For the three-dimensional anisotropic scattering problem the interior transmission problem is defined by

$$
\begin{aligned}
\operatorname{curl} E_1 - i\kappa H_1 &= 0, & \operatorname{curl} H_1 + i\kappa N(x)E_1 &= 0 \\
\operatorname{curl} E_0 - i\kappa H_0 &= 0, & \operatorname{curl} H_0 + i\kappa E_0 &= 0 & \text{in } D \quad (7.4.5)
\end{aligned}
$$

and

$$
\begin{aligned}
\nu \times (E_1 - E_0) &= \nu \times \operatorname{curl}_x q\Phi(x,z), \\
\nu \times (H_1 - H_0) &= \frac{1}{i\kappa}\nu \times \operatorname{curl}_x \operatorname{curl}_x q\Phi(x,z) & \text{on } \partial D, \quad (7.4.6)
\end{aligned}
$$

for the functions E_0, E_1, H_0, H_1 in $C^1(D) \cap C(\bar{D})$ and a given direction q where ν is again the unit outward normal to ∂D. Later we will solve the inverse scattering problem for an anisotropic medium using an integral equation which has a solution if and only if the above interior transmission problem has a solution E_0, E_1, H_0, H_1 such that E_0, H_0 is an electromagnetic Herglotz pair. This is true only in very special cases. Thus we will need to look for approximate solutions of the integral equation. To prove their existence we use a weak form of (7.4.5), (7.4.6). To motivate the weak form of the interior transmission problem we use Green's first vector theorem (see for example (6.2) of [8]) and follow Section 2.6 to obtain after some calculations that for every solution of (7.4.5), (7.4.6) the functions $w := (I - \frac{1}{3}M)E_1$ and E_0 satisfy the system of strongly singular integral equations

$$
(I + (T + \kappa^2 V)\mathcal{M})w = E_0
$$

and the boundary condition

$$
-\nu(x) \times \left((T + \kappa^2 V)\mathcal{M}w\right)(x) = \nu(x) \times \operatorname{curl}_x q\Phi(x,z), \quad x \in \partial B.
$$

This motivates the following definition of a weak solution to (7.4.5), (7.4.6) where Z^0 is the space of all entire solutions to the Maxwell equations restricted to D, Z is the closure of Z^0 in $L^2_{\Gamma}(D)$ and B is a ball with $\overline{D} \subset B$.

DEFINITION 7.4.1 *A pair w, v with $w \in L^2_{\Gamma}(D)$ and $v \in Z$ is said to be a solution of the weak interior transmission problem with source point $z \in D$ and direction q if w, v satisfy the integral equation*

$$(I + (T + \kappa^2 V)\mathcal{M})w := v \qquad (7.4.7)$$

and the boundary condition

$$-\nu(x) \times \left((T + \kappa^2 V)\mathcal{M}w\right)(x) = \nu(x) \times \text{curl } q\Phi(x, z), \quad x \in \partial B. \quad (7.4.8)$$

Remark. The homogeneous weak interior transmission problem is to find a solution $w \in L^2_{\Gamma}(D), v \in H$ of (7.4.7), (7.4.8) with Φ set equal to zero.

To prove a uniqueness theorem for the weak interior transmission problem we first need to have a closer look at the space Z. We will need the following lemma.

LEMMA 7.4.2 *Let B_R be a ball with radius R centered at the origin and $E, H \in C^1(B_R)$ satisfy Maxwell's equations in B_R. Then E has the expansion*

$$E(x) = \sum_{n=1}^{\infty} \sum_{m=-n}^{n} \left(a_n^m M_n^m(x) + b_n^m \text{curl } M_n^m(x)\right) \qquad (7.4.9)$$

which converges uniformly with all its derivatives on compact subsets of B_R.

Proof. According to Theorem 6.23 of [8] we can expand $\hat{x} \times E$ in terms of vector spherical harmonics, i.e., we have the uniformly convergent expansion

$$\hat{x} \times E(x) = \sum_{n=1}^{\infty} \sum_{m=-n}^{n} \left(\tilde{a}_n^m(|x|)\text{Grad } Y_n^m(\hat{x}) \qquad (7.4.10)\right.$$
$$\left. + \tilde{b}_n^m(|x|)(\hat{x} \times \text{Grad } Y_n^m)(\hat{x})\right).$$

We know that the set

$$\mathcal{R} := (0, R) \setminus \{r \in \mathbb{R}, j_n(\kappa r) = 0 \text{ or } j_n(\kappa r) + \kappa r j_n'(\kappa r) = 0$$
$$\text{for one } n \in \mathbb{N} \text{ or } \kappa \text{ is a Maxwell eigenvalue for } B_r\}$$

is dense in $[0, R]$. Let us choose $R' \in \mathcal{R}$. Then we can write the expansion (7.4.10) for $|x| = R'$ in the form

$$\hat{x} \times E(x) = \sum_{n=1}^{\infty} \sum_{m=-n}^{n} a_n^m \, j_n(\kappa|x|) \, \hat{x} \times \left(\text{Grad } Y_n^m(\hat{x}) \times \hat{x} \right) \text{(7.4.11)}$$

$$+ \sum_{n=1}^{\infty} \sum_{m=-n}^{n} b_n^m \frac{1}{|x|} \left(j_n(\kappa|x| + \kappa|x|j_n'(\kappa|x|)) \right) \hat{x} \times \text{Grad } Y_n^m(\hat{x}).$$

Using the continuous dependence of the solution to the interior Maxwell problem with respect to the tangential component of the boundary values in the maximum norm and

$$j_n(\kappa|x|) \text{Grad } Y_n^m(\hat{x}) \times \hat{x} = M_n^m(x),$$

$$\frac{1}{|x|} \left(j_n(\kappa|x| + \kappa|x|j_n'(\kappa|x|)) \right) \hat{x} \times \text{Grad } Y_n^m(\hat{x}) = \hat{x} \times \text{curl } M_n^m,$$

we obtain the uniform convergence of the series (7.4.9) together with its derivatives on compact subsets of $B_{R'}$.

We would like to show that the series (7.4.9) is also convergent on compact subsets of $B_R \supset B_{R'}$. To this end let us consider $\rho \in \mathcal{R}$ with $\rho > R'$, for which we obtain an analogous sequence with coefficients α_n^m and β_n^m. Computing $\hat{x} \times E(x)$ on $|x| = R'$ for both expansions and using the orthogonality of Grad Y_n^m and $\hat{x} \times$ Grad Y_n^m for $n \in \mathbb{N}, m = -n, ..., n$ we obtain $a_n^m = \alpha_n^m$ and $b_n^m = \beta_n^m$, which ends the proof. $\square$

THEOREM 7.4.3 *Let M be coercive. Then for every source point $z \in D$ and every direction $q \in \mathbb{R}^3$ there exists at most one solution of the weak interior transmission problem.*

Proof. Without loss of generality we can assume that $z = 0$. Let w, v be the difference between two weak solutions of the interior transmission problem, i.e., w, v satisfy the homogeneous weak interior transmission problem and in particular the homogeneous boundary condition

$$-\nu(x) \times \left((T\mathcal{M} + \kappa^2 V \mathcal{M})w \right)(x) = 0, \quad x \in \partial B. \tag{7.4.12}$$

By the uniqueness of the exterior Maxwell problem and analyticity we obtain from (7.4.12) that $\left((T + \kappa^2 V)\mathcal{M}w \right)(x) = 0$, $x \in \mathbb{R}^3 \setminus \overline{D}$. We consider curl $(T + \kappa^2 V)\mathcal{M}w$, use curl grad $\varphi = 0$ for two times continuously differentiable functions φ, the identity

$$\text{curl curl } N_n^m = \kappa^2 N_n^m$$

and the addition formula (7.4.3) to obtain the equations

$$\int_D M_n^m(y) \cdot \overline{Mw(y)} \, dy = 0, \qquad \int_D \text{curl } M_n^m(y) \cdot \overline{Mw(y)} \, dy = 0, \quad (7.4.13)$$

for $n = 1, 2, \ldots$ and $m = -n, \ldots, n$. Since according to Lemma 7.4.2 the functions M_n^m and curl M_n^m are dense in Z, we have the orthogonality relation

$$(h, Mw) = \int_D h(y) \cdot \overline{Mw(y)} \, dy = 0 \qquad (7.4.14)$$

for all $h \in Z$. On the other hand from (7.4.14), taking the curl of $(T + \kappa^2 V)Mw$, we can deduce that curl $(T + \kappa^2 V)Mw$ vanishes outside B and hence by Maxwell's equations (7.4.12) is valid. We have proven that (7.4.12) is equivalent to $(h, Mw) = 0$ for all $h \in H$, i.e., $Mw \in H^\perp$.

Now let $(v_j) \subset H^0$ be a sequence with $v_j \to v$, $j \to \infty$, in $L_{\Gamma}^2(D)$ and note that $(I + (T + \kappa^2 V)M)^{-1}$ exists in $L_{\Gamma}^2(D)$ and is bounded. Then for

$$w_j := (I + (T + \kappa^2 V)M)^{-1} v_j$$

we also have the convergence $w_j \to w, j \to \infty$, in $L_{\Gamma}^2(D)$. For $x \notin D$ we define $w_j(x)$ by

$$w_j(x) := v_j(x) - \left((T + \kappa^2 V)Mw_j \right)(x), \quad x \in \mathbb{R}^3 \setminus D. \qquad (7.4.15)$$

The incident fields v_j solve the free space Maxwell equations in $\mathbb{R}^3$ and are thus divergence free. The functions

$$E_j(x) := (I - \tfrac{1}{3} M(x))^{-1} \left(v_j - ((T + \kappa^2 V)Mw_j) \right)(x), \quad x \in \mathbb{R}^3, \quad (7.4.16)$$

satisfy the Maxwell equations in B and we have $ME_j = Mw_j$. Note that ME_j converges to Mw in $L_{\Gamma}^2(D)$. From Green's first vector theorem applied to E_j and Mw_j in B we obtain by straightforward calculation that

$$\int_B \left\{ |\text{curl } (E_j - Mw_j)|^2 - \kappa^2 |E_j - Mw_j|^2 \right.$$

$$+ \kappa^2 \overline{E_j} \cdot ME_j - \kappa^2 \overline{Mw_j} \cdot ME_j \Big\} \, dy$$

$$= \int_{\partial B} (\nu \times \overline{(T + \kappa^2 V)ME_j}) \cdot \text{curl } ((T + \kappa^2 V)ME_j) ds. \qquad (7.4.17)$$

We first observe that from $Mw \in H^\perp$ and with the help of the addition formula (7.4.3) we obtain

$$\text{curl } \left((T + \kappa^2 V)Mw \right)(x) = 0, \quad x \in \partial B.$$

Using the Cauchy-Schwarz inequality, this yields

$$\int_{\partial B} (\nu \times (\overline{(T + \kappa^2 V) M E_j})) \cdot \text{curl } ((T + \kappa^2 V) M E_j) ds$$

$$\overset{j \to \infty}{\longrightarrow} \int_{\partial B} (\nu \times (\overline{(T + \kappa^2 V) \mathcal{M} w})) \cdot \text{curl } ((T + \kappa^2 V) \mathcal{M} w) ds = 0.$$

Since $\mathcal{M} w \in H^\perp$ and $w \in H$ we have

$$\int_B \overline{w(y)} \cdot \mathcal{M} w(y) \, dy = 0$$

and thus

$$\kappa^2 \int_B \overline{\mathcal{M} w_j(y)} \cdot M(y) E_j(y) \, dy \overset{j \to \infty}{\longrightarrow} 0. \tag{7.4.18}$$

We now take the limit $j \to \infty$ for the imaginary part of (7.4.17) to obtain

$$\int_D \text{Im}\left(\overline{E(y)} \cdot M(y) E(y)\right) dy = 0. \tag{7.4.19}$$

Since M is coercive this yields $E = 0$, i.e., $w = \mathcal{M}^{-1} M E = 0$, and hence $v = (I + (T + \kappa^2 V) \mathcal{M}) w = 0$. The proof of the theorem is now complete. □

THEOREM 7.4.4 *Let* $\|\Gamma(x)\|_2 \leq \gamma$ *for all* $x \in D$ *with a sufficiently small constant* γ. *Then for every source point* z *and direction* q *there exists a solution to the weak interior transmission problem.*

Proof. Without loss of generality we can assume that $z = 0$ and $q = (1, 0, 0)$. In this case we have

$$\text{curl } q\Phi(x, 0) = \frac{i\kappa^2}{\sqrt{12\pi}} N_1^0(x).$$

We consider the space

$$Z_1^0 := \text{span}\{\text{curl } M_1^0, M_n^m : \ n = 1, m = \pm 1 \text{ or } n = 2, 3, ..., m = -n, ..., n\} \tag{7.4.20}$$

and let Z_1 be the closure of Z_1^0 in L_Γ^2. Then $L_\Gamma^2(D) = Z^{\perp_\bullet} \oplus Z$ where

$$Z^{\perp_\bullet} := \{a \in L_\Gamma^2(D) : \langle h, a \rangle = 0 \text{ for all } h \in Z\}.$$

The space $Z_1^{\perp_\bullet} \cap Z$ is not empty. Let ψ be a unit normal vector in $Z_1^{\perp_\bullet} \cap Z$ and define c_ψ by

$$c_\psi := \frac{\kappa}{\sqrt{3\pi} \langle \psi, M_1^0 \rangle}.$$

Then for every element $w_{\perp_*} \in Z^{\perp_*}$ using the addition formula (7.4.3) we obtain

$$\text{curl } (T + \kappa^2 V)\mathcal{M}c_\psi(w_{\perp_*} - \psi) \;=\; \text{curl } q\Phi(x, 0)$$

outside some ball B with $\overline{D} \subset B$. By unique continuation, this holds for $x \in \mathbb{R}^3 \setminus D$. Since $(T + \kappa^2 V)\mathcal{M}c_\psi(w_{\perp_*} - \psi)$ solves the Maxwell equations in $\mathbb{R}^3 \setminus D$ this implies that $(T + \kappa^2 V)\mathcal{M}w = q\Phi(x, 0)$ for $w := c_\psi(w_{\perp_*} - \psi)$, i.e., that (7.4.8) is satisfied.

Let P be the projection operator in $L_\Gamma^2(D)$ onto $Z^{\perp_*}$ defined analogously to the orthotropic case of Theorem 7.3.4 and consider the integral equation

$$(I + PT_0\mathcal{M} + P(T - T_0)\mathcal{M} + P\kappa^2 V\mathcal{M})w_{\perp_*} \;=\; P(T + \kappa^2 V)\mathcal{M}\psi \quad (7.4.21)$$

in $L_\Gamma^2(D)$. The operators $P(T - T_0)\mathcal{M}$ and $P\kappa^2 V\mathcal{M}$ are compact. We can now proceed exactly in the same way as in the proof of Theorem 7.3.5 to show the existence of a solution to equation (7.4.21) and to complete the proof.   □

Let us now investigate the three-dimensional anisotropic medium. We assume that the electric far field patterns $E_{pl}^\infty(\hat{x}, d, p), \hat{x} \in \Omega$, of the scattered fields for incident plane waves with direction of incidence d and polarization p are given for all $d, p \in \Omega$. We use $E_{q,true}^\infty(\hat{x}, z), H_{q,true}^\infty(\hat{x}, z), \hat{x} \in \Omega$, for the far field pattern of the electromagnetic field

$$E_q(x, z) := \text{curl }_z q\Phi(x, z), \quad H_q(x, z) := \frac{1}{i\kappa}\text{curl }_z E_q(x, z), \quad (7.4.22)$$

$x \in \mathbb{R}^3 \setminus \{z\}$. By $L_{tan}^2(\Omega)$ we denote the space of all square integrable tangential vector fields on the unit sphere. To solve the inverse electromagnetic scattering problem described above we look for special approximate solutions $a(., z) \in L_{tan}^2(\Omega)$ of the equation

$$\int_\Omega E_{pl}^\infty(\hat{x}, d, a(d, z)) \, ds(d) \;=\; E_{q,true}^\infty(\hat{x}, z), \quad \hat{x} \in \Omega \quad (7.4.23)$$

which satisfy the crucial property

$$\lim_{z \to \partial D} \|a(., z)\|_{L^2(\Omega)} = \infty, \qquad \lim_{z \to \partial D} \|v_a(., z)\|_{L^2(D)} = \infty \quad (7.4.24)$$

where $v_a, \frac{1}{i\kappa}\text{curl } v_a$ is an electromagnetic Herglotz pair with density $a \in L_{tan}^2(\Omega)$, i.e., $v_a, \frac{1}{i\kappa}\text{curl } v_a$ is a solution of the Maxwell equations with v_a having the form

$$v_a(x) := \int_\Omega e^{i\kappa x \cdot d} a(d) ds(d), x \in \mathbb{R}^3 \quad (7.4.25)$$

with $a \in L^2_{tan}(\Omega)$. We note that (7.4.23) has a solution if and only if the interior transmission problem (7.4.5), (7.4.6) with source point $z \in D$ and direction q has a solution E_0, E_1, H_0, H_1 such that E_0, H_0 is an electromagnetic Herglotz pair.

To prove the existence of approximate solutions of (7.4.23) satisfying (7.4.24) we first need the following lemma where, as previously defined, Z is the closure in $L^2_\Gamma(D)$ of the space of all entire solutions to the Maxwell equations.

LEMMA 7.4.5 *The space of electromagnetic Herglotz wave functions*

$$v_a(x) := \int_\Omega e^{i\kappa x \cdot d} a(d) ds(d), \quad x \in \mathbb{R}^3 \qquad (7.4.26)$$

with $a \in L^2_{tan}(\Omega)$ is a dense subspace of Z.

Proof. Let us consider $v \in Z$ and an approximation $v^\epsilon \in Z^0$ to v with respect to the norm in $L^2_\Gamma(D)$. Using Lemma 7.4.2 we can expand v^ϵ with respect to the basis functions $M^m_n(x)$ and curl $M^m_n(x)$ with $n = 1, 2, \ldots$ and $m = -n, \ldots, n$. The expansion also converges in the space $L^2_\Gamma(D)$. We will complete the proof by showing that the functions $M^m_n(x)$ and curl $M^m_n(x)$ are Herglotz wave functions, i.e., that the electromagnetic fields $M^m_n(x)$, $\frac{1}{i\kappa}$curl $M^m_n(x)$ and curl $M^m_n(x)$, $\frac{1}{i\kappa}$curl curl $M^m_n(x)$ for $n = 1, 2, \ldots, m = -n, \ldots, n$ are electromagnetic Herglotz pairs. An entire solution (E, H) to the Maxwell equations is an electromagnetic Herglotz pair if and only if it possesses the growth property

$$\sup_{R>0} \frac{1}{R} \int_{|x| \le R} \left\{ |E(x)|^2 + |H(x)|^2 \right\} dx \ < \ \infty \qquad (7.4.27)$$

(see Theorem 6.30 of [8]). For $M^m_n(x) = j_n(\kappa|x|)$Grad $Y^m_n(\hat{x}) \times \hat{x}$ we have

$$\text{curl } M^m_n(x) \ = \ j_n(\kappa|x|) \text{ Grad } \times \left(\text{Grad } Y^m_n(\hat{x}) \times \hat{x} \right) \qquad (7.4.28)$$
$$+ \kappa^2 \hat{x} j'_n(\kappa|x|) \times \left(\text{Grad } Y^m_n(\hat{x}) \times \hat{x} \right)$$

and curl curl $M^m_n(x) = \kappa^2 M^m_n(x)$. For the electromagnetic fields defined above, (7.4.27) now follows from the growth properties of the Bessel functions j_n (see (3.44) of [8]). $\qquad \square$

We now proceed to the proof of the existence of an approximate solution of (7.4.23) satisfying (7.4.24).

THEOREM 7.4.6 *For every $\epsilon > 0$ and every $z \in D$ there is a solution $a(.,z) \in L^2_{tan}(\Omega)$ of the inequality*

$$\left\| \int_\Omega E^\infty_{pl}(\hat{x},d,a(d,z))\, ds(d)\; -\; E^\infty_{q,true}(\hat{x},z) \right\|_{L^2(\Omega)} \leq \epsilon \qquad (7.4.29)$$

such that

$$\lim_{z \to \partial D} \|a(.,z)\|_{L^2(\Omega)} = \infty \qquad \text{and} \qquad \lim_{z \to \partial D} \|v_a(.,z)(.)\|_{L^2(D)} = \infty \qquad (7.4.30)$$

where v_a is an electromagnetic Herglotz wave function with density $a(.)$.

Proof. Let $\{w(.,z), v(.,z)\}$ be the weak solution to the interior transmission problem with polarization q and source point z. We first use Lemma 7.4.5 to approximate $v(.,z)$ by a Herglotz wave function $v_a(.,z)$ with tangential density $a(.,z)$, i.e., for given $\tilde{\epsilon} > 0$ we can find $a \in L^2_{tan}(\Omega)$ such that

$$\|v(.,z) - v_a(.,z)\|_{L^2_t(D)} \leq \tilde{\epsilon}. \qquad (7.4.31)$$

We now define $w_a = w_a(.,z)$ by

$$w_a := (I + (T + \kappa^2 V)\mathcal{M})^{-1} v_a.$$

Using the boundedness of $(I + (T + \kappa^2 V)\mathcal{M})^{-1}$, the continuity of $(T^\infty + \kappa^2 V^\infty)\mathcal{M}$ and

$$(T^\infty + \kappa^2 V^\infty)\mathcal{M}w(.,z) = E^\infty_{q,true}(.,z)$$

we obtain

$$\|(T^\infty + \kappa^2 V^\infty)\mathcal{M}w_a(.,z) - E^\infty_{q,true}(.,z)\|_{L^2(\Omega)} \leq c''\tilde{\epsilon} \qquad (7.4.32)$$

with some positive constant c'' where $T^\infty \mathcal{M}w$ and $\kappa^2 V^\infty \mathcal{M}w$ denote the far field patterns of the functions $T\mathcal{M}w$ and $\kappa^2 V \mathcal{M}w$, respectively. We now interchange the order of integration to obtain

$$\left((T^\infty + \kappa^2 V^\infty)\mathcal{M}w_a(.,z) \right)(\hat{x}) = \int_\Omega E^\infty_{pl}(\hat{x},d,a(d))ds(d), \quad \hat{x} \in \Omega, \qquad (7.4.33)$$

i.e., $a(.,z)$ satisfies (7.4.29).

For the proof of (7.4.30) we have to proceed in a different way than for the proof of (7.3.48), since the operator T now maps $L^2(D)$ continuously into $L^2(D)$, but not into $H^1(D)$. We first observe that the field curl $q\Phi(x,z)$ is not square integrable on any ball $B_r(z)$ with radius $r > 0$ and center z, i.e., we have

$$\lim_{z \to \partial D} \|\text{curl } q\Phi(.,z)\|_{L^2(B \setminus \overline{D})} = \infty \qquad (7.4.34)$$

where B is a ball containing $\overline{D}$ in its interior. We also have the estimates

$$\|\text{curl } q\Phi(.,z)\|_{L^2(B \setminus \overline{D})} = \|(T + \kappa^2 V)\mathcal{M}w(.,z)\|_{L^2(B \setminus \overline{D})}$$
$$\leq c\|w(.,z)\|_{L^2_\Gamma(D)} = \|(I - (T + \kappa^2 V)\mathcal{M})^{-1}v(.,z)\|_{L^2_\Gamma(D)}$$
$$\leq c\|v(.,z)\|_{L^2_\Gamma(D)}. \qquad (7.4.35)$$

From (7.4.31) we obtain

$$\|v(.,z)\|_{L^2_\Gamma(D)} = \|v_a(.,z) + (v(.,z) - v_a(.,z))\|_{L^2_\Gamma(D)}$$
$$\leq \|v_a(.,z)\|_{L^2_\Gamma(D)} + \tilde{\epsilon}. \qquad (7.4.36)$$

Combining (7.4.34) - (7.4.36) and $\|.\|_{L^2_\Gamma(D)} \leq c\|.\|_{L^2(D)}$, with the help of the Cauchy-Schwartz inequality we obtain (7.4.30). This ends the proof.□

Given the results of Theorems 7.3.6 and 7.4.6, respectively, the support $\overline{D}$ of M is now reconstructed by finding a (regularized) solution g or a of the far field equation (7.4.23) for z on some lattice $\mathcal{L}$ containing $\overline{D}$ and then determining those points $z \in \mathcal{L}$ where $\|g\|_{L^2(\Omega)}$ or $\|a\|_{L^2(\Omega)}$ achieves its maximum.

References

[1] Berger, M.S.: *Nonlinearity and Functional Analysis*. Academic Press 1977.

[2] Bruno, O.P. and Reitich, F.: Calculation of Electromagnetic Scattering via Boundary Variations and Analytic Continuation. Preprint.

[3] Chadan, K. and Sabatier, P.C.: *Inverse Problems in Quantum Scattering Theory*. Springer-Verlag, Berlin Heidelberg New York 1989.

[4] Colton, D. and Erbe, C.: *Spectral Theory for the Magnetic Far Field Operator in an Orthotropic Medium*. Nonlinear Problems in Applied Mathematics, T.S. Angell, et al., editors, SIAM Publications, Philadelphia, 1996, 96-103.

[5] Colton, D., Giebermann, K. and Monk, P.: A regularized sampling method for solving three dimensional inverse scattering problems. SIAM J. Scientific Computation (to appear).

[6] Colton, D. and Kirsch, A.: A simple method for solving inverse scattering problems in the resonance region. Inverse Problems 12, 383-393 (1996).

[7] Colton, D. and Kress, R.: *Integral Equation Methods in Scattering Theory*. John Wiley and Sons 1983.

[8] Colton, D. and Kress, R.: *Inverse Acoustic and Electromagnetic Scattering Theory*. 2nd Ed., Springer-Verlag 1998.

[9] Colton, D. and Kress, R.: Eigenvalues of the Far Field Operator and Inverse Scattering Theory. SIAM J. Math. Anal. 26, 601-615 (1995).

[10] Colton, D. and Monk, P.: A linear sampling method for the detection of leukemia using microwaves. SIAM J. Appl. Math. 58, 926-941 (1998).

[11] Colton, D. and Päivärinta, L.: Far-field patterns and the inverse scattering problem for electromagnetic waves in an inhomogeneous medium. Math. Proc. Camb. Phil. Soc. 103, 561-575 (1990).

[12] Colton, D. and Päivärinta, L.: The uniqueness of a solution to an inverse scattering problem for electromagnetic waves. Arch. Rational Mech. Anal. 119, 59-70 (1992).

[13] Colton, D., Piana, M. and Potthast, R.: A simple method using Morozov's discrepancy principle for solving inverse scattering problems. Inverse Problems 13, 1477-1493 (1997).

[14] Colton, D. and Potthast, R.: The inverse electromagnetic scattering problem for an anisotropic medium. Q. Jl. Mech. Appl. Math. 52 (3), 349-372 (1999).

[15] Colton, D. and Sleeman, B.D.: Uniqueness theorems for the inverse problem of acoustic scattering. IMA J. Appl. Math. 31, 253-259 (1983).

[16] Engl, H., Hanke, M. and Neubauer, A.: *Regularization of Inverse Problems*. Kluwer 1996.

[17] Garabedian, P. R.: *Partial Differential Equations*. Chelsa Pub. Comp. 1986.

[18] Gilbarg, D. and Trudinger, N.S.: *Elliptic Partial Differential Equations of Second Order*. Springer-Verlag 1977.

[19] Groetsch, C.W.: *The Theory of Tikhonov Regularization for Fredholm Equations of the First Kind*. Pitman, Boston 1984.

[20] Günther, N.M.: *Potential Theory*, Frederick Ungar, New York 1967.

[21] Giebermann, K. and Potthast, R.: An efficient method in three-dimensional inverse obstacle scattering (in preparation).

[22] Hadamard, J.: *Lectures on Cauchy's Problem in Linear Partial Differential Equations*. Yale University Press, New Haven 1923.

[23] Hähner, P.: A periodic Faddeev-type solution operator. Jour. of Differential Equations 128, 300-308 (1996).

[24] Hähner, P.: *On Acoustic, Electromagnetic and Elastic Scattering Problems in Inhomogeneous Media*. Habilitation thesis, Göttingen 1998.

[25] Hähner, P. and Hohage, T.: New stability estimates for the inverse acoustic inhomogeneous medium problem and applications. Preprint.

[26] Hanke, M., Hettlich, F. and Scherzer, O.: A convergence analysis for the Landweber iteration for nonlinear ill-posed problems. Numer. Math. 72, 21-37 (1995).

[27] Hettlich, F.: An iterative method for the inverse scattering problem from sound-hard obstacles. In: *Proceedings of the ICIAM 95, Vol. II, Applied Analysis* (Mahrenholz and Mennicken, eds). Akademie Verlag, Berlin (1996).

[28] Hettlich, F.: *The Domain Derivative in Inverse Obstacle Problems*. Habilitation Thesis, Erlangen 1999.

[29] Heuser, H.: *Funktionalanalysis*, Teubner 1986.

[30] Hohage, T.: Logarithmic convergence rates of the iteratively regularized Gauss-Newton method for an inverse potential and an inverse
scattering problem. Inverse Problems 13, 1279-1299 (1997).

[31] Hohage, T.: Iterative Methods in Inverse Obstacle Scattering: Regularization Theory of Linear and Nonlinear Exponentially Ill-Posed
Problems. PhD-Thesis, University of Linz 1999.

[32] Hörmander, L.: *The Analysis of Linear Partial Differential Operators
III.* Springer-Verlag 1985.

[33] Isakov, V.: On uniqueness in the inverse transmission scattering problem. Comm. Part. Diff. Equa. 15, 1565-642 (1990).

[34] Isakov, V.: Stability estimates for obstacles in inverse scattering. J.
Comp. Appl. Math. 42, 79-89 (1992).

[35] Isakov, V.: New stability results for soft obstacles in inverse scattering.
Inverse Problems 9, 535-543 (1993).

[36] Isakov, V.: *Inverse Problems for Partial Differential Equations.*
Springer-Verlag, New York 1998.

[37] John, F.: Continuous dependence on data for solutions of partial differential equations with a prescribed bound. Comm. Pure Appl. Math.
Vol. XIII, 551-585 (1960).

[38] Jones, D.S.: *Acoustic and Electromagnetic Waves.* Clarendon Press,
Oxford 1986.

[39] Kersten, H.: Grenz- und Sprungrelationen für Potentiale mit quadrat-
summierbarer Dichte. Resultate d. Math. 3, 17-24 (1980).

[40] Kirsch, A.: New Characterization of Solutions in Inverse Scattering
Theory. Preprint.

[41] Kirsch, A.: Factorization of the far field operator for the inhomogeneous medium case and an application in inverse scattering theory,
Inverse Problems 15, 413-429 (1999).

[42] Kirsch, A.: Characterization of the shape of the scattering obstacle
using the spectral data of the far field operator, Inverse Problems 14,
1489-1512 (1998).

[43] Kirsch, A.: *An Introduction to the Mathematical Theory of Inverse
Problems.* Springer-Verlag, New York 1996.

[44] Kirsch, A.: The domain derivative and two applications in inverse scattering. Inverse Problems 9, 81-96 (1993).

[45] Kirsch, A.: Numerical algorithm in inverse scattering theory. In: *Ordinary and Partial Differential Equations, Vol. IV* (Jarvis and Sleeman, eds). Pitman Research Notes in Mathematics 289, Longman, London 93-111 (1993).

[46] Kirsch, A.: Surface gradients and continuity properties for some integral operators in classical scattering theory. Math. Meth. in the Appl. Sci. 11, 789-804 (1989).

[47] Kirsch, A. and Kress, R.: Uniqueness in inverse obstacle scattering. Inverse Problems 9, 285-299 (1993).

[48] Kirsch, A. and Monk, P.: A finite element/spectral method for approximating the time-harmonic Maxwell system in $I\!R^3$. SIAM J. Appl. Math. 55, 1324-1344 (1995).

[49] Kirsch, A. and Ritter, S.: A Linear Sampling Method for Inverse Scattering from an Open Arc. Preprint.

[50] Kress, R.: Integral equation methods in inverse acoustic and electromagnetic scattering. In: *Boundary Integral Formulations for Inverse Analysis* (Ingham and Wrobel, eds). Computational Mechanics Publications, Southampton, 67-92 (1997).

[51] Kreß, R.: *Linear Integral Equations.* 2nd Ed. Springer-Verlag 1999.

[52] Kreß, R.: A Newton method in inverse obstacle scattering. In: *Inverse Problems in Engineering Mechanics* (Bui et al., eds). Balkema, Rotterdam, 425-432 (1994).

[53] Kreß, R.: Integral equation methods in inverse obstacle scattering. Engineering Anal. with Boundary Elements 15, 171-179 (1995).

[54] Kress, R. and Rundell, W.: A quasi-Newton method in inverse obstacle scattering. Inverse Problems 10, 1145-1157 (1994).

[55] Kress, R. and Rundell, W.: Inverse obstacle scattering with modulu of the far field pattern as data. In: *Inverse Problems in Medical Imaging and Nondestructive Testing* (Engl et al., eds). Springer-Verlag, Wien, 75-92 (1997).

[56] Kress, R. and Rundell, W.: Inverse obstacle scattering using reduced data. SIAM J. Appl. Math. 59, 442-454 (1998).

[57] Lax, P.D. and Phillips, R.S.: *Scattering Theory*. Academic Press, New York 1967.

[58] Leis, R.: *Initial Boundary Value Problems in Mathematical Physics*. John Wiley, New York 1986.

[59] Louis, A. and Maaß, P.: A mollifier method for linear operator equations of the first kind. Inverse Problems 6, 427-440 (1990).

[60] Michlin, S.G. and Prößdorf, S.: *Singuläre Integraloperatoren*, Akademie Verlag, Berlin (1980).

[61] Mönch, L.: A Newton method for solving the inverse scattering problem for a sound-hard obstacle. Inverse Problems 12, 309-323 (1996).

[62] Müller, C.: *Foundations of the Mathematical Theory of Electromagnetic Waves*. Springer-Verlag, Berlin Heidelberg New York 1969.

[63] Müller, C.: Radiation patterns and radiation fields. J. Rat. Mech. Anal. 4, 235-246 (1955).

[64] Murch, R.D., Tan, D.G.H. and Wall, D.J.N.: Newton-Kantorovich method applied to two-dimensional inverse scattering for an exterior Helmholtz problem. Inverse Problems 4, 1117-1128 (1988).

[65] Nachman, A.: Reconstructions from boundary measurements. Annals of Math. 128, 531-576 (1988).

[66] Newton, R.G.: *Scattering Theory of Waves and Particles*. Springer-Verlag, Berlin Heidelberg New York 1982.

[67] Novikov, R.: Multidimensional inverse spectral problems for the equation $-\Delta\psi + (v(x) - Eu(x))\psi - 0$. Translations in Func. Anal. and its Appl. 22, 263-272 (1988).

[68] Ola, P., Päivärinta, L. and Somersalo, E.: An inverse boundary value problem in electrodynamics. Duke Math. Jour. 70, 617-653 (1993).

[69] Potthast, R.: On the convergence of Newton's method in inverse scattering. Preprint.

[70] Potthast, R.: Stability estimates and reconstructions in inverse scattering using singular sources. J. Comp. Appl. Math. 114, 247-274 (2000).

[71] Potthast, R.: Integral equation methods in electromagnetic scattering from anisotropic media. Mathematical Methods in the Applied Sciences 23, 1145-1159 (2000).

[72] Potthast, R.: Electromagnetic scattering from an orthotropic medium. J. Int. Eq. Appl. 11 No. 2, 197-215 (1999).

[73] Potthast, R.: On a concept of uniqueness in inverse scattering for a finite number of incident waves. SIAM J. Appl. Math. 58, 666-682 (1998).

[74] Potthast, R.: A fast new method in inverse scattering. In: M. Tanaka and G.S. Dulikravich (eds): *Inverse Problems in Engineering Mechanics*, International Symposium on Inverse Problems in Engineering Mechanics 1998 (ISIP'98), Nagano, Japan, Elsevier Science/UK 1998.

[75] Potthast, R.: A point-source method method for inverse acoustic and electromagnetic obstacle scattering problems. IMA Jour. Appl. Math. 61, 119-140 (1998).

[76] Potthast, R.: A fast new method to solve inverse scattering problems. Inverse Problems 12, 731-742 (1996).

[77] Potthast, R.: Domain derivatives in electromagnetic scattering. Math. Meth. in the Appl. Sci. 19, 1157-1175 (1996).

[78] Potthast, R.: Fréchet differentiability of boundary integral operators in inverse acoustic scattering. Inverse Problems 10, 431-477 (1994).

[79] Potthast, R.: *Fréchet Differenzierbarkeit von Randintegraloperatoren und Randwertproblemen zur Helmholtzgleichung und den zeitharmonischen Maxwellgleichungen.* Dissertation Göttingen 1994.

[80] Ramm, A.G.: *Scattering by Obstacles.* D. Reidel Publishing Company, Dortrecht 1986.

[81] Ramm, A.G.: Recovery of the potential from fixed energy scattering data. Inverse Problems 4, 877-886 (1988).

[82] Roger, A.: Newton Kantorovich algorithm applied to an electromagnetic inverse problem. IEEE Trans. Ant. Pro. AP-29, 232-238 (1981).

[83] Reed, M. and Simon, B.: *Scattering Theory.* Academic Press, New York 1979.

[84] Simon, J.: Differentiation with respect to the domain in boundary value problems. Num. Func. Anal. Opt. No. 2 (1980).

[85] Stefanov, P.: Stability of the inverse problem in potential scattering at fixed energy. Phys. Rev. 56, 99-107 (1990).

[86] Sun, Z. and Uhlmann, G.: Recovery of singularities for formally determined inverse problems. Commun. Math. Phys. 153, 431-445 (1993).

[87] Tikhonov, A. N., Goncharsky, A. V. Stepanov V. V. and Yagola A. G.: *Numerical Methods for the Solution of Ill-posed Problems*. Kluver, Dordrecht 1995.

[88] Tobocman, W.: Inverse acoustic wave scattering in two dimensiona from impenetrable targets. Inverse Problems 5, 1131-1144 (1989).

[89] Wang, S.L. and Chen, Y.M.: An efficient numerical method for exterior and interior inverse scattering problems. Inter. Jour. Imaging Systems and Technology 1, 100-108 (1989).

[90] Wilcox, C.H.: *Scattering Theory for the d'Alembert Equation in Exterior Domains*. Springer-Verlag Lecture Notes in Mathematics 442, Berlin Heidelberg New York, 1975.

Index